It'll Buff Out

A Private's Tales of War and Shenanigans with the 10th Mountain Division in Afghanistan: 2001-2005

Daniel Pace

GZE publications

Dedication

This book is dedicated to all of our country's wartime infantry privates. Whether they were called joes, doughboys, GI's, grunts, knuckle-draggers, ground-pounders, or dog faces, this motley crew of citizen-soldiers, ne'er-do-wells, and occasional heroes has always borne the brunt of our nation's wars, and for that reason, they will forever hold a special place in my heart.

Of those infantrymen, the book is most especially dedicated to the three who bore the brunt of our own time at war in Afghanistan with the 1st Battalion, 87th Infantry Regiment.

Chad Edward Fuller, Scout platoon, KIA, August 31st, 2003
Adam Lee Thomas, Scout platoon, KIA, August 31st, 2003
Evan William O'Neill, Alpha Company, KIA, September 29th, 2003

May they forever be remembered for their sacrifice.

The book is also dedicated to Michelle Bowens, for reasons she would understand if she were still with us.

Glossary

Ate Up (Ayt'-up) **adj. slang.** Bad. (e.g. Your uniform is ate up, Private. Go fix that shit before I put my foot in your ass.)

Bag of Dicks (bag-ov-diks') **n. slang.** A common Army simile for something that isn't executed well or looks improper. (e.g. That guy landed like a bag of dicks. Also. Your rucksack looks like a bag of dicks; repack that shit!) (see also: football bat and soup sandwich)

Barracks Lawyer (Ber-icks-loi'-yer) **n. slang.** A lower enlisted soldier who claims to have certain knowledge of one or more Army systems and processes. Barracks lawyers know the best way to negotiate larger bonuses, the most effective method of securing promotion, and the ins and outs of the uniform code of military justice. Strangely, despite their extensive knowledge, barracks lawyers rarely seem to enjoy the fruits of the systems they claim to know so well, which suggests their knowledge is mostly bullshit.

Blue Falcon (Bloo'-fowl-kin) **n. slang.** A Buddy fucker. One who stabs another (figuratively) in the back for his own benefit.

Broke Dick (brok'-dik) **n. slang.** An injured, or pretending to be injured, soldier.

Cherry (chair'-ee) **n. slang.** A new guy. See also: Cherry-pot-pie, cherry-ass-fuck, or FNG (Fucking New Guy).

Chow Hall (Chow'-hall) **n. slang.** The Army term for a cafeteria. Chow halls vary widely in quality and visual appeal, ranging from Soviet-esque slop lines that specialize in such army classics as SOS (shit on a shingle) to gleaming (usually Air Force) nutritional Meccas.

CIB (See'-eye-bee) **n. acronym.** (Combat Infantryman's Badge) The CIB is the award given to infantrymen for participating in ground combat. There is no other award in the military that I know of (except perhaps for the Purple Heart) that encourages as much hair splitting and armchair quarterbacking as the CIB, and many staff officers have fought fiercely to ensure that their being within 200 meters of an artillery round that may have been fired at or around them counts as 'participating in ground combat.'

Cock Cushion (cok'-kush-in) **n. slang.** One of many Army terms for a mustache. See also: Dick duster.

Cock Holster (cok'-hol-ster) **n. slang.** An informal term for a soldier's mouth. See also: trap, hole, and suck.

Dick Beaters (Dik'-Beet-urs) **n. slang.** An informal term for a soldier's hands.

E-Tool (Ee'-tul) **n. abbreviation (entrenching tool).** A foldable shovel that is carried by every infantryman on his rucksack. The E-tool has myriad uses, from providing a sort of stool to balance on when relieving one's self in the woods, to digging holes, to serving as an improvised weapon. I've heard many far-fetched accounts of soldiers slaying enemies with their E-tools, although none of them sounded credible.

EIB (Ee'-eye-bee) **n. acronym** (Expert Infantryman's Badge). The EIB is an award provided to infantrymen who complete a wide variety of tactical, medical, and weapon-based tasks. EIB training is conducted in a round robin format with tightly controlled standards and evaluation criteria. The EIB is a major discriminator in the infantry community for promotion and professional success, and units put a good amount of work into preparing their soldiers to succeed at the EIB tests.

ETS (Ee'-tee-ess) **n. acronym** (Expiration Term of Service). The much-anticipated and widely celebrated end date of a soldier's time in the Army.

Fart Sack (Fart'-sak) **n. slang.** A sleeping bag.

FOB (Fahb) **n. acronym.** (Forward Operating Base). A temporary military base. A FOB is generally larger than a combat outpost (COP) or firebase, although the terms are sometimes used interchangeably.

Footbread (foot' bred) **n.** The flat, pita-like bread commonly served in Afghanistan. It was called footbread because the bakers kneaded with their feet. It was delicious and best enjoyed while actively avoiding thinking about where the baker's feet had been earlier in the day.

Guidon (Gy'-don) **n.** A small flag used in the Army to mark a unit commander's position. In our unit, each company had a guidon. The guidon is carried by a particularly hooah soldier whose primary qualities are generally motivation, squared-awayness, and the ability to not fall out of formation runs while carrying the guidon.

GWOT (Gee'-wot) **n. acronym** (Global War on Terror). The collection of military actions undertaken by the United States following the World Trade Center attacks in 2001 and formally ended by President Biden in August 2021.

HESCO Barrier (Hess'-ko-bear-ee-er). A collapsible metal frame lined with woven cloth. When unfolded and filled with dirt, HESCOs form an explosion and impact-resistant brick that can be used to build a wall. HESCO walls are used to build FOBs around the world.

High Speed (Hi'-speed) **adj.** good. The term is used equally often to sarcastically describe something that is not good. (see also: squared away)

HMMWV (Hum'-vee) **n. acronym** (High mobility, multipurpose, wheeled vehicle). A diesel powered, military vehicle originally produced to transport light equipment and personnel around secured areas of the battlefield and to serve in other non-combat functions. During the Global War on Terror, the HMMWV instead became one of the primary combat vehicles employed by the U.S. military in both Iraq and Afghanistan for most ground operations until it was eventually replaced (in large part) by more capable vehicles. Often written as Humvee.

Hooah (Hoo'-ahh) **interj.** An army interjection that can mean almost anything. As a loud, group response, it generally indicates mandatorily enthusiastic support. As a quieter, individual response, it can either mean 'I understand' or 'fuck you,' depending on the tone with which it is uttered. As a noun, it can serve as a slang term for a soldier. (e.g. Go grab a couple of those young hooahs over there so we can put them to work.) Finally, as an adjective, it can be used sarcastically to express fondness for an activity. (e.g. This crap-burning detail is some extra hooah shit. I need to thank my recruiter if I ever see his ass.)

Hooch (Hooch') **n. slang.** The term specifically refers to a sort of tent-like shelter infantrymen commonly set up using the standard issue Army poncho. In common usage, the term is used to refer to any sort of lodging one might be staying in. (e.g. Let's get back to the hooch. It's cold as hell out here.)

Hump (Hump) **v. slang.** To carry something. (e.g. It's your day to hump the .50 cal. Pack that shit up and be ready to move out in five mikes.)

IED (I'-ee-Dee) **n. acronym** (Improvised explosive device). a homemade bomb.

IVO (In-vi-sin'-it-ee-ov) **adj. acronym** (in-vicinity-of). Near a known object. (e.g. The terrorist is IVO the 4th street mosque).

Joe (Jo') **n. slang.** A lower enlisted soldier in the Army.

Joe Rumor (Jo-roo'-mer) **n. slang.** The informal network of information that permeates the lower enlisted world. Joe rumor is generally apocryphal, although sometimes it turns out to be more accurate than the officially provided material. (See also: RUMINT)

Haji (Ha'-jee) **n. slang.** A term American soldiers used for any resident of Iraq or Afghanistan. The term derives from the Arabic term for a person who has made the *hajj*, the once-in-a-lifetime trip to Mecca required by Islamic law. Depending on the tone and context in which it was used, Haji could be either derogatory or just generally descriptive. The term was frequently shortened to 'Haj.'

K-Pot (Kay'-pot) **n. slang.** An informal term for the kevlar helmet issued to soldiers in the Army during the 80s and 90s. This helmet weighed between 3.1 and 4.2 pounds, and it was supported on the soldier's head by a complex set of strings and leather bands that were notoriously difficult to assemble, uncomfortable to wear, and surprisingly foul smelling after long use. The kevlar helmet is also sometimes referred to as the dome of obedience.

LACE Report (Layss' ree-port) **n. acronym (**Liquid, Ammunition, Casualties, Equipment**)** The LACE report is the standard report sent by infantrymen to quickly report their status to higher. At higher echelons, it's replaced with more complicated reports that have language like "intent" and "tactical task" in them, but at the individual infantryman's level, the LACE report–supplemented with a healthy dose of profanity–is sufficient to communicate one's entire status to one's boss.

LCE (El'-see-ee) **n. acronym.** An acronym for 'load carrying equipment,' the LCE is a belt-and-suspenders system worn by soldiers in the 80s and 90s. The LCE was a modular system that could be outfitted with a wide variety of pouches for ammunition, canteens, medical supplies, and other military necessities. It was replaced during the GWOT with similar systems that included integrated armor.

Lum (Loom) **n. abbreviation.** Short for illumination and expressed as a percentage, lum is the term used to describe how bright a given night is. (e.g. "Lum is good tonight. 80%) (See also: Illum)

M249 (Em'-too-for-nine) **n.** The M249 Squad Automatic Weapon, or SAW, is a belt-fed 5.56 caliber light machine gun carried by infantry squads in the U.S. Army. During the GWOT, it was commonly mounted on vehicles alongside the MK-19 automatic grenade launcher, but because of the latter weapon's unsuitability for urban warfare, the SAW often became the truck's primary weapon.

M4 (Em-for') **n.** The M4 carbine was the standard weapon of almost all active-duty soldiers during the GWOT. The weapon is 5.56 caliber, which gives it almost no recoil, and is usually fed by a 30-round magazine. Accessorizing one's M4 was a favorite pastime of many soldiers during the GWOT, and the variety of sights, slings, grips and lights that could be found on M4's across the theater was mind boggling.

Ma Deuce (Maw'-Duce) **n. slang.** The slang term for the M2 .50 caliber machine gun, a venerable machine gun that's seen service (in one form or another) since 1923. Its distinctly loud report and ability to render its targets extra-definitely-dead ensure its continued popularity a hundred years later.

MEDEVAC (Med'-i-vac) **n. acronym (**medical evacuation). the transportation of a casualty from the point of injury to a medical treatment facility in combat by a dedicated air or ground evacuation asset. It's like calling an ambulance in combat. Also, **v. acronym** (medical evacuation). The act of conducting the above activity (e.g. Joe's hit; MEDEVAC his ass!)

Mikes (miks) **n. slang.** minutes

Meat Rocket (Meet'-rok-et) **n. slang.** One of a wide variety of terms of endearment used by senior enlisted personnel to refer to lower enlisted personnel. (see also: studs, hammer-heads, dirt-darts, and hooahs)

NCO (En'-see-oh) **n. acronym.** (Non-commissioned Officer) The official term for enlisted personnel between the rank of corporal and sergeant major. When addressed, NCOs below the rank of first sergeant are referred to as 'sergeants.'

NVGs (en'-vee-geez) **n. acronym.** (Night Vision Goggles) A generally helmet mounted device used to see better in the dark. During this period of the GWOT, most infantry units used a combination of PVS-7s and PVS-14s. Both of the systems functioned by amplifying ambient light to produce a grainy green image of the environment. Each of the systems had distinct drawbacks, though. 7s were binocular but eliminated the user's depth perception. 14s were monocular, but because only one eye was amplified, they tended to give the user splitting headaches after extended use.

PCCs (Pee'-see-sees) **n. acronym.** (Pre-combat Checks) Checks conducted by leaders on their soldiers prior to executing a mission.

POG or **Pogue** (Pohg') **n. acronym.** (People Other than Grunts) The term of derision used by the infantry for everyone else in the military. That includes (much to their chagrin) cavalry scouts, attack aviation pilots, tank drivers, and artillerymen, although, notably, it doesn't include Special Forces guys.

Poguey Bait (Pohgee'-bayt) **n. slang.** Treats brought from home to supplement an infantryman's diet. Common examples of poguey bait are beef jerky, candy of the non-melting variety, and booze.

PT (pee'-tee) **n. acronym.** (Physical Training) The Army's word for exercise. Each unit in the Army runs its PT program in a different manner, with varying degrees of quality and effectiveness, although almost all programs incorporate a block time in the morning for PT to be conducted. At the time this book was set, PT was conducted every morning on Fort Drum from 0630 to 0800. Our unit's program was usually led at the squad leader level and heavily emphasized formation running, group calisthenics, and regular ruck marches. Woe to the soldier who fell out of a unit formation run.

QRF (cue'-ar-eff) **n. acronym.** (Quick Reaction Force) A mission for a given unit, during which that unit remains on alert at a designated location and ready to roll out quickly to respond to a problem. In Afghanistan at this time, whichever unit wasn't patrolling was generally expected to serve as a QRF for whichever unit was.

Ranger Panties (Rayn-jer' pan'-tees) **n.** Also known as 'silkies,' Ranger Panties are the ultra-short nylon shorts worn by infantrymen during this period. They are notoriously immodest, particularly when wet.

RIP (Rip') **v. Acronym.** (Relieve in Place) The act of taking over someone else's military mission. During the GWOT, units usually RIP'd with one another when one unit arrived in country and the other redeployed. (Note: RIP is also commonly used as a noun. e.g. We conducted a RIP with the 82nd.)

RTO (arr'-tee-oh) **n. Acronym.** (Radio Telephone Operator) The guy in an infantry unit responsible for carrying and operating the unit's radio. The RTO usually accompanies the commander of the unit and is responsible for maintaining communication with whatever higher element that unit is reporting to.

Rucksack (Ruk'-sak) **n. German.** The German word for backpack, which became the Army's *de facto* term for any of the various backpacks issued to soldiers. The term is often abbreviated to 'ruck' during everyday use.

Sandbag (Sand'-bag) **v. slang.** To avoid work by either making yourself hard to find, or by creating the appearance of doing work while actually accomplishing nothing. (see also: sham)

Smaj (Smaj') **n. abbreviation.** A *very* informal abbreviation of the sergeant major rank. The word smaj is used only in conversation, and even then, only when the smaj isn't listening.

Smoke (Smok') **v. slang.** To force someone to do exercise, usually for punitive reasons. e.g. "Jones was late to formation so Sergeant Smith smoked his ass for an hour."

Snatch Patch (Snach'-pach) **n. slang.** A style of Army haircut where the barber places the palm of his hand on the recipient's forelock and shaves the rest of his hair to the skin. It's a sort-of anti-tonsure.

Snivel Gear (Sni'-vul-geer) **n. slang.** Any clothing designed to keep one comfortable. A person wearing snivel gear is generally both derided and envied by one who is not.

SOG (ess'-oh-gee) **n. acronym. Sergeant of the Guard.** The sergeant of the guard is the NCO in charge of a given guard shift. His job is to check up on the various guard positions his team is responsible for to make sure the guys are doing what they're supposed to be doing.

SOP (ess'-oh-pee) **n. acronym. Standard Operating Procedure.** The Army term for the approved way of doing something. Units have SOPs for everything from how to spraypaint the vehicle number on a Humvee's bumper to how long the lanyard should be that tethers a canteen to a canteen pouch. Sometimes SOPs are codified in a sort of Army bible called the TACSOP, (Tactical Standard Operating Procedures) but in our battalion, they were more commonly found in printed and stapled sheafs of paper, each of which was carefully secured in a filing cabinet so that it could be referenced when the Sergeant Major came calling for an inspection. Woe to the private who didn't know the answer to one of his SOP-related questions.

Squared Away (Skward'-a-way) **adj. slang.** Good. (e.g. That haircut is squared away, Private.)

TOC (Tok') **acronym. Tactical Operations Center.** The building, tent, vehicle, or poncho that the headquarters element of a given unit sets up to command and control operations. TOCs can be small affairs, with nothing more than a map and a radio, or they can be large, elaborate, and computerful. TOCs are often full of pogues (see above).

Voluntold (Vol-un-told') **v. slang.** The act of being unofficially ordered to do a nominally voluntary activity. (e.g. "The unit is collecting donations for the Army Emergency Relief fund. Participation in the program is voluntary. However, if we have 100% participation in this donation campaign, the unit will receive a three-day pass this weekend. If we do not, the unit will conduct area beautification of Range 37B this weekend.")

Yardbird (Yard'-burd) **n. slang.** An untrained or inept private, or a guy on a menial work detail.

(The) Z monster (zee'-mon-stir) **n. slang.** The cryptozoological nemesis of people on guard duty. The Z monster makes you fall asleep, even when you don't want to. His methods are so insidious that even after waking, his victims are often unaware they ever fell asleep at all, at least until they see the dicks their buddies drew on their foreheads when they look in the mirror the next morning.

Forward

We aren't no thin red 'eroes, nor we aren't no blackguards too,
But single men in barricks, most remarkable like you;
An' if sometimes our conduck isn't all your fancy paints,
Why, single men in barricks don't grow into plaster saints;
While it's Tommy this, an' Tommy that, an' Tommy, fall be'ind,"
But it's "Please to walk in front, sir," when there's trouble in the wind
There's trouble in the wind, my boys, there's trouble in the wind,
O it's "Please to walk in front, sir," when there's trouble in the wind.

Rudyard Kipling, "Tommy"

"A thing may happen and be a total lie; another thing may not happen and be truer than the truth."

Tim O'Brien, "The Things They Carried"

This book is vulgar, and by the end of the first part, you're going to hate Specialist Pace. These are both necessary, though, so bear with me. The book is vulgar because junior enlisted life in the Army is mostly a nasty, low business. So telling a story about it requires describing the sorts of things most of us would rather avoid thinking about and maybe don't even believe occur. You'll hate Specialist Pace because at twenty-one I was a sarcastic prick who relentlessly criticized and bitched about every facet of Army life, and it took a good deal of life experience to change my point of view.

I tell you these things up front so that as you delve into the story of my life with the 1st Battalion, 87th Infantry and our time in Afghanistan, you will be emotionally prepared to read about the sorts of things that would make Jerry Springer blush. I assure you, there's no way that you could be any more shocked than my wife, Alycia, and I were when we left the campus of Texas A&M University and moved into enlisted housing at Fort Drum, New York.

Once you look past the vulgarity and bitching, though, junior enlisted life has a great deal of beauty to it that its gritty surface would never suggest exists. It's exciting, and it's animated. It's packed full of crazy schemes and wild dreams, and it's just bursting with the sort of *chutzpah* that only eighteen to twenty-five-year-olds possess. Yes, many of those young hooahs turn out to be criminals and miscreants, but the lower enlisted ranks also produce some of the best people in the world–Jackie Robinson and Charles Schultz to name a few–and the camaraderie that exists between the members of this community of uprooted young people from across the country is refreshingly free of every sort of cultural, racial, and socio-economic influence. There's also incredible candor and transparency. Never since then have I seen the complete lack of superficiality that I saw among privates in the Army. Everyone knows everything about everyone, so nobody hesitates to say anything. Conflict is worked out directly and quickly, and everyone moves on. And there's wonderful generosity, the sort of open-handed willingness to share everything with a buddy that can only exist in an environment where everyone has nothing.

While it certainly has a dark side, lower enlisted life is a fascinating and often beautiful thing, well worth studying in its own right.

Aside from that, and perhaps most importantly since this is a military history book, we should never forget that these are the people who *actually fight wars.* Most military accounts focus on the actions and decisions of senior leaders, but junior enlisted personnel–not generals–make up the vast majority of actual combat formations. Privates clear the houses, throw the grenades, and take the prisoners. So, how can we understand what a given war was like if we don't understand the individuals who fought it? If we avoid generalization and focus on the Global War on Terror (GWOT) specifically, this assertion becomes even more apt, because the privates described here were literally the people who fought the GWOT for twenty years, and they carried all the generational and societal baggage they brought with them to the Army (or had instilled in them by it) over that two-decade period.

Our battalion produced future sergeants major, Silver Star recipients, intelligence officers, Special Forces lieutenant colonels, and even (although they weren't privates in this story) a couple of three and four-star generals. Our battalion's junior enlisted soldiers grew up to be the late-GWOT's senior leaders. So, understanding their formative experiences in the Army can help us better understand our military's current leaders in turn. Just as the platoon leaders and privates of World War One became the generals and sergeants major of World War Two, everything I accomplished in Iraq as a platoon leader was influenced by my time in Afghanistan as a private and sergeant.

Likewise, looking past the events and people portrayed here toward larger organizations and trends, we can get a sense of what sort of organization the Army was as it entered the GWOT. In many ways, it was very much still the Army of the 1990s, trained and equipped to achieve Cold War-era foreign policy objectives. Its equipment, tactics, and experience drove it toward the conduct of maneuver warfare, clearing trenches, destroying tanks, and attacking defensive positions. Obviously, twenty years of counterinsurgency changed all of that. But that change came neither quickly nor easily, and understanding how the Army operated in 2003 and what sort of place it was can help us understand why it was so hard for the organization to adapt, and why the Army struggled to find its footing during the early years of Afghanistan and in post-invasion Iraq.

This deeper look also helps bring to light other important truths about the Army, such as how perishable readiness can be in a community that turns people over as quickly as the Army does, or the challenge of balancing readiness and capability against personnel burnout, or how difficult it can be to build and hold onto institutional knowledge. *Street Without Joy,* which chronicles Bernard Fall's observations about the French units operating in Indochina, drips with lessons that are universally applicable to militaries conducting limited warfare in other times and places. Fall's book reveals how analyzing the microcosm of a small unit and life at the small unit level can broaden our understanding of major military activities. Similarly, in this book, I hope to shed new light on the GWOT (and warfare in general) by homing in on a 10th Mountain Division infantry platoon in the early 2000s.

Oh, and as you work through this story, please remember that this book *is* set in the early 2000s, and that the world then, as many of you may remember, was a bit different. For one thing, it was a bigger and more mysterious place. There were fewer cell phones, there was no social media or *YouTube*, and the Internet was in its larval stage, so people had less access to information about just about everything than they do now. This was made starkly evident as I reread my letters from this time period to fact-check myself while writing this book. I'm shocked by some of the things I encountered for the first time in Afghanistan, ranging from the Afghans' choice of headgear (the *pakol)* to the existence of what I now know to be a chai latte (tea with cream, sugar, and spices); things that are now so commonplace that it's hard to imagine not knowing about them. At that time, though, they were novel and noteworthy enough to me that I described them in depth to my wife back home, and she was equally struck by their novelty. Whenever possible, I've tried to capture that sense of wonder and balance my naivete with what experience taught me after many more years of exposure to Afghan and Islamic culture.

Another key difference between now and the early 2000's is that back then, everyone was in constant competition to shock everyone else. I cannot overstate this. As we, the last members of Generation X, reached adulthood, songs like the Bloodhound Gang's *Lift Your Head Up High and Blow Your Brains Out* were still popular. The rawest sorts of racially charged humor were ubiquitous. And any indication that you might be offended only provoked further ridicule from your peers, *particularly in the Army.* It was an altogether grittier and less sensitive time. Also, a number of societal barriers which no longer exist were still firmly in place in the military. For instance, there were no women serving in the combat arms; any sort of transexual anything would have gotten you immediately discharged for deviance; and everything related to homosexuality was firmly in the closet. I literally had to sign a sworn statement attesting to the fact that I'd never participated in any sort of homosexual activity before I could join the military. That's how different things were.

So, when you read about the way we talked and the things we said to one another all those years ago, try to remember that, once upon a time, all the information in the world wasn't as accessible as it is now. Also try to remember that as scandalized as you might be by some of the things we said, they weren't all that remarkable at the time. That being said, this book does not seek to glamorize that time or those events, but rather to portray them honestly and accurately. Pretending we were other than we actually were would do a disservice to the unit, its members, and to the study of history itself.

Because in the end, of course, *it was what it was.*

Prologue

College Station, Texas; Fall, 2000

He who makes a beast of himself gets rid of the pain of being a man.

Samuel Johnson

The doctor said, "What'cha been thinking about?" Bob said, "That's the point. I wasn't thinkin about nothin. Now I gotta do something else"

NOFX, "Bob"

"See, Dan, they have these things called port calls. And when there's a port call, everyone on the ship gets off at whatever badass tropical location they happen to be at and gets hammer drunk."

The hot Texas sun beat down on Brad and me as he explained what our lives would be like after we joined the Navy.

"At these ports, there are hot chicks everywhere, and they love sailors. You know why they love sailors, Dan? Because sailors have money, and because sailors look badass in uniform. That's why."

I nodded along silently as Brad's comments slowly wound their way into my brain. It was blazing hot in the parking lot we were standing in–easily 103 or 104 degrees–and Brad's words weren't penetrating my consciousness very quickly. The sluggish comprehension might have been caused by the heat, or it might have been caused by the fact that I was six beers and a joint into the day–who could say? Regardless, I liked the overall flavor of what he was saying, and I was enthusiastic about taking the step we were about to take.

We were standing outside the Navy Recruiter's office at the Post Oak Mall. We'd arrived there after one of our usual long mornings of drinking and arguing about books, video games, and childhood T.V. shows had led us to conclude that we needed more adventure in our lives. Adventure and money. And women. At the moment, we didn't have any of the three. We were both seniors in college, getting ready to graduate in December, and neither of us had any idea what to do with our lives.

To say Brad and I were slackers would be a wild understatement. I was getting ready to graduate with a B.A. in Philosophy and a 2.78 GPA. Brad was on his third or fourth major and had lost more credits transferring between universities than most people needed to finish their degree. It's not that either of us was dumb–at least not in the mental processing power sense–but we were drunk most of the time, high or frying on drugs the rest of the time, and we didn't give a flying shit about anything. We were exactly the sort of wasted space Tipper Gore cautioned the mothers of the country their kids would turn into if the music industry didn't immediately add 'explicit lyrics' labels to rock and rap album covers.[1]

One quality we did have, though, was impetuousness, and that was the quality that led us to the Navy recruiter. I wish I could say we were standing there thanks to long-term planning, passion, patriotism, or at least an understanding of what the hell the military was or did. But that wouldn't be accurate. On the contrary, we were armed with nothing but impetuousness and a complete lack of concern about the consequences of our decisions.

"Chicks, boats, booze, and tropical sun," I thought. "That sounds amazing. I like all of those things."

"Let's go, Brad. It's time to join the Navy."

We walked across the parking lot to the door to the recruiting office, pausing for a moment to admire the cool pictures of fighter jets, guys in rubber boats, and aircraft carriers on the posters that adorned its windows, then pulled the handle. It didn't budge. The door was locked.

"What the fuck?"

My eyes wandered down the door to a small placard which read "Closed for lunch. Back at 1300."

I checked my watch. It was 12:15. Our dreams were shattered. No chicks. No booze.

1. And who knows, maybe she was right?

No nothing. Just forty-five minutes of sweating in a parking lot. My stomach rumbled. I turned to Brad.

"Dude, there's no way I'm waiting 45 minutes for this guy to show back up. I'm hungry, and I've got to get to work at 3:00."[2]

Brad was equally frustrated. "Yeah, fuck the Navy. Let's get out of here."

We turned to leave, but something in my brain stopped me. I looked back over my shoulder and saw the word 'Army' in big, green block letters over the next door down from the Navy office.

I'd never given a single thought to the Army. I had vague ideas about what Army people did from books and movies, but nothing I could specifically describe. For some reason, though, at that moment a quote from Hunter S. Thompson floated through my head–"Don't take any guff from these swine."

Should a young man in search of adventure abandon a promising chance for excitement just because some guy was out to lunch? No. Certainly not. I stopped and turned to Brad.

"Wait. Let's check out the Army instead. I'm sure they do cool shit, too, and it looks like their office is open."

Brad paused for a moment, taking a drag off the cigarette he was smoking. Then shook his head. "No way. My dad was in the Army in 'Nam, and he says it's full of nothing but criminals and idiots."

"How bad could it be?" I thought. "Besides, what the hell else am I going to do when I graduate?"

To Brad, "I'm going to check it out anyway. Fuck it, it's probably cool."

"You're on your own then, Dan. I'm going to walk around the mall, and I'll meet you at the truck."

I nodded, and turned back toward the row of recruiting offices. I walked across to the Army office's door and pulled the handle. It was open, so I walked in, pausing a moment to feel the wave of frigid air conditioning pleasantly wash over me and to let my eyes adjust to the relatively dim office lighting.

---

2. I worked evenings at the Freebirds World Burrito across from campus. The job didn't pay much, but since I closed the place most nights, I could trade my free employee burrito for a pitcher of beer at the bar next door.

The office was quiet, and nobody immediately greeted me as I stood looking around. I wasn't used to what I would later come to know as the 'government-*chic*' style of decor. The room I was standing in was predominantly gray. Gray cubicles, gray desks, and dated-looking gray computer terminals. The overhead lighting–provided by a couple of fluorescent tube bulbs–flickered slightly, bathing the gray room in a sort of bright, overly clear, but somehow sickly light. The gray was offset by brightly colored posters that adorned all the walls and cubicles. "What some of the most successful women this year are wearing" proclaimed one–its fearsome, helmeted, female visage glaring at me aggressively. "Army. Be all that you can be. 1-800-USA-ARMY" announced another. A third depicted two guys who looked vaguely like Crockett and Tubbs from *Miami Vice* poking out from the top of a tank. "19K. Armor." the poster said.

"What the hell is a 19K?" I wondered.

My sight seeing was interrupted when a gentleman stood up from behind one of the cubicles. His head was entirely shaved except for a small patch of hair on top, and he was wearing dark green slacks, black patent-leather shoes, and a light green button-down shirt with brass emblems on the collar, a colorful square of cloth above the left breast pocket, and a name tag that read "Santiago". He looked about 30, and he was smiling as he walked over to greet me.

"Hi, how are you today, son? My name is Staff Sergeant Santiago. Are you thinking about joining the Army?"

To his great credit as a recruiter, Santiago registered no disapproval as he reached out to shake my hand, even though I probably looked and smelled like a drunk bum.

I was dressed in a plaid shirt, torn up old jeans, and grease-stained, indoor soccer shoes. My hair was almost shoulder length and half-dyed blonde, and my glasses had one broken lens from a bottle that had hit me in the face at Mardi Gras in New Orleans that spring.[3] I looked something like Shaggy from *Scooby-Doo,* and I probably smelled like him too, except that instead of being covered in the scent of Great Dane and a Volkswagen van, I stank of beer, pot, and the sort of rotten body odor one acquires in

3. The New Orleans EMTs were remarkably unsympathetic when I staggered up to one of their checkpoints on Bourbon Street with a bleeding face. "Get the fuck out of here, shithead," was all the medical advice they offered. If I'd been as sober as they were that night, I'm sure I'd have said the same thing to me. Being a Mardi Gras EMT is probably a pretty thankless job.

Texas when living without either air conditioning or money for laundry.

I shook the recruiter's hand and greeted him back; "I'm Dan, and I think I am interested in joining. I graduate from A&M in December, and I want to get out of here and do something awesome."

The recruiter nodded, smiling, and said, "I think we can help you with that, Dan. Why don't you come over to my desk, and we'll work through a few forms. Do you have a little time to take a practice test? I need to see what options you're eligible for."

"Sure, I have an hour or so. Is that enough?"

"Absolutely"

Santiago escorted me toward a section of unoccupied cubicles and clicked a few keys on the computer sitting on the desk inside one of them.

"Have a seat and take this test, and let me know when you're done. Here's scratch paper and a pencil if you need it."

He left a legal pad and a number-two yellow pencil next to me and walked back to his desk. I sat down and looked at the computer monitor, which displayed an old, Windows 3.1-style operating system. A bright blue, pixelated block in the middle of the screen said "Practice Armed Services Vocational Aptitude Battery (ASVAB). Click 'start' to begin."

I clicked 'start', and the first of a series of multiple choice questions appeared on the screen. "If A=3 and B=2, then A + B = ?"

Bear in mind that at this point I was a senior in college. The week prior, I had taken the GRE to get into graduate school. Slacker though I was, this multiple choice question seemed ridiculous. I looked around to see if someone was screwing with me. Nobody appeared to be assessing my reactions to the test. On the contrary, nobody seemed the least bit interested in what I was doing.

Returning my attention to the monitor, I clicked the answer, and then looked at the next question.

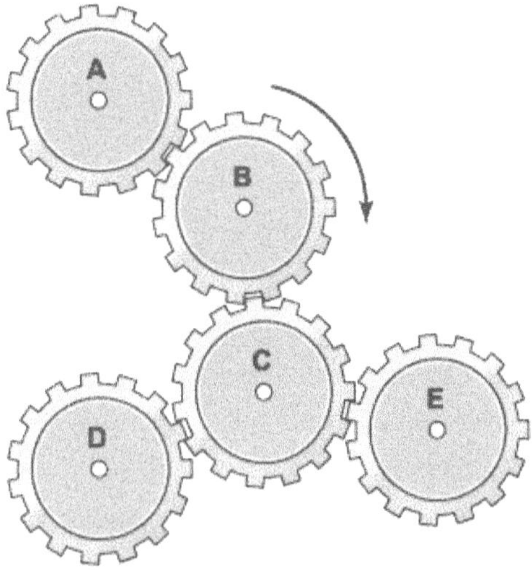

"Which way does gear 'E' move when 'B' rotates in the direction indicated by the arrow?"

Again, I'm no MacGyver, but I couldn't imagine anyone struggling to answer this question. I clicked the appropriate answer and kept at it for the next hundred or so questions. Most of them were similarly silly. As I finished the test, I stood up and stretched. What was the point of administering a test that nobody could fail?

I walked over to Santiago's desk. He looked up, surprised, and said, "Do you need something? Was there a glitch in the system?"

"No, I'm finished with the test."

He looked at his watch. "Really? In ten minutes?"

"Yeah, it said I was done."

He stood up, walked over to the computer, and clicked a few buttons.

"Yeah, it looks like you finished. Great. Let's see how you did."

A dot-matrix printer across the room started buzzing and spat out a piece of paper. He tore it off the printer, and we walked back to his desk and sat down.

"How'd I do?"

His eyes scanned the paperwork for a minute. "You did great. It was only the practice test, but if you do that well on the real test, you'll be qualified to do anything you'd like to do in the Army."

"Oh, awesome. What sorts of things does the Army do?"

"Well, there are a lot of positions available. I'm a logistics sergeant. There are also intelligence positions, combat arms jobs, personnel jobs..."

He paused. My face must have registered my total lack of comprehension. I had no idea what any of the words he was saying meant. What was combat arms? Doesn't everyone in the Army fight in combat? What's an intelligence position?

Santiago reached under his desk for a binder and handed it to me. "Here, Dan. Take a look through here, and I think what I'm saying will make more sense."

I took the binder and opened it as Santiago turned to type on his computer.

The binder was full of colorful leaflets tucked into plastic sleeves. Each leaflet depicted a guy or girl in camouflage doing some sort of activity, and below each picture was a blurb that described what they were doing and what the position was called. I thumbed through the binder and thought to myself.

"Cook? Nope, not doing that. To hell with cooking."

"Counter-intelligence agent? That sounds like some cool James Bond shit." I marked it with my finger, continuing.

"Supply? Nope. Boring."

"Adjutant? No idea what that word means."

I flipped several more pages, seeing nothing that caught my eye, until I saw a photo of a bunch of soldiers parachuting into a field. In the foreground was a guy with his face entirely covered in camouflage kneeling on the ground. He was wearing a backpack and carrying a grenade launcher in his hands. To a guy raised on movies like *Predator* and *Commando*, he looked super bad-ass. "Infantry," the leaflet said.

I got Santiago's attention and pointed at the book. "That's what I want to do. I want to walk around and do cool Army stuff like this dude."

Santiago's eyebrows went up, and he nodded. "Okay, I think we can get you an infantry contract. We can even get you to Airborne School if you want. That means you get to jump out of planes like the guys in the picture. One of the other recruiters used to do that." He turned toward another guy with a severe haircut and a beat-up face across the office. "Hey, Sergeant Jones, you were an airborne ranger, right?"

The guy looked at us and replied. "Yeah, until my knees gave out. Now I'm here."

"Dan here wants to join the infantry."

Jones looked at me and grinned. I grinned blankly and nodded back. At the time, the moment meant nothing to me. Looking back, though, I've thought about it a good bit. I wonder what Jones, with his ruined knees and his twenty-year infantry career, thought about this dirty idiot standing in the recruiting office that hot afternoon. To this day, I can't picture him without thinking of the scene from *Starship Troopers* when the legless guy says to young Johnny Rico, "The mobile infantry made me the man I am today."

I turned to Santiago and finished up the initial paperwork. I had to sign a big pile of papers, swearing a wide variety of things, like to never having used drugs, not wetting the bed or sleepwalking, and never having committed any homosexual activity. The last page indicated that if I graduated from one station unit training (whatever that was), I'd get a $9,000 bonus and $50,000 for college, and all I owed the Army was four years of time.

As I shook Santiago's hand and walked out of the office, I couldn't believe my good fortune.[4] To a guy in my situation, nine grand was an impossibly large sum of money. I'd be rich! And I'd get to spend four years in the Army having awesome adventures, parachuting and shooting stuff.

Brad was waiting in the parking lot.

"What took you so long?"

"I joined the Army, man. I'm getting $9,000 to jump out of airplanes. You missed out."

He seemed impressed as he took a drag from his cigarette. "Awesome. That's some good shit. Congratulations, Dan. They give you any of that cash now? We got beers to drink."

"Nah, they sure didn't. Spot me?"

"Course I will, you broke motherfucker."

Brad and I walked back to his truck and drove back toward our dorm. That evening, I shaved my head and walked across the Texas A&M campus to Northgate with Brad and a few other friends to get blind drunk celebrating my upcoming career move.

This shit was awesome. I was going to avoid real life for a while, make great money, and have some motherfucking adventures! To hell with grad school, I'm joining the Army!

---

4. Santiago probably felt similarly. For a recruiter, whose career is driven by the number and quality of people he gets to join the Army, a college graduate with no criminal record who required no work to land was unicorn material.

Part 1

Through the Looking Glass

They say that in the Army, the chow is mighty fine.

They say that in the Army, the chow is mighty fine.

A chicken fell off the table, and started marking time.

A chicken fell off the table, and started marking time.

Oh Lord, I want to go-oh.

Oh Lord, I want to go-oh.

But they won't let me go-oh.
But they won't let me go-oh oooooooooooooome, hey!

Army marching cadence

"When I use a word," Humpty Dumpty said in rather a scornful tone, "it means just what I choose it to mean — neither more nor less."

"The question is," said Alice, "whether you can make words mean so many different things."

"The question is," said Humpty Dumpty, "which is to be master — that's all."

Lewis Carroll, "Through the Looking Glass and What Alice Found There"

I

March, 2001; Fort Benning, Georgia

"*What makes the grass grow green?!*"

"*BLOOD! BLOOD MAKES THE GRASS GROW GREEN, DRILL SERGEANT!*"

Charlie Rock's response was thunderous and instantaneous. It had better be.

Charlie Rock, or properly, Charlie Company, 1st Battalion, 19th Infantry Regiment, was arrayed in an extended rectangular formation in a pit of soggy mulch across from the Battalion headquarters in the Sand Hill area of Fort Benning, Georgia. We were wearing identical camouflage battle-dress uniforms, or BDUs, our heads were shaved bald, and we were holding heavy, fake, rubber assault rifles, each of which was equipped with a heavy, real, steel bayonet.

"*Whirl right!*"

We whirled, striking imaginary enemy soldiers with our bayonets and shouting, "*Yahhhhh*" for emphasis as we did so.

"*Whirl left!*"

"*Yahhhh*" Again, we whirled.

Even in March, the Georgia morning was warm, and as we swung our bayonets about and shouted, we sweated heavily in the sticky southeastern air. My Army-issued glasses were fogged up enough that I couldn't see much of what I was swinging at, but it didn't really matter.[1] What mattered was to make plenty of noise. Insufficient noise angered Drill Sergeant Goelich, and when he was angry, Charlie Rock suffered.

And Charlie Rock had already suffered much that morning.

The morning had started at 0500, as all our mornings did except Sunday, with the lights being flipped on and a great deal of yelling as the residents of our barracks tripped over one another to shit, shower, shave and get downstairs to assemble in the open area under our building. Unfortunately, this morning Charlie Rock hadn't assembled quickly enough, so people were still straggling downstairs when the commander showed up to give us our morning instructions. Nothing much came of it while the commander was talking to us. But as soon as he left for his office, Goelich emerged from the shadows, face glowing red with rage, and he smoked the shit out of us, berating us loudly as he made us do pushups and side-straddle-hops for wasting the commander's time.[2]

"Do you dumb sons of bitches think the commander has nothing better to do than stand around waiting for you to wake up and make your way downstairs?! Do you think the enemy is going to wait for you to wake up and put on your make-up?! Hell no! He's going to kill you while you're still trying to figure out what the hell you're going to wear for the day! You privates will figure out how to show up on time for things! You'll figure it out quick, or you're going to get *really goddamn strong*!"

After the smoke session, Charlie Rock was herded upstairs to change from the physical training clothes we usually wore in the morning to our battle dress uniforms. Then we were herded back downstairs to collect rubber rifles and bayonets. We'd screwed that up too, though, failing to line up in any sort of organized fashion quickly enough to placate Goelich, and so we earned ourselves another healthy dose of smoke.

Somehow, after three or four more mishaps, we finally made it to the mulch pit, where we were now trying to appease Goelich by demonstrating our ferocity and motivation.

"*Bayonet sequence one...Execute*!"

"*Hiyah*!" In unison with my comrades, I swung the butt of my rifle up and across my body, cracking an imaginary communist in the jaw. "*Yaaah*!" I swing the butt back

---

1. Army issued glasses are rugged, plastic, and cheap. They are nearly impossible to break and fully-functional, but they also make their wearer look like an idiot, particularly when the wearer has a shaved head. This last bit has led them to be affectionately referred to as birth-control glasses, or BCGs, because of the well-established fact that in the history of the Army, no private wearing BCG's has ever impregnated, or been impregnated by, anyone.

2. The side-straddle-hop is Army-speak for a jumping jack.

down, simultaneously bringing the bayonet across my body at an angle, slashing the bad guy's chest open. "*Raahh*!" I completed the sequence, stabbing the bayonet forward and killing my fictional opponent.

"*Recover*"

I resumed my original position, rifle held in both hands and ready to attack again on command.

"What the hell am I doing right now?" I wondered. Here I was, a theoretically-college-educated twenty-first century adult, standing in a pit full of wood chips, pretending to stab imaginary people with a rubber rifle with a knife on the end of it. In spite of the physical strain of the activity and the perpetually looming threat of smoke from Goelich, it was hard not to crack a smile and laugh at the ridiculousness of the situation. However, the thought had no more than crossed my mind when Goelich appeared at my side, radiating fury and placing the rim of his brown, round Drill Sergeant's hat about an inch from my right eye. Shit. Either my mind had wandered off for a second and I'd missed something, my face had betrayed my inner monologue, *or Goelich could hear my thoughts.*

"Is something fucking funny, Pace?!"

Goelich was about my height but much bigger. He had the sort of palid, pinkish, heavily freckled complexion that some folks of German descent who spend their lives in the sun have, and his eyes were pale blue, red-rimmed, and utterly without levity.[3]

"No, Drill Sergeant."

"Then why the fuck are you smiling?! You're supposed to be training to be a killer, not grinning into space like a goddamn idiot?! Get the fuck down and start knocking out pushups."

"Roger, Drill Sergeant." I laid my weapon down in the mulch and got down to start doing pushups.

"Don't lay your weapon in the mud, goddamnit! How the fuck do you think you're going to kill the enemy if you get mud in the chamber?!"

---

3. As it turns out, Drill Sergeant Goelich and I grew up in the same hometown, San Angelo, Texas. I didn't find this out until the end of the course, when he became a bit more approachable. Normally, you'd imagine this would improve our relationship. But because I had a college degree, Goelich binned me into the "officer-material" category. In his world view, there was nothing San Angelo could produce that was lower or more loathsome than a lieutenant, and thus my degree doomed any potential friendship we might have had.

"With the bayonet, Drill Sergeant?"

The words exited my mouth before my brain had a chance to exert any control over my wayward pie-hole.[4]

"Oh, so we've got a goddamn smart-ass on our hands, do we?! Drill Sergeant Trevino, would you mind assisting Pace here while I continue the drill with the platoon? He seems to need a bit of additional instruction on the proper care of his assigned weapon and military bearing."

As I worked to adjust my rifle so I could do pushups with it laid across my hands (thus keeping it out of the mulch) Trevino walked up and stood in front of me. He squatted down, and I looked up from the top of the pushup position; he locked eyes with me. Drill Sergeant Trevino was one of our other drill sergeants. He was shorter than Goelich and more understated. On the surface, he was the less terrifying of the two. But when I looked into his black eyes, Robert Shaw's speech from Jaws popped into my head. "*Sometimes that shark looks right at ya. Right into your eyes. And the thing about a shark is he's got lifeless eyes. Black eyes. Like a doll's eyes. When he comes at ya, he doesn't even seem to be livin'...*"

Trevino's mouth smiled a bit as he squatted there, but the smile never reached those black eyes.

"Oh, so we have a comedian on our hands, do we? I hear comedians love attention. Come on comedian, let's go get you a bit of extra attention."[5]

Trevino ordered me to my feet and led me to the back of the formation. For the next twenty minutes, as my peers continued their bayonet training, he smoked the bejesus out of me. He ordered me to do pushups and six-inches until I was covered in brown mulch. He had me work through air squats and side-staddle-hops until my quads and calves were screaming. He even ran me through the more esoteric exercises like monkey-fuckers and front-back-goes. By the end of my tour-de-smoke with Drill Sergeant Trevino, I had experienced the breadth and depth of the Army's repertoire of punitive exercises, and I was a soaking groggy mess as I fell back into formation to continue training.

---

4. A constant problem in my Army career, and in my life as a whole.

5. If all of this sounds like something out of *Full Metal Jacket*, there's probably a reason. That movie captured the overall feel of basic training better than any other movie I've ever seen. But given its popularity, I suspect it probably impacted the way drill sergeants themselves came to play their roles. Life imitating art imitating life and all.

"Stupid brain. Stupid mouth," I thought, "Why'd y'all have to get me in trouble?"

Neither provided any response as I continued to swing and stab.

"Assemble to the right, March!"

Goelich's last command meant we were done with this event. Upon hearing that, we collapsed forward and to the right into a compressed rectangle; four ranks by around nine columns–the Army's standard formation. After some jostling–particularly since we all still had our bayonets fixed–we got into position.

"Right *face*!"

We all turned to the right.

"Port *arms*!"

We all picked up our rifles and held them in front of us.

"Forward, *march*!"

We started walking forward.

"Double time, *march*!"

The whole rectangle started jogging, our weapons bobbing in front of us. I was in the second rank, so the bayonet of the guy next to me in the third rank was uncomfortably close to my face.

"Where are we going?" I mouthed the question as quietly as I could to my battle buddy, Powers, who was jogging in front of me.

"No idea, man. We're going the wrong way to get to the ranges, and we don't have real rifles anyway. It's too early for lunch, though. Who knows?"

Infantry One Station Unit Training, OSUT for short, wasn't the sort of place that encouraged communication. We did what we were told as quickly as possible, and we learned not to ask questions. That tendency had been beaten out of us early.

The formation jogged back to the road that led between the mulch pit and our battalion, the 1st Battalion, 19th Infantry. As we ran, Drill Sergeant Goelich called a running cadence to keep us in step. We repeated each line as he shouted it.

"C-130 rolling down the strip!"

"C-130 rolling down the strip!"

"Sixty-four troopers on a one-way trip!"

"Sixty-four troopers on a one-way trip!"

"Mission top secret, destination unknown!"

"Mission top secret, destination unknown!"

"Don't even know if they're ever coming home!"

"Don't even know if they're ever coming home!"

"'Cause they're hard core."

"Hard core."

"All day!"

"All day!"

"Every day!"

"Every day!"

As it always did, the singing took our minds off the sweat running into our eyes that jogging in the morning heat caused, and after a few minutes, we reached a range sign that read "bayonet assault course"

The drill sergeants slowed us to a walk and maneuvered the formation off the street, lining us up in a huffing and puffing rectangle near the beginning of what looked like a long obstacle course.

A sergeant we didn't know, dressed in BDUs and immaculately polished boots but without a drill sergeant's hat, walked smartly to the front of our formation and shouted, "Good morning! This is the bayonet assault course! Your task this morning is to negotiate the bayonet assault course. The conditions are: number one, given nine lanes on a 300-meter bayonet assault course over irregular terrain with four types of targets: thrust; parry thrust target; parry, butt stroke to the groin target; and parry, butt stroke to head target. Number two, given all targets are marked with a sign to indicate the required attack. Number three, given seven types of obstacles. Number four, given a soldier in battle dress uniform with load-carrying equipment and a rifle with a fixed bayonet. The standard is that the course must be successfully negotiated by each soldier, with each soldier obtaining kills on 75 percent of the total targets in his lane. The course must be negotiated in 5 minutes or less. Do I have any questions?!"[6]

Even though few of us had comprehended or retained any of what the instructor said, there were no questions. There were never any questions. Questions invited attention, and in basic training all attention was negative. All I understood was that I needed to do 75 percent of whatever I was about to do correctly, or I'd get the crap smoked out of

6. This is how the instructors at Army training commands speak. The wording is unnatural, as with so many things in the Army, but you get used to it. I, of course, was in no way used to it that morning in 2001, and so I had no idea what the instructor was telling us to do. In the interests of recording the experience accurately, I looked up the tasks, conditions, and standards of the bayonet assault course on armymstudyguide.com to ensure I recounted the instructor's words correctly.

me.

The instructor continued. "Demonstrator, *post*!"

"*AHHHHH*" A young soldier in BDUs yelled loudly as he ran from behind the formation to join the instructor. He was wearing 'private first class' rank, which meant he wasn't a sergeant, but it also meant he was a real soldier–one who had already completed basic training–and therefore was infinitely better than any of us.

"Private First Class Jones will now demonstrate how to correctly execute the bayonet assault course. You will follow along to ensure you are also able to successfully execute the bayonet assault course!"

Goelich stepped in. "When Private First Class Jones runs this course, you will follow me as we run alongside the course and watch his execution. Then, you will run the course yourselves, and we'll see whether you learned anything during bayonet training this morning. Ground your gear!"

We all took off our K-pots, LCEs and weapons, and we set them down in the mandatory fashion: K-pot on the ground, facing forward, LCE wrapped around the K-pot with the canteens on either side and the harness draped over the top, weapon balanced on the K-pot to keep the bolt out of the mud.[7]

Once we had finished, Goelich took off at a run, leading us in a line behind him like a string of grunting baby ducks. When we arrived at the first obstacle, a six-foot-high wall made of horizontally-stacked logs, we heard Jones yell from the starting line as he sprang into action. I watched in fascination as he charged toward the wall, yelling the whole time. Powers tapped my elbow, distracting me from the scene.

"Hey Pace, you got any food, man? All this shouting has got me hungry as hell."

I looked at him out of the corner of my eye and murmured my reply, trying to look like I was still paying attention to the demonstration. "Nah man, I got nothing."

"Fuck. man, this is some bullshit." He glanced over at me. "At least we got to take off our K-pots. That thing is uncomfortable as hell, and when I'm wearing it, I look like an idiot."

---

7. The K-pot, or kevlar, was the helmet of the 1990s Army. The LCE, or load bearing equipment, was a belt-and-suspenders rig that had pouches for ammunition and canteens. It, too, was a standard piece of gear for a soldier at that time.

"Yeah. You're right on both counts."[8]

Our conversation hadn't been noticed by the drill sergeants, but, unfortunately, by the time I looked back at the obstacle course, Jones had already made it over the wall. "Shit," I thought, "I didn't see how he climbed it."

Goelich led us off at a run again to the next obstacle, a series of hurdles, and again we watched Jones loudly execute. We continued this pattern, sprinting ahead of Jones, then watching him demonstrate how to execute each obstacle, until we made it all the way around the circular course and back to where we'd begun. Between muttered conversations with Powers, rearranging the crap in my pockets, and trying to take in our surroundings, I only managed to see about half of the demonstrations by the time we made it back to the starting point. As soon as we stopped, Goelich gave us our next instructions.

"Alright, Privates. Fall back in on your gear and drink water. Drill Sergeant Trevino will call you up one squad at a time to execute this course. First Squad, you don't get a break. Follow him, time now. The rest of you better reach your little dick-beaters into your cargo pockets, pull out your Soldier Handbooks, and study. If I catch you bullshitting over here, it won't go well for you."

I was in Third Squad, so I had a few minutes. I picked up my plastic canteen and took a drink of water, then pulled my small white Soldier's Manual out of my pocket and looked at it. The booklet was pocket-sized and bound across the top. Its front cover depicted a fearsome looking Army specialist's frowning visage. He was camouflaged and helmeted, and he looked like the sort of guy who ate bayonet assault courses for breakfast. I thumbed open the book to a random page and started reading.

"1. React to a nuclear attack with no warning (a brilliant flash of light).

a. Close your eyes immediately.

b. Drop to the ground in a prone, head-on position. If in the hatch of an armored vehicle, immediately drop down inside the track.

c. Keep your head and face down and your helmet on.

d. Stay down until the blast wave passes and debris stops falling.

---

8. Because of the Army's one-size-fits-none approach to helmetry, Powers' observation was indeed true. There were only a few helmet sizes available to soldiers, so guys with long skinny heads like me ended up getting extra large helmets and looked like Dark Helmet off *Spaceballs*. On the other hand, people with round, but relatively small heads like Powers ended up with tiny helmets that left them vaguely resembling walking penises.

e. Check for casualties and damaged equipment."

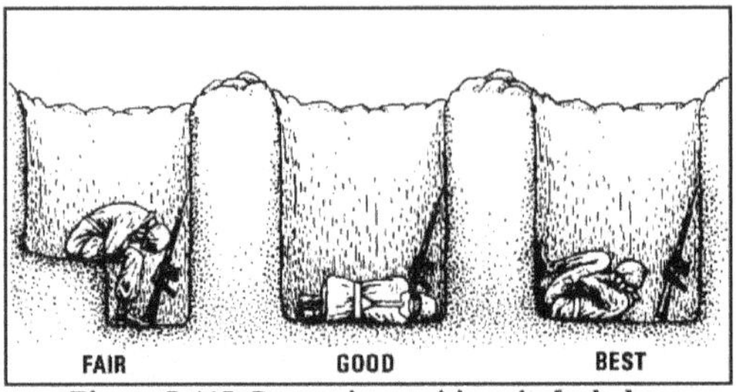

Figure 5-115. Protective positions in foxholes.

The black and white illustration below the text depicted a soldier wearing a chemical suit lying in a hole. He was curled up in the fetal position and lying on his side. Had he survived the attack, or was he dying? The picture was unclear.[9]

"What in the fuck did I get myself into?" I wondered aloud again.

My eyes flicked left and right nervously. Fortunately, none of the drill sergeants had heard my comment. I turned my attention back to the book and flipped through it at random, pausing to inspect the illustrations. There was a bewildering array of Army ranks and insignia, descriptions of the Army's phonetic alphabet, and sketches of how to give hand and arm signals during an infantry patrol. Even two months into basic training, the purpose of most of the material in the book remained utterly lost on me. That didn't mean I didn't need to know it, though. Heaven help the poor sod who was asked a question about something in the Soldier's Manual and didn't provide the answer promptly and with confidence.

I heard voices from my left and looked down the squad to see what was going on. Root, a cheerful storage unit burglar from the rust belt, who'd joined the Army to avoid jail, was arguing about something with Thomas, an urban gang-banger from the Northeast. I couldn't exactly hear what they were saying, but still nudged the guy to my right and quietly muttered. "Tell them to shut up before the drill sergeants hear that shit."

Too late.

"*What the fuck are we having a conversation about, ladies?!*"

Shit. Shit. Shit.

---

9. When we conducted this training later, the drill sergeants added that when we were lying on the ground with our heads down to protect us from the blast wave of the nuclear attack, we should also cup our balls in our hands and kiss our asses goodbye.

Instead of getting to wait our turn to run the bayonet assault course, Third Squad now had to spend that time grunting and sweating in the front-leaning-rest for Drill Sergeant Goelich.[10]

Eventually though, our turn came, and we were ordered to stand up, don our equipment, and jog over to the start of the course. I waited apprehensively until the instructor from earlier turned to me and shouted, "*Go!*"

"YAHHHHH!" I sprinted down the lane in front of me toward the first obstacle, the six-foot log wall. As I arrived, an instructor shouted "*Over the wall!*" I reached the wall at a sprint, set my weapon on top of it, hauled myself over, and grabbed my rifle to continue running. I negotiated the next two obstacles, a set of hurdles and a three-foot-ditch that had to be jumped, hollering all the while.

I turned a corner and saw a sort-of rubber, man-sized dummy in front of me. It had a hinged pole sticking out of the front of it. As I ran toward the target, an instructor shouted "*parry, butt-stroke to the head!*" This morning's training kicked in. I stopped just short of the target, swung my rifle to knock the hinged pole out of my way, then struck the dummy in the crotch with the butt of the rifle.

Take that, commie!

"Did that look like his head, Private?"

I stood confused for a second. What was he upset about? Shit. I'd hit the dummy in the groin instead of the head. What do I do now? Do I take another swing?

"Continue the course, Private!"

I continued running around the course, pausing to slay the targets that appeared before me with bayonet stabs, butt-strokes to the head, or groin-strikes as dictated by the cadre.

For the rest of the three-hundred-yard-long course, I crawled under barbed wire, jumped over fences, and slogged through muddy ditches. By the end, I was covered with sweat, struggling to catch my breath, and my BCGs were fogged to near opacity. But in spite of all the ridiculousness and apparent lunacy of the event, *I felt good.* I felt like I'd just killed the bad guys, saved the princess, and scored the final run. Knock, knock, motherfucking commies. Guess who's here with a bayonet? This guy, that's who.

My moment of triumph was interrupted by Drill Sergeant Trevino.

---

10. The front-leaning-rest is an overly complicated Army word for laying on the ground, arms extended, at the top position of the pushup exercise.

"*Private Pace, why is the sky blue?!*"

I thought, momentarily stumped as my brain tried to sift through the haze of exhaustion. What was the response to this one? Oh yeah, that's right!

"Because God loves the infantry, Drill Sergeant!"[11]

"Correct, Private. Fall in on your gear."

I ran back to my position in the formation and drank water, watching as my battle-buddies finished the course and fell in around me. I'd gotten the answer right. I knew a thing. I'd stabbed fake bad guys and not fucked it up. Score! I still had no idea what any of this meant, but at least I hadn't gotten the shit smoked out of me again. Not yet.

I looked around nervously, wondering which drill sergeant would catch me doing something wrong next.

"*Matlock!*"

"*Here, Drill Sergeant!*"

A pile of letters sailed across the room and scattered to the floor around Matlock's bunk, where he scrambled to pick them up.

"*Nash!*"

"*Here, Drill Sergeant!*"

Another pile of letters. Another scrambling private.

"*Pace*"

"*Here, Drill Sergeant!*"

My pile was small, only two letters, and they fell to the floor in the 'kill zone' in front of my bunk. The kill zone, an area of highly-polished floor in the center of the barracks denoted by a long rectangle of 100-mile-per-hour tape, was off limits to lowly privates. So, without touching the floor, I held onto the end of my metal bunk bed, leaned out

---

11. This is actually a more nuanced answer than I originally thought. On the surface, the required answer to the question "Why is the sky blue?" refers to the similarity between the color of a clear sky and that of the baby-blue color of the Army infantry's standards and uniform insignia. God made the sky blue because he loves his infantrymen. Years later, though, it occurred to me that since there's nothing worse for an infantryman than a miserable, rainy day, the blessing of a clear, blue sky suggests that God *must* love the infantry, because if he didn't, it would be raining.

as far as I could, and pulled the letters in with my socked foot.

Mail! Contact with the outside world!

One letter was official-looking. The return address was marked Defense Finance and Accounting Service. My Leave and Earnings Statement, (LES) I thought. Sweet.

The other was from a friend back home, Alycia.

Even better.

As the drill sergeant continued the nightly mail call, I tore open the first letter and pulled out the folded paper inside. It was my monthly pay stub, and in three long columns it depicted how much I made in March, $1,288.80, how much the government took back, $520.60, and what was left over, $767.70.

That meant I'd been paid $383.50 for the last two weeks of running, rucking, shooting, shouting, and getting smoked.

Holy shit, I was rich! The Army kinda sucked, but it certainly paid well.[12] Especially when that bonus money finally came in, I'd practically be a millionaire.

My pay stub also showed that I had accumulated *five paid days off* in the last two months. We got paid time off? I couldn't believe it. Why didn't everyone join the Army? Sure, I couldn't smoke pot anymore, and I had to keep my hair cut short, but that was a small price to pay for this kind of affluence.

Riding the wave of my misguided economic high, I picked up Alycia's letter. Alycia was my on-again-off-again friend/love interest from back home. We'd known each other since we were eleven, and we still saw each other sometimes at college. She was smart, lovely, and well-read–pretty much the opposite of everyone surrounding me here at Fort Benning–and her letters were a much-appreciated connection with the world back home. I tore open the envelope and unfolded the flower-patterned stationery.

January 23, 2001
Bryan, TX

Dear Daniel,

12. As ludicrous as it sounds, I actually thought this. Assuming twelve hour days, which is on the short end of what we worked in basic, that meant we were earning around $2.80 an hour. Man, oh, man, how one's perspective can change.

I haven't written letters in so long–I'm really out of practice. I'm so used to using the computer, on which I can proofread everything. I'm sorry if sections of my letters don't make sense.[13]

I was excited to receive your third letter this afternoon, and it sounds like you're having quite the experience. It surprises me that boot camp is so much like what we see in movies – I never would have expected that. It sounds quite entertaining. I wish I could be a fly on the wall there. I'm also surprised they gave you guys penicillin shots if they're really trying to treat gonorrhea. The CDC no longer recommends this because of all the penicillin-resistant strains out there now. Maybe they were just bullshitting you guys. Or they could just be really cheap bastards.

If you really want to hear about my daily life, I can comply. It seems pretty boring to me, though, so if it's too much, let me know. For now, I'll assume you're so bored that you want a lot of info. I'm pretty much never home. I worked over 30 hours this week and I'm taking 15 hours of classes. I essentially just shower and sleep here. I get up every day at 6:00 to work in the lab for a few hours before class. After class, I've been working until around 1900 (like my military time?).

I'm not in any way motivated this semester – I'd rather work at the lab than study. I've already failed two quizzes–probably because my books were still in the shrinkwrap. Strangely, I couldn't care less. I think I can get a few C's and still graduate summa cum laude. In fact, I could make a 0.0 this semester and my GPA would be a 3.47. Hence the apathy.

The two classes I'm most interested in now are Molecular Genetics and Recombinant DNA and Biotechnology. So far, they're cool as shit, and I've already learned some cloning techniques I can use at work. They'll be good for me since I'll be doing a lot of research over the next few years in grad school. As far as work goes, my goal is to get a clone before I leave. (You're probably picturing something way sexier than what this actually entails since I'm working with cyanobacteria instead of animals, ha!).

This Saturday I got a call from the University of Wisconsin – Madison, which is my top choice for grad school! The guy who called kept telling me, "Congratulations, this is a very tough program to get into." They want me to come for an interview the first weekend in March, but from what I understand, they don't spend the money to fly you

---

13. During the 2001-2005 period, any time Alycia and I were separated, we wrote to each other frequently. This letter, as with all the other letters that appear in this book, is a compilation of direct quotes from Alycia's letters to me.

up there unless they already want you. I guess they want to lay eyes on the applicants at least once just to make sure we're not assholes. From what I've heard, grad school interviews are more like weekend-long parties they use to try to convince us to pick their school over others we might be considering. Since I have several other such interviews scheduled this spring, my last college semester should be an exciting one!

I guess that's it for now because I have some shit to do before I go out tonight. Take care of yourself - I want you to come back in one piece. Keep your chin up, Private!

Love,
Alycia

As I finished reading the letter, I sat with my eyes closed for a second and thought about how nice it would be to be with Alycia right now. Oh the things we'd do.

"Hey douchebag, what are you smiling about? You get a letter from your boyfriend or something?"

Powers' interruption snapped me out of the moment. I wasn't in College Station, or a bar, or Narnia, or anywhere else pleasant. I was at Fort Benning, Georgia, in the Charlie Company Barracks, and I was surrounded by savages, miscreants, and screwballs. They were my people now, though, and I was one of them.

"Nah, it's from your mom. She's thanking me for the good time we had last fall and asking if she can see me during graduation. You cool with us fucking in the other bed in y'all's hotel room?"

Powers laughed in reply, and I laughed too.

Powers was the closest thing to a friend I had in basic training. His last name's alphabetical position ensured our bunks were next to each other, which meant he was my battle buddy–the Army term for your assigned friend. Your battle buddy did everything with you. If you went somewhere, he went somewhere. If you screwed up, it was his fault for not fixing you, and vice versa. Powers was a good battle buddy, and we got along well. He was from Las Vegas, and he was a smooth-talking hustler who'd finished high school and been short the money he needed for college, so he'd joined the Army for the Montgomery G.I. Bill. Like me, he'd had no association with the Army prior to arriving at Fort Benning. So, also like me, he was frequently caught off guard by the

Army's many idiosyncrasies.[14]

I stuffed Alycia's letter back in its envelope. Mail call was wrapping up, and the drill sergeants were getting ready to give us a class on packing a ruck, so we stowed our letters in our wall lockers–locking them carefully to avoid the penalty for having an unsecured locker–and made our way to the front of the bay to join the rest of our platoon.

We all sat down in a semicircle, watching as Goelich brought a large, green nylon sack with an attached metal frame–a rucksack–and stood in front of us.[15]

"All of you get close so you can see." He held up the ruck. "Today I'm going to teach you how to properly pack a ruck for a combat patrol."

Goelich's demeanor was more relaxed during these evening training sessions. In spite of his often ruthless approach to training our platoon, he was a well-meaning guy who took his job as a drill sergeant seriously. He'd been to war, and he knew how quickly an idiot could get himself killed in combat, so he wanted to impart the necessary knowledge to us to ensure we avoided that fate.

"See, if you ever pass this course, you privates are going to be infantrymen, and someday, God willing, you'll go to war. And when you go to war as infantrymen, you're frequently going to find yourselves in situations where the only things you have in the entire world to solve the problems you're facing are in you and your battle buddy's rucksacks.

"When I was a long-range reconnaissance scout in Desert Storm, my team and I ran patrols deep into the desert. We'd dig hide sites out there ahead of friendly forces, conceal ourselves, and report on enemy locations so our boys could kill them. Sometimes those patrols lasted for days, and during those days, we couldn't leave our holes for anything or we'd be seen and killed, so the only things we had were in our rucksacks. Food. Water. Batteries. Hell, we even had to shit in bags, tape them up, and store those in our rucksacks too. Anyone have a guess why?"

An intense guy who'd flown in from Texas with me raised his hand.

"Yeah, Vaughn. Why?"

"The smell, drill sergeant?"

Goelich nodded. "That's right, Vaughn, the smell. See, if we just shit wherever, then

---

14. After graduation, I never saw Powers again. That happens all the time in the Army.

15. A rucksack is a backpack with an external frame. The term is a German word, *rucksack*, that made its way into Army vernacular.

the Iraqis would have smelled it and come looking for us, and then we'd have been fucked. Even though we packed it up tight, sometimes dogs would still sniff around and find us, but as long as we killed them quietly, that wasn't too big a deal. Most of the time.

"Once, a dog did bark before we killed it, and a guy came out of nowhere to see what the noise was all about, but I caught him from behind and stabbed him through the chest. Damn knife was too long though, and it went right through the skinny son of a bitch and into my leg. We didn't get compromised though.

"Anyway, enough on that. Here's what you need in your ruck and how you pack it so you can find all your shit in the dark. First, you pack the outside pockets with your wet weather jacket, poncho, and wet weather pants. If you don't stuff those in first, you won't be able to fit them in later, once the main bag is full. Next, line the main bag with a wet weather bag to keep your shit dry, then stuff all the crap you'll never use like your dry clothes and extra boots way down at the bottom. Keep the shit neat while you pack it, or you'll run out of space for all the gear you'll need to reach quickly, like ammo or squad gear, and pack it exactly the same way every time, so after a year or two, you can find it all by feel."

Goelich continued for around an hour, carefully showing us how to fit as much stuff as humanly possible in a ruck, and embellishing his instruction with tales of war, wounds, and womanizing. We were enthralled.

By the time he was finished it was late, and most of us were nodding off where we sat. Goelich looked around at his zoning audience and grinned. He stood up and put the ruck away.

"Get in your racks, privates. Don't let me catch you up here bullshitting, or I'll smoke the shit out of you. Turn out the lights and get some sleep so you don't fuck shit up tomorrow the way you fucked it up today."

We scattered across the room to our racks, breaking the carefully folded hospital corners of our green wool blankets to climb in. Goelich turned out the lights and left, and the men of Third Platoon–Turd Platoon, Worst-Platoon-in-the-history-of-the-Army, Shitheads, Worthless Platoon–Charlie Company went to sleep.

As I nodded off, I thought back to what Goelich had said about stabbing an Iraqi through the chest. Where the fuck was I right now? Was this for real? I'd been in basic for two months with one month to go, and the Army still felt like some sort of psychotic version of scout camp. Would we really be going to war? Did all of the crazy shit they made us do here actually prepare us for combat? What the hell was combat anyway? Why would I want to go kill anyone? Would I have to? I guess I was cool with killing people. How big a deal could it be? Everything Goelich had described sounded fun in a sort-of-theoretical but hard-to-picture sense. But none of it seemed very plausible.

"You ready to stab some people to death with our bayonet skills, Powers?"

Powers didn't respond. He was already asleep in the rack above me, probably dreaming about strippers or croupiers or whatever else people from Las Vegas dream about.

Ah well, there was plenty of time to figure everything out later. For now, at least we were getting paid well.

"Alright, privates. Today is the big day, and your families are all here in the stands watching, so don't look like a bunch of fuck-tards."

Charlie Rock was formed up in our company area wearing our Class B uniforms, which consisted of dark green polyester pants, a light green short sleeve shirt, and a dark green hat that was affectionately referred to as a 'cunt cap' because it supposedly resembled female genitalia. Each of our right shoulders was adorned with a light blue cord, a prestigious token of the ordeal we'd just completed that marked us as that most highly honored and frequently shat upon type of soldier–the infantryman.

In true, Army, hurry-up-and-wait fashion, we'd been standing here for three hours while the commander walked through the barracks, the drill sergeants handed out orders, and the admin clerks sorted people's last minute issues. Our belongings were packed into green duffle bags and piled up behind us, the barracks were cleaned and closed out, and morale was high. We'd completed every hurdle the drill sergeants placed in front of us, from the five-mile 'eagle run,' to enduring the teargas chamber, to the twenty-five mile 'bayonet' ruck march, and today was our last day at basic.

All that remained was our graduation ceremony and our drill sergeants' final words of wisdom, which were being dispensed to us now.

"And once the ceremony is over, and you're back with your families and Mary Jane Rotten Crotch and whoever the fuck else came to see you today, don't go doing anything stupid. Remember, you're soldiers now, and that means you need to carry yourselves with some dignity. Respect the uniform, and respect those who've worn it before you."

Trevino added a bit more practical advice. "And don't forget that you're subject to

UCMJ now. Don't think that just because you graduated you can do whatever you want. Your units are tracking you now, and if you do something dumb before you get there, you'll pay for it later. Some of you *might* actually go on to have pretty successful careers, so don't go screwing it up now."

Someone snickered from the formation, and Trevino turned toward him. "I wasn't talking about you, Nash. There sure as hell isn't a successful career in your future; you're lucky to be standing here at all."

We all laughed. Nash wasn't alone. We were probably all lucky to be standing here.

Goelich chimed in again. "Someday, somehow, I might get out of this chickenshit training job and find my way back to a combat unit, and on that day, I might actually bump into one of you again. So help me God, if you're fucked up on that day and anyone else in the unit finds out I'm the one who trained you, I'll kill you myself."

We all laughed, but I noticed Goelich didn't. "Shit," I thought, "Maybe he would kill us, who could say?"[16]

Trevino: "I know some of you feel rich right now, with all that big time Army money burning a hole in your pocket. Shit, some of you have probably never seen that much money before. Don't go spending it all on stupid shit like gold chains, new clothes, or a car from the first dealership you find outside the gate on Victory Drive. Those guys prey on dumbasses like you, and you'll walk out of there with a piece of shit car and a twenty-five percent loan if you're not careful. I know you're not going to listen to me, but you shouldn't buy a car at all. You're moving into the barracks, and you don't need that shit anyway. Save your money, privates."[17]

Goelich continued. "And one last thing. For God's sake, don't get married to Mary Jane Fucking Rotten Crotch from back home while you're on leave and show up to your unit with all of her kids and credit card debt."

Powers and I gave each other the side eye and smirked. What kind of idiot would do that?

"Yeah, I see you dumbasses smirking. All of you are thinking, 'What's he talking about?' and 'What kind of idiot would do that?,' but I guarantee at least one of you idiots is going to get married in the next two weeks and ruin your fucking life. Divorce

---

16. I haven't bumped into him since. I guess he still might.

17. Trevino's advice fell on deaf ears. I didn't buy a new car, but within two weeks, I'd filled a bathtub entirely with beer and got a huge, overpriced, ridiculous tattoo.

ain't cheap, and if you're married more than ten years, she gets half of your retirement forever. Remember that."

Whatever else Drill Sergeant Goelich might have wanted to tell us that day was cut short, as the First Sergeant shouted the order for our platoons to march to the parade field for the graduation ceremony. We did so. And after a few speeches, some marching, and a strained rendition of the Army song, the men of Charlie Rock were released. We'd entered Fort Benning fourteen weeks previously as a ragtag collection of whoevers from across the country. But we were leaving the base today as infantrymen.

2

May, 2001; College Station, Texas

"With this ring, I be wed."[1]

I slid the ring onto Alycia's finger, and we looked into each other's eyes and smiled.

I'd left Fort Benning two days ago, after proposing to Alycia over dinner with a bunch of my classmates at the Longhorn Steakhouse in Columbus, Georgia. It wasn't the most romantic setting, what with the shouting bunch of drunken privates, and the smell of cigarettes and beer mixed with medium-rare beef. As with all late-night drunken escapades, when we woke up the next morning at the La Quinta Inn, we weren't totally sure the conversation counted. But after a bit of hesitant exploration, we each confirmed that the other was still interested, and Alycia accepted my renewed proposal.[2]

Now we were standing together–me in my pickle suit and Alycia in a lovely white wedding dress–under a taxidermied bear in a smoke-filled bar called Duddley's Draw in College Station Texas. We'd spent a good bit of time drinking and playing trivia at Duddley's Draw during college, so it was a natural choice for our wedding venue.

1. I later found out that this should have been, "I thee wed." But despite my misstep, the J.P. signed our marriage certificate. I blame my ignorance on a childhood spent watching *Nightcourt* and *Mr. Ed* reruns instead of doing anything useful.

2. We weren't certain about the proposal, but we *were* sure our friend Root had arrived in the middle of the night and drunkenly asked if he could sleep on our floor, because he was still lying there when we woke up. I'd answered the door drunk and naked–basic training eliminates modesty entirely–we'd given him an extra blanket, and we'd all gone happily back to sleep.

The justice of the peace, a guy who was improbably but appropriately named Truelove, completed the ceremony, and our gathered friends clapped and whooped as Alycia and I kissed to seal the deal. The moment was further punctuated when a druggie burn-out who'd frequented the bar as long as any of us could remember wandered in through the front door, staring at the wedding party for a bit before leaving in confusion.

It was a great day.

Not only had I ignored Trevino's advice and wasted much of the money I'd earned in basic on tattoos and booze, but I'd also ignored Goelich's advice and immediately gotten married. Ah well, what did they know? I had a week of leave before I was scheduled to report to the 10th Mountain Division in Fort Drum, New York, and I was going to live it up.

After the ceremony, Alycia and I grabbed burritos from the restaurant next door where I'd previously worked, then posed for pictures on the porch and headed home to change out of our formal wear before we returned to Duddley's for the reception.[3]

The reception was a drunken blur. Jose Cuervo and Natural Light flowed like water, and all our various old friends from college drank, laughed, shouted, and snapped photos of each other using the disposable cameras Alycia had placed on every table in the bar for that purpose. The grocery store cake was cut, served, and eaten off paper plates. Alycia's brother got so drunk that he inexplicably decided to walk back to his house. He wandered for an hour before he was spotted and recovered at a stop light by the maid of honor and brought home. In short, good times were had by all.

The next few days passed similarly, with me staying in Alycia's apartment and both of us spending the days partying with old friends. The whole week after basic felt just about like being in college again, except that, for once, I wasn't worried about the size of the tab when we left the bar for the night.

When the day finally came for me to fly to Fort Drum, Alycia and I said our goodbyes. She had graduated two weeks prior, and our plan was for me to get to Drum, figure out what the real Army was all about and where we'd live, and then have her join me. We were both excited about moving to New York. Aside from basic training, I'd never lived outside of Texas, and the thought of being stationed so close to New York City seemed

---

3. That photo is still hanging across from the bar at Duddley's Draw today. Next time you're grabbing a chugger of Natural Light, look in the glass case for the picture of the beautiful lady in white standing next to the chump in a green suit with a bad haircut.

pretty appealing.[4]

When I left College Station for the airport, I had a single duffle bag that contained my worldly belongings–mostly uniforms and a few books–a bank account with a few hundred dollars, a new wife, and a four-year obligation to serve in the infantry.

I couldn't wait to see what the Army was like. All of Goelich's combat stories still floated through my head daily, and I was excited to see what being in the *real* infantry was like. Would I have a grenade launcher? Where would I live? Would we go to combat? What sort of people would be in my platoon? Would we ride in helicopters and slide down ropes onto rooftops? Or would we sneak around in the woods and ambush bad guys?

Head swimming with fantasy and speculation, I boarded the airplane to Syracuse, looked out the window to say farewell to my home state, and left the past for an unknown future.

---

4. This was another instance of my phenomenal ignorance. Fort Drum is in no way close to New York City.

3

June, 2001; Fort Drum, New York

"Here's your can of paint. Come get me when you need another one. And don't drip any on the floor, or the First Sergeant will have your ass. You need to get all the doors on this side of the hallway done by lunch."

"Roger, Sergeant."

"Hooah, Sergeant."

"You have any questions?"

"No, Sergeant."

"Negative, Sergeant."

Having finished providing us with instructions, the armorer stalked back to the arms room to repair weapons. He was a stocky older guy who was several inches shorter than me but much heavier, and he wore a perpetual scowl that made it clear what he thought of managing a couple of dipshit privates like Brock and me. After he was safely out of earshot, I turned to Brock.

"This is some pretty hooah shit, huh?"

"Yeah, it's badass. Some good army shit for sure, right here."

The two of us were standing in the Headquarters Company barracks of the 1st Battalion, 87th Infantry Regiment. The barracks was a three story building. Each floor was arranged in an extended 'L' shape, with long white hallways running in either direction from an apex where the CQ desk sat.[1] The hallway had gray doors spaced evenly down its length, and its tile floor gleamed with well-buffed paste wax. Brock and I were currently about half-way down the left leg of the 'L' with a couple of brushes and a can of paint.

Brock had arrived at the unit the same week I had, and his room was down the hall from mine in the barracks. That's where I was staying while the Army sorted out where Alycia and I would live when she arrived. Brock was a cheerful redneck from Florida. He had a shaved head, the sort-of naturally sloppy appearance that drew the ire of sergeants everywhere, and a sense of humor that incorporated frequent impersonations of Mary's brother from *Something About Mary*. We got along well.

Neither of us was very thrilled about painting. But while the rest of the company was down at West Point training cadets for the summer, those of us who were too new to deploy stayed at Drum where we were put to work on area beautification.

I hefted my brush and popped open the paint can. "Fuck it, these doors aren't going to paint themselves. We'd better get to work."

So we did, painting each heavy metal door along the barracks hall with a fresh coat of identical gray paint and trying not to get any paint on our uniforms so we didn't get smoked for looking like dirtbags.

Brock noticed me painting too quickly and gave me an irritated look. "Don't go too fast, Pace, or they'll just give us more shit to do. We gotta stretch this out until lunch at least."

We hadn't been at the unit more than a few weeks, but it was already clear that one of the keys to being a successful private in the infantry was to avoid doing more shit-work than necessary while still looking like you were staying busy. After all, the work involved in making the company area more beautiful was endless, but the duty day was not, and the sergeants in charge of us weren't all that interested in working too hard themselves–unless you provoked one of them with outright defiance.

And so we painted slowly, wielding our brushes with the sort of discernment and

1. The CQ desk, or Charge of Quarters desk, was the spot where the day's designated sergeant sat after duty hours to keep the barracks from devolving into a monkey shit fight. The CQ was usually on duty for twenty-four hours, and his responsibilities were to perform hourly checks on all doors, call the battalion if anything serious transpired, and to discipline unruly privates when they did things like drunkenly try to shit in the company mop bucket.

attention to detail that would have made Michaelangelo proud.[2] I thought back to my Classical Philosophy class, where the professor had lectured us on the Stoics and the secret of happiness. What had he said? Something about the key to happiness being understanding what you could control and what you couldn't?[3] Well that was easy in the Army, because I understood very clearly that I didn't control jack shit. However, I must have misunderstood the point the philosopher was trying to make about happiness, because painting doors wasn't exactly making me glow with delight.

As 1130 neared, footsteps from down the hall caught our attention. We sped up our painting as the armorer approached.

"You two done with those doors yet?"

We both snapped to the position of parade rest, feet spread and hands locked behind our backs. This was the mandatory position for a private to assume when speaking to a sergeant, unless that private wanted his ass smoked.

I responded. "No, Sergeant, but we're getting there."

He eyed the doors we'd finished, then looked at those remaining.

"None too quickly I see. You two bullshitting?"

"No, Sergeant."

He stared at us skeptically.

"Well, it's lunch time. Stuff some food down your gullets, then get back to work."

"Roger, Sergeant. Oh, and my wife arrives this afternoon. We have an appointment to sign for housing at 1300."

The armorer's scowl deepened. "You married for real, Pace, or is this some bullshit arrangement just so you can get out of the barracks?"

"No, Sergeant. We're actually married. She's moving here from Texas."

"Yeah. Well, we'll see about that. Go do your housing bullshit then. Brockmiller, you don't get to sham out of painting though, so get back to work after chow."

"Roger, Sergeant."

The armorer continued to stand there, eyeballing us both.

"And both of you need to shine those boots better and iron your uniforms. You look

---

2. The painter. Not the turtle.

3. "*The chief task in life is simply this: to identify and separate matters so that I can say clearly to myself which are externals not under my control, and which have to do with the choices I actually control.*" –Epictetus, Discourses, 2.5.4–5

like bags of ass."

Brock chimed in. "Roger, Sergeant. Bag of ass, Sergeant."

The armorer's eyes snapped toward Brock. "You think you're funny there, Brockmiller? Why don't you knock out twenty push ups for being funny?"

"Roger, Sergeant."

"And why don't you join him? You're his battle buddy, right?"

"Roger, Sergeant."

Brock and I knocked out the required twenty pushups while the armorer loomed over us. As I pushed, I noticed that his boots were immaculately polished–so well polished that I could see the reflection of my face going up and down in them. How the hell did he get his boots to look like that?

We stood up after completing the pushups and resumed the position of parade rest.

The armorer glowered at us a bit longer, then muttered, "Carry on," and turned to walk back up the hall.

Once he was out of earshot, I turned to Brock and grinned.

"Man, I'm super jealous you get to come back and paint doors this afternoon. All I get to do is meet my wife and go sign for a house, which means I won't get to live here anymore."

"Yeah, fuck you. While you get to go fuck off and get some ass, I'll be painting doors and cleaning random shit for Sergeant Cheerful."

"Good luck, man!"

Alycia was supposed to meet me at noon in the battalion parking lot so we could get some food and head to the housing office to meet the representative who would take us to our new place. I hadn't spoken to her since early that morning when she left her motel in Pennsylvania, so I wasn't entirely sure she'd be on time. Still, standing around in the parking lot would be better than hanging out in the barracks. Privates who looked like they weren't doing anything were liable to be picked up for whatever shit detail a passing sergeant happened to need them for. So it was always best to avoid high-traffic areas.

Brock and I put the lids on the paint cans, pushed them against the wall, and set the brushes on top. We'd been painting in our t-shirts, so we put our uniform tops back on and walked out of the building. We split up on the sidewalk outside. Brock went to the chow hall for lunch, and I went to the parking lot to wait for Alycia.

"And then she said, 'At first I thought you were a stuck up bitch and a slut because you were reading in the front yard and hitting on my man. But then after I talked to you, I realized you were cool.'"

Alycia was recounting her first encounter with Misty, the lady who lived across James Street from us. We'd been in our house for a few weeks now and meeting the neighbors had been proving interesting.

The house was a two bedroom, one bath, vinyl-sided duplex. The brown wall-to-wall carpet smelled clean, the linoleum floor in the kitchen was mostly spotless, the Formica countertops were almost entirely scratch-free, and the walls between our house and the one next door were so thick you almost couldn't smell the cigarette smoke through them. The house didn't have an air conditioner, but the heat worked fine, and it even had its own washer and dryer hook-ups. In short, it was so far beyond the housing Alycia and I had lived in during college that we were blown away by how nice it was.[4]

As nice as the house was, though, we didn't have anything to put in it. As described earlier, I flew to Fort Drum with a single duffle bag of clothes and books. Alycia had arrived with a bit more–a 1997 Nissan Altima packed with two cats, some blankets, and her own wardrobe. But that was it. Before she'd left Texas, a moving truck had arrived to pack the rest of our stuff, but even that was limited to a couch Alycia's mom had given us and some of her childhood furniture, and we had no idea when the truck would arrive. Our house was empty.

Also, I hadn't actually spent *all* the money I'd earned in basic. So, between what was left from that and the few hundred dollars Alycia had in the bank, we set out to equip ourselves for lives as adults. The first stop, naturally, was the liquor store. We picked up a hundred dollars worth of the basics–vodka, gin, Frangelico, Bailey's– then swung by Pier One Imports so we'd have a few glasses to pour it into. Fortunately, Alycia had a bit more sense than I did, so our next stop was to head to Sears to get a washer and dryer. A desktop computer, two lawn chairs, a litter box, and some kitchen stuff rounded out our purchases, and we spent a night on the town in Syracuse to top off the spree.

---

4. A friend from high school and I had shared a non-air conditioned, single-room dorm with a communal restroom. As you might imagine, when August rolled around, the Texas heat made our room smell pretty amazing.

Now we'd spent all the money I'd earned in basic, and, as it turns out, you don't actually get your Army enlistment bonus right when you join the Army. You get most of it after you've been in long enough to prove you're not going to get thrown out, which meant we were pretty much dead broke. Which was why Alycia was recounting her encounter with Misty to me from a lawn chair in the front yard.

"You remember when I saw Tyrone walking Rocky down the street and walked over to say 'hi' to him? That's when she thought I was hitting on her husband. I mean, the guy's walking a 130-pound Rottweiler. I just wanted him and the dog to know I was friendly. I didn't think anyone would interpret it as flirtation."

Alycia continued.

"And then, do you remember when we were sitting out here last week reading in the yard? Misty said she saw us outside and immediately walked over to Brandy's cousin Nadine's house to gossip about how 'the two stuck-up new neighbors were *reading* in the front yard.' As though reading is stuck up!"

I laughed. So far, our basic system of values and social skills seemed to be translating poorly to James Street.

James Street was technically mixed-rank housing, which meant it housed both enlisted soldiers and officers. But in reality, the only officer was one unfortunate captain who lived a few houses up from us with his family. The rest of the street was populated by lower and mid-ranking enlisted folks from all over the country. Open and friendly Southerners lived side-by-side with xenophobic mountain people, deliberately abrasive New Yorkers, and California surfers. The only thing we all had in common was that we were in the Army.

And of all the Army people, Alycia and I were the newest on the street, which meant we didn't understand much of anything, particularly since we'd just spent the last four years with our heads up our asses in college.

"Doesn't anyone else here read?" I wondered aloud.

"I don't know. I imagine they do. And surely adult men and women talk to one another, without trying to sleep with each other, right?"

"No idea. I don't work with any women. I know how the guys at work talk *about* women, and it doesn't suggest they care much about their conversational skills, if you get my drift."

Our conversation was interrupted when the neighbor's door opened and an angelic looking golden-haired toddler wearing nothing but underpants wandered out. Alycia and I were sitting in our lawn chairs, and, as we watched, the little boy walked across his yard, picked up an empty can, and started playing with it at the end of his house's driveway. He was no more than a foot from the street.

"How old do you think that kid is?" I asked.

"I don't know, maybe three or so? His name is Tucker. I heard his mom calling him the other day."

"Oh, do you know her?"

"A little. Her name is Brandy. She doesn't come outside much, but we talked for a few minutes earlier this week, and she said she works nights."

"Maybe she's a nurse or something."

"I dunno, maybe."

The kid continued playing with the can, rolling it back and forth in the driveway.

We watched him play for a few minutes. Neither of us had spent much time living anywhere around young kids, so seeing him outside and playing was novel.[5]

"Should he be there playing alone?"

Alycia shrugged. "I don't know. I don't think so, but who knows? Maybe that's normal here."

As a car drove by, slowing noticeably as it neared the boy playing near the street, his playing there didn't seem normal.

"Is his dad there?"

"No, Brandy said he was 'in the field' for a few weeks. I think that means he's not home."

"Should we do something?"

"I'm not doing shit. You think the neighbors think we're weird and stuck up, now? Just wait until we call CPS on their kid."

"Yeah, good point. And you said he's a sergeant, right? He'd probably smoke the shit out of me if I got into his business."

Having made our decision, we continued reading quietly in the front yard. Alycia was picking her way through an old favorite–*Anne of Avonlea*–and I was reading a new Chuck Palahniuk book, *Invisible Monsters.* The kid continued to play quietly with the can, no adults in sight.

But then our reading was again interrupted, as Misty and Tyrone's door popped

5. We have five now. It's no longer novel.

open. Tyrone and Rocky emerged in a hurry, with Misty's shrill shouts following him until the door slammed shut, cutting off the sound. Tyrone shook his head for a second, muttering something under his breath that we couldn't hear given where we were. Alycia and I waved as he led Rocky down the street past our house.

"Hey Tyrone, how's the day treating you?"

"All good. Not too bad at all. Just taking this guy for a walk. Y'all getting settled okay?"

"We sure are. Just trying to enjoy this beautiful summer day."

Tyrone was a friendly guy from Mississippi, and we'd had nothing but positive interactions. We watched him walk past us, with Rocky pausing to sniff the toddler playing in the neighbor's yard. Tucker was delighted and patted the giant Rottweiler's head. Tyrone led the big dog away, gesturing to Tucker to stay in his yard when the boy tried to follow.

It was evening now, and the light was getting too dim to read outside, so Alycia and I decided to go inside. We looked next door, then at each other.

"Are you *sure* we shouldn't do anything about that? I feel weird leaving that kid alone by the street."

"I have no idea. I really don't want the neighbors to hate us, though."

"Yeah, I know what you mean."

Fortunately, we were saved from having to make a decision when Tucker stood up and toddled back into his house. I'd still seen no signs of parental life, but at least the kid was inside now. I guess that was good enough.

Alycia and I walked back into our house, still not quite sure about the nature of our new environment. We had dinner on the fold-out card table and chairs we'd borrowed from the Army lending closet and went to bed on the pallet of blankets Alycia had put together on the floor of our bedroom.

Alycia snuggled under my arm and laid her head on my chest.

"Well, babe, the neighborhood is weird, and we don't have any stuff, but I feel pretty good about things. Can you believe how nice this house is?"

"Nope, it's amazing. I wasn't sure what the Army would set us up with, but I didn't think it would be anything this awesome."

"I wouldn't mind having a bed though."

"Yeah. We've got the one from my mom in the stuff arriving from Texas. I can't imagine it'll take more than a few weeks. There are a table and a couch in there, too."[6]

We lay there for a bit, and I could clearly hear the neighbors laughing loudly as they watched something on television. The smell of cigarette smoke faintly reached my nose.

"Oh, I forgot to tell you, I finally finished *Black Hawk Down* today. It was horrible. Everyone in the book either got injured or killed."

"Yeah, it looked rough, for sure. Remember those guys in the turrets of the Humvees? The gunners? Rumor has it I'm going to be one of them when the platoon gets back."

That woke Alycia up.

"Are you kidding?! I think every one of them got killed!"

I shrugged in the darkness. "Who knows? It's not like we're at war or anything. I'm sure it'll be fine."

That seemed to calm her down.

"Yeah, you're right. It's not like we're at war or anything."

6. Our stuff actually took 53 days to arrive. Around six weeks after I got to Drum, my platoon sergeant asked how my family was getting settled. When I told him we were fine, but that we didn't have any furniture yet, he blinked in surprise. "Why didn't you say something sooner?" I hadn't said anything sooner because it had never occurred to me that anything was wrong. Was six weeks a long time to go without furniture? Also, if there's one thing the Army had taught me so far, it was that you didn't ask questions.

4

September 11, 2001; Fort Drum, New York

"Shut the fuck up and get to work you two."

"Roger, Sergeant Mac," we replied.

Brockmiller and I strung the thick nylon cables across the motor pool parking lot near our storage bay. Each cable was around 20 feet long, with a nylon loop on one end, and a length of heavy chain on the other. The cables came in a set of four, and the 52 pound set was joined together by a thick metal 'U' that made the whole thing resemble a sort of octopus when stretched out. The cables were called sling legs, and the sling leg set was designed to be hooked up to a Humvee so that the truck could be lifted off the ground by a helicopter and flown to wherever it needed to go.

Years later in Afghanistan, sling legs would prove to be invaluable, but this morning, as the September sun struggled to take the chill out of the New York air, they were just one more heavy, rust-smelling piece of motor pool equipment that the platoon had to lay out and inventory.

Once the legs were stretched out, Brockmiller, Page, Kelsch, Lussi, and I started on the next item, the platoon's snow chains. The fire extinguishers followed, as did the OE-254 radio antennae, the tool sets, the gun mounts, the arctic tents, and all the rest of the platoon's equipment.

Brock and I sweated as we lugged out a 60 pound bag of camouflage netting. Brock loosened the nylon strap so we could empty it. Then, he carefully laid out the various components, while I picked up the technical manual and called off parts while Brock held them up.

"Stake, camouflage, aluminum–24 each."

"Got em."

"Poles, support, woodland–12 each."

"Looks like we're missing one."

I noted the missing component on a piece of paper, stuck it in with the pile of three-foot pole sections, then continued.

"Shape disruptor–6 each."

Today was layout day, and our platoon, the anti-tank, or AT, platoon, was laying out our equipment so it could be cleaned and inventoried. The full Battalion was back from West Point, and we were preparing to deploy on a training exercise. Staff Sergeant Mac was supervising our squad's role in the event, and Sergeant Moore had several of the other guys in the platoon cranking up our Humvees and working through the maintenance checks on them.

Our equipment had to be reviewed by the company executive officer around lunch time, which meant our platoon leaders wanted to review it before then, which meant our squad leaders insisted on reviewing it before that, which meant our team leaders wanted us there inventorying even earlier. All of these 'before thats' meant that we AT worker bees were in the motor pool lugging around equipment before 0800.[1]

As the boom-box we had set up in the storage bay played The Red Hot Chili Peppers' "*Otherside,*" Brock and I continued our work.

"This is some hooah shit, eh?"

"Yeah. Real fucking hooah. Give me that bag, man."

Staff Sergeant Mac, the first squad leader and a notorious hard-ass, overheard our discussion.

"What are you two bitching about now?"

"Nothing, Sergeant. We were just talking about how hooah this layout is. We're really enjoying it."

---

1. This process is known in the Army as the "ten-minutes prior" effect. If a battalion is holding a formation at 0900, the fact that there are five echelons of command between the battalion commander and the lowest privates means the privates will be in formation no later than 0810 (five echelons times ten minutes equals fifty minutes) and most likely even earlier than that. This system may seem ludicrous, but it makes sense when you consider that in the typical Army leader's estimation, a private's time has a value of zero, while the risk that the battalion commander might find the formation to be in a less than perfectly prepared state has an inestimable cost.

"That's good to hear. I'm glad I can make your fucking day a little bit better."

Brock didn't let the moment pass.

"That's wonderful, Sergeant. I had no idea you cared so much about making us happy."

Sergeant Mac's irritation was palpable.

"You know what, why don't you just get the fuck down and start pushing, knuckle-head."

Sergeant Mac turned to me.

"And why don't you join him, Pace. You two bitch enough for a whole goddamn platoon. Maybe some pushups will shut you the fuck up."

Brock and I pushed. Anthony Kiedis sang. Sergeant Mac lectured.

The smoke session continued for a couple of minutes as Brock and I paid the price for never learning to shut our stupid mouths. We sweated our way through pushups, side straddle hops, and lunges until Sergeant Mac was convinced we'd learned our lesson.

As Brock and I brushed ourselves off and got ready to get back to work, we looked around and noticed that a couple of guys had stopped what they were doing and were standing around the radio.

Brock and I walked over to join them. The music had stopped, and the radio broadcaster was saying something about a plane crash.

"What gives? Why'd they kill the tunes?"

"I don't know. I guess a plane hit a bank or trade center or something down in the city."

"Who gives a shit about a plane crash? Planes crash all the time. Why is this one worth stopping the music over?"

Sergeant Mac noticed our lack of activity and chimed in to the conversation.

"Quit lollygagging over there, and get back to work."

We did, wandering away from the radio and continuing our layout and inventory work.

But the music didn't return. Instead, we kept hearing more about the plane crash. Now the announcer was saying something about *another* crash. I didn't know anything about New York City, so the locations the announcer was mentioning meant nothing to me. But he did seem rather excited about the whole situation, which was strange. Brock and I speculated to one another as we continued our task.

"What the fuck is going on with airplanes today?"

"No idea, man. Maybe there was a big pilot party last night, and they're all hungover."

Lussi and Page jumped in on the speculation. They, too, were clearly more focused on the news than the inventory they were doing.

"How hard a job could it be, flying an airplane? Is it really that different than driving a bus?"

"There must be something to it, otherwise they wouldn't get paid so well. I'd trade places with a pilot in a heartbeat."

"When are they going to quit talking about this bullshit and put the music back on?"

We continued our half-assed work and divided attention for a while, but by the time the announcer said a *third* plane had crashed a half-hour later, even Sergeant Mac stopped glowering at us and walked over to the radio. We all took this as tacit approval to stop working and do the same.

Of course, anyone reading this book knows the story of what happened that day. But as hard as it is to imagine, in the 1-87 motor pool, *we really had no idea.* Why would we? There was no television in the motorpool. There wasn't even a land-line, and nobody in the platoon had a cell phone. All we had was a radio.

We also didn't know much of anything about New York City or the National Capital Region. None of us had grown up in either area, so the names and places being mentioned didn't mean anything.

It wasn't until the announcer started describing people leaping from the burning towers; huge, gaping fiery holes in downtown New York; an immediate military lock-down; and the president being notified, that the gravity of what was happening struck us.

And then we heard that fateful word that had been so distant a term to all of us, but was soon to become central to our lives–*terrorism.* The radio made it sound like this might have been an act of terrorism, whatever that meant.

As we jostled each other and speculated about what was going on, a runner arrived from the company headquarters.

"Everyone get back up top. First sergeant says formation in thirty minutes. The battalion commander has some information to put out."

The *battalion* commander? Now? We *never* saw the commander. He was a mythical figure. A rarely-spotted man who wielded almost complete power over our lives. He could grant a four-day pass on a whim. He could take a man's pay for 45 days, restrict him to the barracks, and take his civilian clothing away.[2] To a bunch of privates, the battalion commander was like an occasionally benevolent but sometimes wrathful god–a figure to appease, but someone whom it was generally best to avoid. So, why would he want to talk to us on this random Tuesday morning? Was it related to the stuff on the radio? How could it be?

Sergeant Mac quickly translated the runner's message into guidance.

"You heard him, pack all this shit up. We gotta get up top."

We nodded and hastily undid everything it had taken us all morning to do. But this, at least, was well within our expectations. Army life prepared you well for digging holes and then immediately filling them back up.

When everything was repacked and secured in the storage room, the platoon quickly made its way back to the company. The rest of the company–mostly medics, mortarmen, and scouts –were already milling about the CQ desk and starting to make their way into the battalion formation area, a process that was expedited when the first sergeant walked through with all of the company's platoon sergeants.

"Hurry up, heroes. Get your asses out there to formation, *now*."

There was an energy in the air. None of this was normal. The first sergeant was ordering us to do something directly? That never happened. Even my platoon sergeant, Sergeant Wolf, didn't seem to know what was going on. Someone asked the question we all wanted to ask.

"What's up, First Sergeant?"

His reply sounded strained and irritated.

"You'll find out once your ass is in formation. Move the fuck out smartly, knucklehead."

One of the platoon sergeants in the group reinforced the first sergeant's order.

"Alright, you heard the man. Stop asking stupid questions. It don't matter to you what the fuck is going on, anyway. Your job is to get the fuck outside, time now, before I smoke the shit out of all of you."

None of us needed any more reason than that. The herd pushed its way through the door and made its way out to the battalion area, a large blacktop square. Our company's area was along one of the 50 yard sides, and the other companies, Alpha, Bravo, and Charlie, formed up along two of the other sides, leaving the fourth side open for the

---

2. Joe rumor suggested that in extreme wartime cases, he could even have a guy executed.

battalion command team.

As we arrived outside, the other companies' guidon bearers were already in place, and their NCOs were herding the jostling, griping mob of infantrymen into formation. As military protocol directed, each company's first sergeant joined his company's guidon bearer in front of the company. Soon, the entire battalion was formed up in a more or less organized fashion quietly whispering rumors to one another while we waited for the commander to arrive.

We didn't wait long. A few minutes after we'd settled down, the command team of Lieutenant Colonel LaCamera and Sergeant Major Grippe walked out to the center of the formation area.

Grippe, a huge bear of a soldier with a severe haircut and a booming voice, broke the silence.

"Hooah there, mountain men! Fall in around me and the commander!"

The quad resounded with six hundred shouted "*ahhhhs*" as we did so.[3]

LaCamera nodded to his sergeant major and stepped forward. I'd never seen him up close before. He was around 40, of medium height and build, and with a haircut that would have gotten me smoked by my squad leader but must have been within officer norms.

"Gentlemen, as some of you may be aware, this morning multiple commercial planes hit the World Trade Center in New York City and the Pentagon in Washington D.C. We don't know much yet, but there is some reason to believe this may have been an attack by terrorists."

The crowd around the commander began murmuring as people who hadn't heard the news expressed their disbelief to those around them. LaCamera paused as the murmurs grew louder.

"At ease!" Grippe's voice ended the murmuring immediately. LaCamera continued.[4] "As you might imagine, the situation in New York City isn't good. We don't know exactly what role the Army will play in helping out with the response, but I suspect the National Guard will be activated, and there's the potential we might be called to help as well. I know some of you have families in the city, and I can appreciate how difficult this is. But right now, I need you to be patient while we sort through what happened

---

3. This was the expected response when falling in around the commander.

4. When used this way, 'at ease" is the formal Army way of saying 'shut up.'

and determine what's next.

"For now, you are all restricted to the battalion area until further notice. The base commander has closed the gates, and no one is authorized to enter or leave. Additionally, I'd ask that you exercise discretion when you talk to family members on the phone. There are going to be rumors flying everywhere, and I don't want us to make things worse. Do I have any questions?"

One guy raised his hand.

LaCamera raised an eyebrow and nodded to him. "Go ahead, soldier."

"Sir, are we at war?"

LaCamera paused for a minute and seemed to collect his thoughts.

"I don't know, son. I don't think anyone knows yet."

He looked around to see if anyone else had questions. Nobody did.

LaCamera nodded and said, "Okay, then. Fall back into your company formations."

We did so at the run, leaving LaCamera and Grippe standing alone in the center. Once we were again formed up, LaCamera turned the formation over to the company commanders, who turned the formations over to the first sergeants, who turned us over to our platoon sergeants, who turned us over to our squad leaders, who released us from formation and told us to get back to the company area. There, Sergeant Mac directed us to 'stand fast' in the barracks hallway by the CQ desk while he and the rest of the NCOs went to a meeting.

The soldiers of AT platoon–and pretty much everyone else below the rank of E-5–congregated around the CQ desk and waited. The rumor mill started immediately.

A mortarman piped up, "You heard what the commander said. We're going to NYC to help clean shit up."

Someone responded, "That's not what he said at all. He said the National Guard was doing that shit. Cleanup work is for fucking POGs."

A Brooklyn native chimed in, "The fuck it is. I'd be all over getting back home for some cleanup work. Bitches would be throwing themselves at us."

That response polled well with the group, and there were many enthusiastic volunteers for 'clean up and get bitches' duty.

A medic interrupted the cheerful speculation. "What did he say about terrorism though? What the fuck is that? Like airplane hijackers and shit?"

One barracks lawyer replied. "Nah, he doesn't know what he's talking about. Terrorists don't crash planes. Why would they do that? They land them out of the country and get ransom money. My uncle was a cop, and he told me all about that shit."

Our speculation was interrupted when one of the mortar NCOs walked into the area.

"At ease!" someone shouted.[5]

Silence fell as we all quickly stood and put our hands behind our backs.

The mortarman looked around the room, clearly irritated. "We're trying to have a goddamn meeting, but we can't hear each other because all of you Sallies are out here having a jaw-jacking session. So, First Sergeant sent me out here to put your asses to work. Now, you spec-fours out here take charge of this shit and go clean up your common areas. If I gotta come back out here again, it's going to be ugly."

He walked away shouting, "Carry on!" as he did so.

We grumbled, shuffled, then broke up into our individual platoons to get to work.

Brock and I grabbed mops and walked to our end of the hall to get to work.

I swished my mop in the dirty cold water and stuck the head in the yellow plastic wringer to squeeze it dry.

"Man, you heard all of that shit. What do you think's going on?"

"I'm from fucking Florida, and I don't know shit about New York City except that it's not anywhere near here and the beer is expensive as hell. What I *do* know is that we're locked down in this barracks building, which means until they let us go, we're going to be doing every dumb ass thing any NCO in the company can think of."

I nodded. That was definitely true. What else was there to do but continue mopping? The whole east coast could burn to the ground, but at the moment, all that mattered to me and Brock was that we were locked down, and I was sure nobody was in a hurry to tell us what the hell was happening.

"Grab a couple of sets of DCUs, make sure they fit, and then cut the old patches off and pack them up."

The last three weeks had flown by. After September 11th, the entire tone of the

5. When used in *this* manner, 'at ease' means 'everyone stand up, get to parade rest, and shut your mouths." 'At ease' is always called when an NCO who outranks everyone else in the room walks in. Sergeants call it for staff sergeants, platoon sergeants call it for the first sergeant, and everyone calls it for the sergeant major. Woe to the private who forgets to call 'at ease' when an NCO enters the room.

battalion had changed. Since I'd joined the unit, we'd been almost exclusively engaged in meaningless-feeling activity. All of a sudden, instead of boot-and-haircut inspections, private games like 'go find me a box of grid squares,' and drill and ceremony practice, we were doing new and different things.[6]

For instance, one day I'd been ordered to report to the motorpool where a guy in khakis and sunglasses had shown a cross section of privates from across the battalion how to run a satellite radio. In about fifteen minutes, he'd shown us how to unfold a sat com antenna, set it to the correct azimuth and angle of elevation to connect to the nearest satellite, and configure the radio to communicate globally. At the end of his demonstration, he asked if we had any questions. But once again, the 'never ask questions' reinforcement training we'd received throughout our short careers ensured that none of us did. I imagine he walked away from the class confident that he'd imparted critical knowledge on us even though I *know* we walked away with about as much ability to employ a satellite radio as your average chimpanzee.[7]

Another day, we were bused across post to a warehouse where long rows of folding tables were set up along the walls. As we filed around the room from table to table, we were first handed a new never-before-seen type of rucksack, then new cold weather gear, new thermal underwear, new hats, new t-shirts, new socks, new fleece jackets, and, finally, new armor. The last item was hailed as being particularly incredible because none of us had ever before been issued armor of any sort. This meant we had to rework all our gear to fit around it, which caused no end of consternation for the NCO corps, who had to ensure that whatever that rework looked like would pass the sergeant major's inspection.

As for today, we'd arrived for PT formation and found the battalion area covered with square metal aircraft pallets loaded with piles of tan clothing, which turned out to be hundreds of used, desert-pattern camouflage uniforms.

As we walked around the area, guys from across the battalion were holding up shirts and pants to inspect them for rips or stains and taking off the uniforms they were wearing

---

6. A grid square is a one kilometer by one kilometer square drawn onto a map. They don't exist as physical objects, and thus, there are no boxes of them. Similar army snipe hunts include looking for chemlight batteries, Bravo-Alpha-1100-N-S's, Tango, Romeo - double-three's, or a PRC-E-7 (pronounced Prick-ee-seven) that works. All of these, of course, are designed to make the searcher look like an idiot.

7. We were very good at inventorying all of the pieces of the radio set though. Checklists and inventories were our strong suits.

to try on the new ones. Some were the old, chocolate-chip pattern from Desert Storm, while others were in the newer Desert Camouflage Uniform, or DCU, pattern. Most of them still had the rank and name patches of their previous owners still sewn on.

Brock and I got to work finding our own sets.

"Hey Brock, check this one out. It's got an 82nd combat patch on it."

"That must be an extra high-speed uniform then. Maybe if you wear it some of that shit will rub off on you."

"Ha. Not fucking likely. I wonder if I can find one that's already got my name on it, though. It'd save me a couple of bucks at the sew shop."

Pace is a pretty common name, but after five minutes of digging, I couldn't find it on any of the uniforms. I *did* find a bunch of pants that still smelled like the unwashed ass of whichever dude had last worn them.

"Where the hell have these things been for the last decade?"

"Probably in that same warehouse with the Ark of the Covenant from *Raider of the Lost Ark.*"

From across the quad we heard, "Hey you two! Hurry the fuck up! You ain't picking prom dresses. Grab a couple of tops and a couple of bottoms and move the fuck out."

Our efforts to cherry pick the uniforms had caught the eye of one of the battalion's supply sergeants.

"Roger, Sergeant."

I bent down and grabbed a newer DCU top and bottom, held them up to ensure they at least kind of fit, and dug for another ad hoc set. Brock did the same, and we walked back to the CQ desk where the rest of the platoon was assembling.

Guys were comparing the quality of their 'new' uniforms. Some guys had done well–Page looked like he had a brand new matching set. Other guys like Watkins hadn't been as lucky. One of his uniforms looked like it had been used as a shop towel in a former life. Win or lose, though, there wasn't anything we could do about our choices, so we stuffed them into our new rucks with all of the other stuff we'd gotten over the past few days.

There was a lot of speculation about why we were getting so much new gear. We were pretty sure something was going on, but nobody was sure exactly what. Since September 11th, we'd packed our trucks and taken them to the airfield twice, then driven them back and unpacked them both times. At no point in this process had anybody told us where we might be headed or how long we'd be gone. But they'd definitely told us what to bring.

In Rucksack:

BDU top, summer or winter, 2 EA;

BDU bottom, summer or winter, 2 EA;

Underwear, optional, 6 EA;

Poncho, 1 EA;

Canteen, 2-Quart, 1 EA;

Canteen cover, 2-Quart, 1EA;

E-tool, 1 EA;

E-tool cover, 1 EA;

Etc, 1 EA;

The list was six typed pages long, and left no room for interpretation about what we should wear, which pocket a given item should be in, or what color it should be. All of our gear was to fit into a couple of duffle bags, a rucksack, and on our bodies. When we staggered to and from the airfield carrying it, we looked like the squires who trailed along after King Arthur in *Monty Python and the Holy Grail.*

On one of our staggers to the motorpool to load the trucks, the rumor mill was in particularly high gear. Lussi was sure he knew what was going on. He was a gregarious guy from Wyoming who had been in the platoon a few months longer than me.

"So, I heard from one of the scouts that we're heading to the Sinai for a security mission. He said that's where we were scheduled to go before the towers got hit, and that it wasn't possible to change a unit's orders once they're cut."

Kelsch, a fiery guy who was ready for anything, responded. "Fuck that Sinai bullshit. We're going to war. They wouldn't do all this back-and-forth-to-the-airfield bullshit for some stupid Sinai rotation."

Page, an older and more cynical guy who'd been in the platoon for a record-breaking two years, chimed in, "Man, I've been here a long time, and that's all bullshit. This is all just some drill cooked up to demonstrate readiness or something. We aren't going anywhere."

Sergeant Mac settled the debate, "All of you shut the fuck up and keep walking."

We did. In part because of Sergeant Mac's irritation, but also because even when you're in pretty good cardiovascular shape, walking with 150 pounds of shit on your back limits the amount of breath you have for idle speculation.

Once we got to the motorpool, we loaded our Humvees and drove them to the airfield to join the battalion. Our trucks still had taped 'x's' on the windows and mirrors where the Airforce loadmasters had prepped them to be loaded onto the aircraft. We spent the day sitting on our gear, eating MREs, and speculating. But that evening we once again got the word to go back to battalion. Another day wasted on nothing.

On the plus side, I could at least go home for what was left of the evening. The lockdown had been over for a week. So when we got back to the battalion and were released by Sergeant Mac (after we'd cleaned our common areas, of course), I tried to call Alycia from the CQ desk to get a ride home. She didn't answer, so I bummed a ride home with one of the other off-post guys.

When I'd left the house that morning, she'd thought I was leaving for up to eighteen months. So I was sure she'd be happy to see me.

"Oh my gosh, I'm so tired of saying good-bye to you. Are you ever actually leaving?"

It had been after nine o'clock when I'd gotten home from battalion, and Alycia still wasn't home, so I took a shower and lay down to wait for her. When she walked in, I shouted, "*Surprise*!"

She whirled on me, badly startled by my shout. "Why the hell do you always do that?!"

I grinned, "Bad genes, I guess?"

She lashed out and smacked me on the shoulder. "It's not funny, damnit. We've said good-bye twice this week, and every time we do, I have to accept that you're probably leaving for a year to some hellhole *where you might die*. It's really stressful, and I'm sick of it."

Her rage passed, and she wrapped her arms around herself and started to tear up. I walked toward her and pulled her into a hug.

"I'm sorry, babes. I'm an idiot. I know this shit is stressful. It's driving me nuts too."

"Why won't they just tell us what's going on? I can't keep doing this. Do you know anything new?"

"Nope. We got told to be in at 0600 tomorrow in BDUs. I guess we'll find out then."

"How do people live like this? I love that you're back, but I hate seeing you go and having to say good-bye every time. It's turning me into a mess."

She was crying actively now, so I held her quietly for a minute or so, smoothing her hair as she sobbed through the stress.

"Yeah, me too. Me too."

Once she stopped crying, I gave her a tissue and a moment to wipe her nose and dab her dripping mascara.

"Well, if I wasn't a mess before, I definitely am now."

"Nah, you look great. Best thing I've seen all day, I assure you!"

"Ha. Flatterer."

"Just being honest. You've seen the guys I work with. Where were you this evening, anyway? I was surprised you were gone."

"Christi and I got together to have a drink and commiserate. She's not having a great time with all the chaos. It's driving her nuts not knowing if Jay [one of the NCOs in my platoon] is leaving or not."

"That makes about all of us, I think."

"Why won't they tell us what the fuck is going on?"

"OPSEC is the word that's been floating around the platoon. Nobody will tell anyone shit because apparently what we're doing is a secret. I'm not sure it's supposed to be a secret from the people doing it, though."

"It's crazy making."

"No kidding, but I guess if we don't know what's going on, there's no chance anyone else will either."

We finished exchanging updates on the day, made our way to bed, and curled against one another to find a bit of comfort in the chaos. Nobody knew anything, but there were rumors about everything.

Looking back on that time period with 2025 eyes, the chaos all makes sense, of course. Throughout September 2001, plans were being made and courses of action compared at the highest levels. Even after determining that Osama bin Laden was behind the attacks, the U.S. was uncertain about how to handle the situation. Diplomatically, the U.S. was trying to convince the Taliban to turn over bin Laden, and, militarily, everything was being considered, from a full invasion of Afghanistan to a small, C.I.A.-led operation.

When the Taliban turned down the U.S. demands on September 21st, the decision was made for a C.I.A-and-Special-Forces-led invasion. But even then it took a good deal of time for that guidance to trickle down through military channels to the ground pounders at the bottom. The uncertain push and pull we felt at the 1-87 was the same tension that was felt across the Army as the Big Green Machine ground its way inexorably into action.

I woke up the next morning to the alarm clock's harsh chime and rolled over to turn it off. The red numbers read "5:20." I stayed where I was, lying quietly in the dark and turning to where Alycia was next to me. I could tell by her breathing that she wasn't asleep.

"You awake, babe? I gotta go."

"Yeah. I'm awake. I know."

I rolled over and spooned her. We didn't say anything. What was there to say? We'd said it all the last two times. So now we just lay together in misery and wondered if I'd be home again this evening or if we'd ever see each other again.

"It's that time."

I climbed out of bed, got dressed with the clothes I'd laid out the night before, and scratched our cat, Mickey, on the head. He followed me to the front door, blinking as I turned on the entryway light to lace up my boots. I stood up to leave and saw that Alycia had gotten out of bed and was standing in the living room looking at me.

Her face was wet with tears as she walked to give me a final hug and kiss.

"I love you. Come home."

"I love you, too. I will."

5

October, 2001; Camp K-2, Uzbekistan

Our long game of hurry-up-and-wait had finally ended when we left Fort Drum on October 6th. Our C-130 flight lasted three days, during which time we flew from Delaware to Spain, Italy and Turkey before arriving, ready to go to war, in Central Asia. For those unfamiliar with the experience of flying in a military cargo plane, this had been a remarkably long time to spend in one.

We were seated in long rows along the sides of the plane. The seats–called jump seats–were hollow aluminum folding chairs that were bolted onto the sides of the fuselage. Our trucks were parked in the middle of the plane, and our gear was piled in and around the human

and vehicular cargo in an attempt to maximize our comfort without pissing off the air crew too much. The cargo section of the plane wasn't climate controlled or insulated, so the flight was exactly as comfortable as you'd expect sitting on a metal folding chair that was stuck in a dark, cold, noisy tube and taking you to an unknown destination would be.

During our layover in Moron, Spain we'd watched news coverage of the Air Force dropping bombs on Kabul, and although nobody had confirmed it, by the end of the trip we were pretty sure we were part of Uncle Sam's ass-kicking force being sent to hand-deliver the Taliban some payback for September 11th.[1] We were going to war! When the ramp of the C-130 finally dropped at our final destination, we all lowered our

1. Moron is pronounced more-own, not more-on, by the way, as I later learned.

NVGs and got ready to charge off the aircraft into the darkness and meet whatever the Taliban had to throw at us.

Except that there weren't any Taliban there.

Instead, there was an Air Force guy wearing a desert uniform and ear muffs. Over the whine of the C-130's engines, he exchanged shouted words with Sergeant Mac that the rest of us couldn't hear, and we were waved off the aircraft and onto a long patch of runway.

We unloaded our gear and trucks, still confused about where we were, and drove across the airfield to an area where a bulldozer was pushing dirt into a huge, fifteen foot tall wall that stretched into the distance. As we left the high-pitched whine and bright lights of the aircraft-laden airfield, we passed into sporadic darkness. There were stadium-style lights around the area, but the coverage was spotty, so enormous shadows made it difficult to determine whether we could see better with or without our NVGs. We drove for a few minutes, led by a gator driven by a guy in a hard hat, parked our trucks in a line at the base of the wall, and killed the diesel engines.

As our initial confusion wore off, we could see that the base was actually buzzing with frantic activity. Here, a bulldozer and its chemlight-wearing ground crew worked to continue building the berm wall. There, a team of loud support platoon guys were backing huge LMTVs into place and unloading trailers. Across the way, a pile of chemlights marked a spot where a group of field medics were digging a slit trench.[2]

Our status as bystanders didn't last long, and we were quickly tasked with taking our trucks back onto the airfield to serve as fixed machine gun positions around the perimeter of the newly established base.

I was the gunner in our truck, and as the night progressed, the Ma Deuce and I kept watch across the dusty nighttime airfield. It was cold, but since I still had adrenaline pumping through my system from suddenly arriving in a warzone, I didn't notice it much. Brock, the driver in our truck, cranked the vehicle every half hour to let the engine charge the Humvee's battery; the truck's noise joined the occasional roar of Air Force

2. Once it was done, the medics would throw lye into the bottom, and the camp's toilets (at least for the first few nights) would be complete. A slit trench can only handle so much urine, though, and these initial field toilets quickly turned into small lakes of piss. Once full, the medics encircled them with engineer tape to keep people out of the area until the urine soaked into the ground. For the most part this was successful, although we did hear a blood-curdling shriek a few nights into our time at K-2 that was later revealed to have come from an unfortunate female lieutenant who had fallen into one of the trenches.

birds landing or CH-47 helicopters taking off to fly south. Between these events, except for the occasional radio traffic, our portion of the airfield was completely quiet.

"All Killer elements, this is Killer Seven, over."

Sergeant Mac keyed the handmike of the PRC 119 radio that was clipped to his chinstrap with an old MRE spoon.[3]

"This is Killer One, over."

Other trucks chimed in.

"Killer One-Alpha, over."

"Killer Two, over."

"Killer Two-Alpha, over."

"Roger. Send LACE reports, over."

The LACE–or liquid, ammo, casualties, equipment, report–was the infantry's standard update to higher headquarters, designed to convey maximum information with minimal words spoken. This kept the radio network clear. Since only one person could talk on the net at a time, it easily clogged up if folks decided to wax poetic or kept the microphone keyed while they thought about what to say.[4]

Sergeant Mac, being Killer One, was the first to report.

"Killer Seven, this is Killer One. Green, green, green, green. Over."

"Killer One, this is Killer Seven, roger. I copy all green, over."

The rest of the platoon passed their information as well. All the trucks were green. Sergeant Wolf continued, "All Killer Elements, be advised, HHC had one casualty tonight. One support NCO was caught between an LMTV and a trailer while ground guiding the vehicle. His pelvis is crushed, and he'll be MEDEVAC'd immediately, break."[5]

Brock and I looked at each other. Holy shit, that sounded painful.

"All soldiers are advised to use extreme caution while ground guiding. How copy, over?"

---

3. For whatever reason, the clip that the radio handmike came with to attach it to things was useless, so the approved infantry solution was to replace that clip with a plastic MRE spoon and some tape.

4. A habit that would invariably invite a rebuke from the senior NCO on the net. "It's push to talk, not push to think, asshole. Exercise better radio discipline.."

5. Years later, after talking to the special operations medics who were on the ground at K2, I concluded this poor support guy was actually the first casualty of the GWOT (he lived).

"Killer One copies."

"One-alpha copies."

"Two copies."

"Two-Alpha copies."

"Nothing follows, Killer Seven out."

The world lapsed back into silence.

After a few minutes, Brock broke it by thumping his can of dip to pack it properly before taking a pinch to stick in his lower lip.

"Damn. One guy down already. That sucks. Does he get a CIB [Combat Infantryman's Badge] for getting crushed by a truck?"

"Fuck no, he doesn't."

"Man, that's gotta be the shortest wartime deployment in history."

We quietly pondered this for a minute. Five hours into a war and the guy gets his pelvis crushed. What a way to go.

Brock continued, "Hey Sergeant, what country are we in, anyway?"

"Dunno, Brock. Afghanistan maybe?"

"You think so? Where is Afghanistan, anyway?"

"I'm not sure, really. I think it's in the Middle East. You know, Pace? You're the college guy."

"I don't know either, Sergeant. I know Alexander the Great tried to conquer it, and I know he was from Macedonia, which is near Greece, but I don't know how close the two places are."

Brock and Sergeant Mac thought about that.

"So are Afghans Arabs or something? Or Greeks? Bin Laden looked like an Arab."

"Maybe? I don't really know. What I saw on the news about Osama bin Laden said he was an Arab, but all I saw about the Taliban is that they're a bunch of crazy fundamentalists who beat women, rape boys, and blow up old statues."

"Huh. A bunch of pedophiles. Figures."

"Yeah, I still don't get what any of this has to do with September 11th."

"What I saw said the Taliban are hiding bin Laden from us."

"Well that's fucking stupid, isn't it? Now they're gonna get fucked up quick."

"Yeah, it looks that way. Ah well, fuck em. Sounds like they're a bunch of cave-dwelling douchebags anyway. World'll be better off without them. Pass me an MRE, and don't give me that four fingers of death bullshit."

The four fingers of death were the "frankfurters, beef" MRE. MREs, or meals ready to eat, were the standard Army ration in 2001. As far as food served in plastic envelopes goes, they were alright, but with only twenty-four types available, variety was limited, particularly when MREs were the only food available. The beef frankfurter MRE was one of the least popular in our platoon. But, because every other box of MREs contained one of that meal, and because we didn't get another box until everything in the first one was finished, someone in the truck had to eat the frankfurters every few days.

"That's all we got, Sarge. Pace rat-fucked the last good one last night."

"Fucking Pace. Do pushups."

"Roger, Sarge."

I climbed out of the turret and knocked out my pushups on the roof of the Humvee. A bit of smoke was a small price to pay for avoiding the four fingers. When they could be heated, the greasy Vienna sausages were bad enough. But when they had to be eaten ice cold, they were coated with congealed gel and particularly disgusting–especially the tenth or eleventh time in a month.

Sergeant Mac flipped on his red lens flashlight, sliced open his meal, and started to eat.[6]

Even under the faint red glow, the meal looked gross.

"What the fuck is that grease anyway, Sarge?"

"Fucking spermicidal lube. How the fuck should I know?"

"I bet that shit causes cancer."[7]

"I hope so. If I get cancer, I won't have to sit in this truck with you two goddamn idiots anymore. If the smell is any indicator, this all-MRE diet we're on isn't doing anything good for your digestion."

As if on cue, Brock ripped off a long, wet-sounding fart.

Sergeant Mac recoiled instantly, slopping a bit of hot dog jelly down the front of his armored vest.

---

6. The red lens flashlight was the standard light source used by the infantry when training or operating in combat. At night, a white light can be seen from shockingly far away, so the infantry used red to avoid creating a visible signature that would compromise our location. Even red light was avoided when possible, but as we were already sitting in a Humvee on an airfield, nobody worried about that.

7. I have no reason to think the Beef Frankfurter MRE causes cancer, although, as it turns out, K-2 did. According to one Army study, folks who served there have a 500% increased chance of developing certain cancers. Be all that you can be.

"Goddamnit, Brock, I'm *eating*. Take that shit outside, you nasty fuck."

Brock only grinned in response. We were already sitting in a truck on the perimeter of nowhere. What else could our squad leader really do to him?

"On a related note, Sergeant, when do you think we'll get to do laundry? I may have had an accident."

"Shut up, Brock. Scan your fucking sector."

Our vigil continued for the rest of the night.

As the sun came up, the landscape around us came into view, and we got our first look at what would later be known as Camp K-2, short for Karshi-Khanabad Air Base, the site in Uzbekistan from which most of the early combat forces flowed into Afghanistan.

For such an important site, it wasn't much to look at. For starters, everything was brown. The dirt was brown, the asphalt airstrip was dusty enough to be brown, the air was a hazy brown, and even the few buildings we could see across the far side of the airfield looked brown. The only things I could see that weren't brown were the things we'd brought with us.

Our Humvees were Panama-green, the Air Force planes were steely gray, and the Hesco barriers that guys were erecting were the same grayish-white color as the walls in a Department of Motor Vehicles office.[8]

This was my first experience with Hesco barriers, the modular bricks the U.S. military would use over the next twenty years to build bases across the Middle East. Eventually, they became such old hat that I didn't notice them anymore, but in 2001 I found them fascinating. The Hesco barrier is a cubic cloth sack inside a collapsible metal cage. They come stacked flat on aircraft pallets, but when unloaded and unfolded, they become a cube that can be filled with dirt to create a durable barrier. The cubes I saw were usually around five-feet to a side, although, like Legos, Hesco barriers come in a variety of shapes and sizes.

The reason they fascinated me is that Hescos were one of the first things I encountered in the military that seemed designed to save privates from doing manual labor. Almost everything else, from human chains to police calls, seemed premised on privates doing as much work as possible in the least efficient manner possible. In the world of military fortification, nothing epitomized this better than the sandbag.

In the Army, filling a sandbag is generally a three-to-four man job. The first guy holds

8. Eventually, of course, all of these things became brown too. Nothing escaped the dust at K-2 for long.

the bag open and ties it closed when it's full, a second guy shovels dirt into the bag, and another guy or two (usually specialists) offer advice or criticize those doing the actual work while smoking cigarettes. Once the bag is finished, it then either gets hand-carried into position by yet another guy, or passed down a human chain to wherever it needs to go. Depending on the proximity of the nearest NCO, filling and positioning a sandbag can take between two and fifteen minutes. This means that, since a fighting position can require hundreds of bags, sandbag fortifications are time and labor-intensive.

The Hesco embodied the opposite approach. As I watched, one or two guys unfolded the Hescos and positioned them in rows. A third guy used a front end loader to fill the Hesco with dirt. The whole process took about twenty minutes per block. The only other laborer I saw was a guy with a dump truck who dropped off dirt piles for the dozer guy to dig into. To someone who'd spent months doing manual labor in the infantry, the efficiency of this process looked magical.

I eventually tore myself away from watching the engineers and scanned my sector, looking out across the barrel of my machine gun.

Nope, there was still nothing out there.

The terrain was completely flat, so even in the brown, dusty air, if something was coming, I'd see it long before it could do anything bad to us. I took a few minutes to mark down the few terrain features on my notepad that I could use as range references if I needed to shoot at anything. Then I got bored and decided to use my pencil and notepad to work out how much private labor the Hesco team was saving.

After a few minutes of scribbling, I heard, "What the fuck are you doing up there, Pace?" Apparently, Sergeant Mac heard the pencil scratching.

"Math, sergeant."

"What kind of math?"

"The kind of math with numbers. You know, like the shit we used to watch on *The Electric Company* when we were kids. "

"Fucking smart ass. What did you figure out?"

"Well, since a Hesco takes the place of about 150 sandbags, by my calculation, that means a Hesco team of 4 guys fills the equivalent of 450 sandbags in an hour. Four unsupervised privates, on the other hand, fill about 4 bags an hour. That meant that for every hour worked, the Hesco team saves a team of 4 privates 112.5 hours of work."

Sergeant Mac snorted. "Yeah, and what do you think those 4 privates will do with those 112 hours, Professor?"

Brock cut in. "We'd probably smoke a bunch of cigarettes and jerk off. We might even accidentally blow something up or set it on fire."

"Exactly, Brock. That's why we keep you motherfuckers busy. 'Cause you'd be as fucked up as Hogan's Goat if we don't."[9]

I thought about Sergeant Mac's response. I wanted to argue with him, but honestly, he was probably right. There are few things more hazardous to good order and discipline than a bunch of underworked, bored privates. Still, I couldn't let him have the last word.

"Maybe you should write a book on that, Sergeant. You could make a fortune selling the business world on cutting edge Army leadership principles like, 'keep motherfuckers busy so they don't get in trouble'."

Brock snorted with laughter, and Sergeant Mac bristled. "You know, I've had enough of your shit, Pace, why don't you..."

His response was cut short by the rumble of a diesel engine as a Second Squad truck roared up next to us. Kelsch was on the gun, and he gave me a chin-lift-what's-up-greeting. Sergeant Mac and the Second Squad leader talked for a few minutes while I tore off the sheet of paper from my notebook with the terrain features and ranges and handed it to Kelsch. When the NCO's were done, Sergeant Mac got into the truck and directed Brock to park the truck next to a section of the growing berm. The three of us unloaded our rucks and duffles and walked a few hundred meters to a green, GP medium tent.

We pushed our way through the tent's entry flap and dropped our gear on three unoccupied cots. Sergeant Mac talked to our platoon sergeant for a minute, then walked back over to where Brock and I were reorganizing our gear. He was grinning broadly.

I looked at him warily as I stood up to get at parade rest. "What's up, Sergeant?"

"Looks like we'll see how good your math was, Pace."

"What do you mean, Sergeant?"

"We need to get a double row of sandbags put up around this tent to protect it from indirect fire. You and Brock get to work."

Even I had to grin at that one. Fuck. The bastard had gotten me. I guess reorganizing my life would have to wait. "Roger, Sergeant. Thanks for that."

---

9. Hogan's Goat is an anachronistic expression that survived in the Army long after it died out everywhere else. According to wordorigins.com, Hogan's Goat first appeared in a 19th century comic strip called *Hogan's Alley*, where the goat regularly butted people and screwed things up. From there, the phrase passed into common usage as a term for someone who butts into other people's business. In Army parlance, the expression is used for someone who is screwed up.

"Don't thank me, Pace. Thank your fucking recruiter."

Brock and I walked outside, where a pile of empty bags and a shovel awaited us. I picked up a shovel, Brock grabbed a bag, and we got to work. At fifteen minutes a bag, this was going to take a while.

October 10, 2001
Calcium, NY

Dearest Daniel,

I'm doing alright, but I miss you terribly, and I'm worried. I wish I knew where you are and what you're doing. The stress is making food turn to sawdust in my mouth. I couldn't manage to eat lunch yesterday, but I made myself eat salad and a pork chop for dinner (along with three amaretto sours!).

So, you know how you said goodbye to me at home and took the car to work the morning you deployed? When one of our neighbors took me to get the car a few days ago, the guard they had stationed at the parking lot – some private with a rifle – wouldn't let me. I volunteered to show him that the title of the car was in my name, but as usual, there is no room for listening to reason in the army. When Misty found out about this, she was pissed, so she called the staff duty desk, pretending to be a 1-87 spouse, and bitched them out. As a result, I went up there (past several armed guards) and was escorted to my car. If that is what a person has to do to solve a simple problem, I'll be in trouble the whole time you're in the army. I don't think I have it in me.

I want to keep busy right now so I can't dwell on my empty house and bed. As long as I exhaust myself and go to bed late, nights aren't too terrible. It's the mornings that I hate – I have to realize afresh every day that you are gone and will be for some time. It's awful to think that I might not see you again until I'm 24 years old! The FRG told us it could be April 2003 before you come home – 18 months! That's a terrible thing to contemplate, but I keep thinking there's absolutely no way they would deploy soldiers for over a year!

I ironed your BDUs yesterday for no good reason. I just missed you, and that task made you seem closer somehow. I'm leaving your PTs on the guest bed (the ones you wore last) until you come home to pick them up. I'm keeping everything so that it looks like you're still here – toothbrush, washcloth, etc.

I love you so much that I ache for you, sweetheart. I'm going to buy picture frames and put photos of you all over the house so I can see you everywhere I go.

Yours always,
Alycia

P.S. I forgot to tell you that Brandy somehow got Trey to put on a thong the other day, pushed him outside, and locked the door. Then she called several neighbors to look out their windows at him. What a bitch!

"Name, Sir?"

"Lieutenant Colonel Jones, John."

After adjusting my slung rifle to more easily thumb through the stained, weathered pages on the clipboard, I looked for the guy's name on the typed list of names. I found it on page three.

"Lieutenant Colonel John Jones. Yep. I see you right here, Sir. You're cleared to go in."

"Thanks, bud."

The guy clapped me on the shoulder and walked past me through the makeshift plywood door attached to the building I was guarding. The door, which was rigged with a piece of 550 cord and a water bottle hooked over a nail to automatically pull it closed, slammed shut with a vibrating thud.

It was 2:30 in the morning and cold. Cold and dusty. The dry air and omnipresent dust, which everyone called moon dust, had given me a nasty sinus infection and laced my snot and spit with a grayish substance that was constantly running out of my nose.[10] I'd been on guard for four hours. The shift had started on the heels of our twelve hour airfield guard shift, and, once this one was over, I could head to the tent to get some sleep before starting our next airfield shift six hours later.

I'd been at war for about three weeks, and so far, all it consisted of was eating MREs

10. And other pleasant substances too, no doubt, if the congressional commission reports about K-2 being contaminated with uranium and old chemical weapons are to be believed. As the old cadence goes, "It's what the Army did for me; Be all that you can be."

and guarding things.

At this particular guard post, it was at least pretty clear who I was guarding the place against–folks who weren't allowed inside the building. When someone walked up, I checked his name to see if he was on the list. If so, he got to enter. If not, I called some bearded guy in khaki pants with sunglasses on his head to come out and check whether or not the guy could enter. So far, in the week I'd been pulling this shift, everyone had been allowed in.

For the first few hours of the first shift, I'd taken the job pretty seriously and thought about how I'd detain someone who wasn't on the list, watching him carefully in case he was a spy or something. But, by this point, I'd gotten pretty lackadaisical about the whole affair.

Our airfield guard shifts didn't help, since I wasn't really even clear on who I was guarding the things from. For that matter, neither were my NCOs. We didn't have any descriptive information about the sorts of bad guys who might approach, and we weren't really sure what to do if they did, other than the standard infantry procedures we'd learned back in basic training.[11]

This ambiguity led to a few weird moments, like when a civilian helicopter had circled the airfield before coming in low and slow to touch down less than a hundred yards from our position. I'd been in the turret when it initially approached our position.

"Hey Sergeant Mac, is this helicopter authorized to land here?"

"I haven't heard shit about civilian helicopters from anyone."

I unlocked the turret and spun it left until the Ma Deuce was pointed at the landing helicopter, which was now directly facing us. One of the two pilots, who I could see clearly through the glass of the cockpit, began waving his arms and shaking his head frantically.

I shouted over the propwash. "Hey Sergeant, he's waving at me. I don't think he wants me to shoot him."

Sergeant Mac yelled back. "Fuck him. Wait a second. I'll call it in."

---

11. According to my basic training handbook, this procedure was to "Challenge a person by saying, 'Halt! Who goes there?' Once a person answers, if he's on your access list, say, 'Advance to be recognized.' If you're challenging a group, you should allow them to advance only one at a time. If you're not satisfied with a person's identification, you must detain him and call the individual in charge of the watch. When two or more individuals approach from different directions at the same time, challenge each in turn and require each to halt until told to proceed."

The pilot and I continued staring at each another. He was clearly talking to someone on the radio, no doubt trying to let someone in charge know that there was a guy pointing a machine gun at him, while I was staring through the sights of the M2, awaiting approval to finally blast the shit out of something during this war.

The standoff continued for about a minute until Sergeant Mac slapped my pant leg and shouted, “He’s good. Don’t shoot him.”

I nodded, spun the turret back to the right, and offered the pilot a thumbs up. He returned it, grinning and shaking his head, before hopping out of the helicopter and running off toward the base, leaving his copilot to power down the bird.[12]

Other than that single moment of excitement, our endless guard shifts seemed largely meaningless.

“What a waste of fucking life.”

“What’s that, bud?”

Shit. While I’d been daydreaming a guy had walked up to my checkpoint and was standing in front of me. He wore a black fleece coat and khaki pants with unlaced desert boots, and he had a full, vibrantly red beard.

“Nothing. Name, Sir?”

“Redbeard”

That was a new one. I thumbed through the paperwork, checking both the "R’s" and the "B’s."

“I can’t find you on the list, Sir.”

“Oh, they added me yesterday. Check the bottom.”

I flipped to the last page. Sure enough, there at the bottom of the list, someone had written “Redbeard” in big red letters with what looked like a crayon. I thought for a moment about calling SOG, then decided it was pointless. We were on a brand new base in the middle of nowhere. Who was here except for people who were either guarding the building or supposed to be allowed into it?

“Umm. I guess you can go in.”

Redbeard grinned and said, “Thanks, pal” as he walked past me.

The plywood door closed with a slam.

---

12. I never found out who the pilot was or why he was there. To this day, I’d like to grab a beer with him and talk about the weirdness of that moment when I almost machine gunned him out of existence because nobody passed down the information that he was landing. That would have been a hell of an investigation.

Hopefully he was allowed to be in there.

What the hell was going on in there, anyway?

Years later, when I was in the Special Operations community myself and knew some of the guys who'd been serving in the Special Forces in 2001, I found out the building housed one of the command and control centers managing the unconventional warfare campaign that was raging in Afghanistan. Throughout the invasion and for several years after, K-2 served as one of the main hubs for force projection into theater. The 10th Mountain Division had been brought in to secure the airfield during the invasion and to respond if any of the radical Islamic elements in Uzbekistan attacked the base. Our role wasn't glorious, but, in its way, it was important.

That night, though, I had none of this context. I was cold, bored, sick, and irritated at what I felt was a waste of time. So, I did what privates in such situations do best–I bitched to myself.

"Man, fuck this place. Fuck not having a shower for the last three weeks. Fuck stupid MREs for every meal. Fuck this bullshit moondust. Fuck my runny nose. Fuck eighteen hours of guard every day. Fuck Sergeant Mac. Fuck this stupid clipboard. Fuck Red Beard. Fuck smelly ass piss tubes.[13] Fuck the Taliban. Fuck Afghanistan. Fuck Brock's smelly ass. Fuck Uzbekistan. Fuck the cold. Fuck the dark. Fuck these uncomfortable ass NVGs, and fuck my stupid fucking helmet."

I continued bitching for a while, snorting moon-dusty snot back into my head, and wallowing in a bit of solitary, night-time self-pity until I saw someone approaching.

It was Brock, and he was here to relieve me. Awesome.

"What's up, Pace? Having a hooah fucking night?"

"You know it, man. Just one more badass night in the war."

"Yeah, you gotta embrace the suck, you know?"

"Yeah, sure. What's up back at the tent?"

"Fuck if I know. I crawled back into my fart sack for some sleep after guard, and I haven't done shit since. Honestly, I don't think I've ever seen the goddamn tent in the daytime, so I don't even know what it looks like in there."

"Ha. True. Here and the airfield. That's our war."

---

13. By this point, the slit trenches had been replaced with piss tubes. The piss tube is a piece of PVC pipe driven into the ground at a 45 degree angle that you urinate into. From the tube the urine runs down into the ground, where it eventually produces a sort of muddy area behind the tubes drenched in urine, or as we called it 'pickled piss.' It smells just as lovely as you'd imagine.

"Yeah. Fuck it. What can you do?"

"Not much, man. Not much at all."

We nodded together in silence for a minute. Brock thumped his can of dip and pulled out a wad to put in his lower lip. I yawned widely.

"I'm out of here, man. Gotta get some sleep so we can do all this awesomeness again tomorrow."

"Yep. Later. You are hereby properly relieved."

I passed the clipboard to Brock and walked off toward the platoon tent, squishing and sliding through the soft moon dust. I felt better already, and by the time I reached the tent, I was cheerfully thinking of what I'd read in the letter Alycia had sent me before I zipped myself into my sleeping bag for a few hours of sleep. There was nothing quite so motivating as getting relieved from a late night guard shift, even knowing that I'd be right back on the same shift in a few hours.

The life of a wartime private is like that–full of a thousand fleeting emotional highs and lows. Late night guard shifts, with their endless hours of quiet discomfort, epic battles against the Z Monster, crushing boredom, and occasional bursts of life-or-death excitement, try the soul.

From a big picture point of view, guard duty is important, and on some level every guy on night guard recognizes this. The fate of his friends and the mission everyone is working together to accomplish is entirely in his hands, and history is riddled with stories about catastrophes that have occurred when guards were lax or fell asleep.

Despite all of this, the night guard still often struggles to see past the misery of the moment. Cold, wet, interminable nights can lead a person down dark paths that are hard to conjure in more comfortable times. Particularly since the long nights leave plenty of time for introspection. Bad news from back home, fights with friends or coworkers, or even small aches and pains all get magnified during lonely hours of standing and staring into the darkness.

What did she mean by that last sentence in yesterday's letter? Is she cheating on me? Why does my foot *still* hurt? Is it broken? Will I have to get out of the Army? What would I do then? Where would I live? Would I be able to get a job? Has it really only been seven minutes since I last checked my watch? It feels like it's been at least a half hour. What's that movement out there? Is it an enemy? Is it a fucking ghost? Are ghosts even real? Why the fuck did I decide to read *The Shining* last night? When is this goddamn cold going to go away? Man, fuck this place.

Night guard can feel long and hopeless.

Still, for every long dark guard shift of the soul, there's its opposite, and the happiness that results from the simplest things. For instance, on a bad day, opening an MRE and finding it contains M&Ms instead of Charms can be the event that changes a guy's whole outlook.[14] Similarly, a break in the clouds that allows a ray of sunshine to take the edge off a wet, drizzly morning can take a soldier from rock bottom to just fine in seconds.

On this particular night, my simple moment of happiness was getting relieved by Brock so that I could take a piss in a tube, crawl into a dark canvas tent with twenty other farting, smelly infantrymen, read a letter by flashlight, and get five to six hours of sleep on a cot.

When I reached the tent, I pulled open the flap and looked around. It was dark and musty, and everyone else was asleep. Picking my way between the cots and duffle bags, I sat down on my cot, pulled off my boots, and dug the letter out of the shoebox of stuff Alycia had sent me. I tore it open and started to read.

October 27, 2001
Calcium, NY

Dear Daniel,

Whenever anything newsworthy happens in Afghanistan, my phone rings off the hook because various relatives, friends, and acquaintances can't wait to be the first to tell me. I'm not too worried about your safety for the moment, but I'm quite concerned about your overall well-being. I wonder what you are doing right now (hopefully sleeping) and what you are feeling and thinking. I dread the idea of what you might have to face if you do enter Afghanistan. I hate so much the thought of you in combat – I can't help worrying about that possibility.

I went to bed semi-early the night before last, but Misty woke me up at midnight. She was calling to tell me she and Tyrone had "had it out," and she wanted me to take her son, Jordan, to school the next day. I woke up early to get ready, and I waited over

14. According to truetreatscandy.com, Charms are a hard candy that's been in military rations since 1931. The trouble is, Charms are cursed to bring rain down upon those who consume them, so nobody does.

half an hour before she called to tell me that Tyrone was coming home to take Jordan to school after all! I was pretty pissed, but because they've been so wonderful to me, I let it slide without comment.

Case in point: Misty brought me some homemade dinner last night, and Tyrone brought me a chocolate milkshake from Arby's this evening. They're both being really nice and checking up on me regularly. Oddly, Tyrone is very kind to both me and Jordan. He's only an ass to Misty, and only to her occasionally. Bizarrely enough, they get along swimmingly most of the time. Misty's always talking about how sweet he can be, and he's certainly shown me nothing but kindness. I never even feel like he's trying to hit on me. Misty tells me that he thinks I'm the only decent person in the neighborhood and that he feels protective.

I wish Brandy and Trey would stop parking their truck in front of our mailboxes. The mail lady won't deliver to us on days when the boxes are blocked, and the other day, I had to run her down on foot. I'm glad I did, because I got several letters from you.

Speaking of Brandy, she is apparently more than just a stripper. She openly admitted to giving the owner of a local dealership a blowjob as the down payment on her vehicle when he stopped by the club where she strips, and she has a string of guys she calls customers coming and going on evenings when Trey isn't home. This isn't hearsay. She talked about all this when I was hanging out with her and Misty outside last week. Sometimes I feel like I've entered another dimension because it's all so unreal.

Misty just called to invite me over to play cards with her and Tyrone. I guess I'll go in a little while – I don't have anything else to do. As chaotic as they are in many ways, I'm so grateful for their friendship. I'm over there past midnight, playing cards and talking several nights a week. I'd be lonely without them!

Forever yours,
Alycia

P.S. Apparently, my brother's visit last week started the rumor mill churning. All the neighbors except for Misty and Tyrone assumed he was my boyfriend. I'm continuously shocked by the grotesque assumptions people around here make, but then again, I'm also continuously shocked by the grotesque behavior so many of them shamelessly admit to! I haven't seen such avid gossip culture since junior high, and that was never so salacious. This is a whole different world, babe.

As I rested on my shovel for a moment, I watched the lieutenant impale another of the enormous local Uzbek bees on a tiny toothpick-sized stick. He had a whole row of them set up along the edge of the fighting position I was digging, and the bees twitched in the morning sun like a row of tiny captives in a movie about the Zulus.

"What the fuck," I thought, "is that weirdo doing?"

My verbalized word choice was more moderated. He was the platoon leader, after all, which meant he wielded the full punitive authority of the Army. "Sir, what are you doing with those bees?"

He stuck his tongue out at the latest bee, then looked up in response to my voice.

"I'm bored, Pace, so I'm impaling these bees."

"You don't think that's kinda weird, Sir?"

He thought for a moment before replying, "No. They're bees. Who cares?"

"PETA, I imagine, Sir."

He thought again. "PETA isn't here, Pace. It's just me, you, and the bees."

"And me," Brock chimed in. "And Sergeant Mac."

Sergeant Mac, who was dozing in the passenger seat of the truck, pulled the rim of his watch cap up off his eyes and offered his input.

"Fuck the bees."

I shrugged. "Fair enough. Fuck the bees, I guess. It still seems super weird to me."

Sergeant Mac again. "You know what seems weird to me? The fact that I don't hear the sound of a shovel hitting dirt. Keep digging."

"Roger, Sergeant."

Then Brock. "Yeah, Pace. Dig faster you shamming bastard."

That irritated Sergeant Mac. "Shut the fuck up, Brock. Nobody asked you. Do pushups."

"Can't, Sarge. I'm on security, and we're at war. The Taliban might sneak up on us while I'm getting smoked."

Sergeant Mac grudgingly accepted Brock's counterargument, pulled his cap back down to block the light, and resumed dozing.

Brock grinned down at me from the turret with the smugness that only a person who's standing around doing nothing but watching someone else do manual labor can express.

I nodded in reply and got back to work. I was digging a fighting position for our Humvee. It was mid-November now, and our squad had been moved from the airfield to a new position on the south side of the base, called, appropriately, South Watch. Unlike the airfield, South Watch overlooked occupied farmland. So, instead of staring at nothing all day, now we watched Uzbek farmers trudging back and forth along the road across our field of vision. On the whole, it was quite an improvement.

The new eye candy came at a cost, though. First, while we were on the six day-long rotation at South Watch, we couldn't leave the post. So, we lost access to what few amenities K-2 provided, and instead ate, slept, and shit in our thirty-foot square security position. Second, instead of being in the center of a secured area, South Watch was actually on the perimeter of the base. Even worse, it was on a patch of dirt. And in the army of the nineties, dirt meant digging.

This habit was essentially sound. Since improvements made to firearms during the Civil War and the hellish trench battles of World War I, digging in had become a staple of troops in defensive positions because dirt is great at absorbing bullets and explosions.

Normally, however, vehicle fighting positions are dug using engineering equipment. Being light infantrymen, though, we had no such equipment. So instead we dug ours by hand. And, as it turns out, the size of the hole necessary to conceal a Humvee is pretty large–especially when the truck crew only has a single shovel and pick to work with.

For five straight days, Brock and I worked. One of us dug, while the other pulled security from the turret. [15] As we sweated, the LT impaled bees, and Sergeant Mac dozed, the hole grew, and our fighting position began to take shape. Eventually, it started to look something like this:

15. It went without saying that neither Sergeant Mac nor the lieutenant, who came out to the position with us in order to spend some facetime with 'the men,' dug in any meaningful way.

Taken from Field Manual 3-21.71: The Mechanized Infantry Platoon and Squad (Bradley)

The only differences between our position and the one depicted in the field manual were that ours had a truck in it instead of a Bradley, and there were no trees in Uzbekistan.

On Day Five, after a full day of test-fitting the truck and adjustments, Brock leaned on his shovel and decreed the fighting position ready for final inspection. The LT, standing up from what had become a full-fledged, bee-themed reenactment of the last scene of *Spartacus* with Kirk Douglas, walked around the position with an appraising eye. After a few minutes of assessing the position's effectiveness from a variety of angles, he concurred with Brock's assessment.

"Looks good, gents. Now, get the camo nets set up."

From the turret, I blinked in surprise. The camo nets? Our truck did have an issued camouflage net–the same one Brock and I had inventoried a dozen times–but the idea we'd set it up out here seemed ludicrous.

"Sir, the camo net is forest green."

"Yeah, so?"

"If we set it up, our position will look like a ten-foot tall bubble of green in the brownest, flattest place on Earth."

The LT thought about this for a moment. "That's true. But the net is also radar scattering. Doctrine says all fighting positions need to integrate camouflage netting whenever possible."

"Sir, in the instructions, they say that's to protect the position from aircraft or something. I'm pretty sure in this situation, the netting will make us *more* visible, not less."

The LT thought again, scratching his chin.

"I hear what you're saying, but I'm going with doctrine. Let's get the net set up."

I nodded, and Brock shook his head, pulling up the tail of his shirt to wipe the sweat off his face as the LT returned to his bees.

I grinned down at Brock from the turret. "Get back to work you shamming bastard. We won't be safe from Taliban aircraft until you get that net up. I'll pull security for you."

Brock smiled as he started to unstrap the netting from the tailgate of the Humvee. Turnabout was fair play.

Setting up the net took another hour, with Brock doing most of the work and Sergeant Mac and I helping string the pieces of netting into place. We finished around dusk, and I was just getting ready to tag out with Brock on the turret when we felt the first drops of rain.

"It rains here?"

"I guess so."

"Huh. And here I thought it was a desert."

"Ah well, I can't imagine it'll rain for long."

It did though. It rained all night and well into the next morning. Fortunately, the camo net kept a bit of the rain off our heads. What it didn't do, however, was keep it out of the fighting position. By morning, the hole we'd spent the last five days digging was almost knee deep in water, and was in danger of spilling over the lower lip of the Humvee's door and flooding the truck.

Sergeant Mac considered the situation and made a recommendation.

"Hey Sir, we need to back the truck out of the fighting position or it's going to get stuck."

Before the LT could respond, I added my two cents.

"But Sergeant, then the Taliban will be able to see us."

Sergeant Mac glowered at me. "Shut up, Pace."

"On the other hand, I've read that standing around in fighting positions full of water can lead to trench foot, so there's that. Trench foot or Taliban. That's quite a conundrum."

"*Shut the fuck up*, Pace."

The LT weighed in. "Sergeant Mac is right. Let's get the truck out of there."

Brock climbed into the truck, fired it up, and after a bit of slipping and sliding, successfully maneuvered it out of the fighting position. Unfortunately, though, the camo net wasn't big enough to accommodate the truck when it was outside the hole, so we had to forego the net and park ten feet away in a new position with a slightly worse view of the sector we were supposed to scan.

The rain, no longer deflected by the camo net, poured freely onto my head and into the truck through the turret-hole. I looked at the netting. Then I looked at the water-filled slop-hole our fighting position had become.

"You're right, Sir. This is much better. Mission accomplished."

"Pushups, Pace. Go."

I climbed out of the turret, placed my hands on the wet hood of the truck, and pushed. As I did so, water ran down the back of my helmet, soaking my T-shirt and dripping down my arms. At the moment, pushups were a good thing. It was cold and miserable, so the exercise didn't just warm me up, but helped me bite back the sarcastic bitching that was running through my head.

Grumbling to myself, I pushed until I was too tired to push any more. Sweaty and dripping, I climbed back into the turret, brushed my muddy hands off on my muddy pants, and stared back into the wet farmland in front of me.

Was this really war? Was the Army really this dumb? What the fuck was I doing here? We'd been 'at war' for a month and a half, and, so far, I'd done nothing but stare into the distance and do shitwork. Was there a point to any of this? Helicopters were flying in and out of the base daily, so surely someone was doing something, but why weren't we?

Part 2

The Evilest Place on Earth

Went to the store

Where all the kiddies shop

Pulled out my machete

And I began to chop

'Cause Napalm sticks to kids.

Burns the little bastards and it never quits.

White, black, yellow and red.
Napalm burns them until they're dead.

Army running cadence

6

January, 2003; Fort Drum, New York

Hey there, reader. I'll bet when you read "Fort Drum, New York," you had a moment of confusion. "Wait a minute," you probably thought. "We were just in Uzbekistan. Shouldn't we be talking about Afghanistan? This *is* a book about the 10th Mountain in Afghanistan, right?"

Before you ditch the book in fury and zip off to leave a one-star review on Amazon, let me explain, or, to paraphrase Inigo Montoya from *The Princess Bride,* "No. There is too much. Let me sum up."

The 1-87 did end up heading to Afghanistan in the winter of 2001/2002. But I didn't go with them. Instead, I got MEDEVAC'd home Thanksgiving week after breaking my ankle in the least heroic way possible–doing PT.[1]

I was embarrassed, and I felt like an idiot. My only consolation that winter was that I was home with Alycia and that I didn't miss much action anyway. When the battalion finally deployed forward into Afghanistan to secure Bagram Airfield in February, they mostly just pulled security there, too. Even when 200 guys from the battalion deployed to the *Shah-i-Khot* Valley in March for OPERATION ANACONDA, one of the largest set-piece battles in the entire GWOT, AT Platoon, whose trucks and heavy weapons were ill-suited to the helicopter insertions used in the operation, didn't really

1. Alycia was never formally notified that I was coming home. She actually heard about it from her cashier at the Fort Drum Commissary, who was our first sergeant's wife and recognized her from a company function.

participate.[2]

When the battalion arrived home that spring, there was some initial buzz as the first war veterans in a decade returned to the U.S. But after a few months, the platoon settled back into its normal routine, and before 2002 was finished, there were already rumors floating around the battalion that we were about to head back to Afghanistan.

This time we'd be going back with a new crew, though. Our platoon sergeant rotated out. We were told we'd get a replacement down range. Our platoon leader was gone too, replaced by a new LT. Sergeant Mac and most of the other sergeants left as well, replaced by Sergeant Stearns and the newly-promoted Sergeant Lussi. Even my battle buddy, Brock, left. His term of service was up, and he had no desire to stick around the Army any longer.

The rest of the platoon was filled with a bunch of new guys: Hogu, Sobon, The Hawk, Dickey, Raimond, Nevin, Horn, Williams, and Mikkelsen, and new NCOs like Staff Sergeant Collazo. Some, like Horn, Nevin, Dickey, Mikkelsen, and Sobon, were transfers from other units or parts of the battalion. The rest were brand new, cherry-ass buck privates, and as rumors of our deployment flew about, the new battalion leadership worked to get the newly remanned organization into shape.

What sort of shape, though? As a retired officer, now looking back on the time between our 2001 and 2003 rotations, I often think about that and what I would have done if I'd been in the new Battalion Commander–Lieutenant Colonel Mike Howard's–shoes.[3]

As I touched on earlier, in 2001 the Army still very much trained to face a Cold War enemy. Our light infantry tactics were focused on defeating peer-level forces, and our equipment–from radar scattering camo nets, to unarmored Humvees, to anti-tank missiles with thermal sights–reflected this focus. Although the opponent hadn't been a Soviet rifle company during OPERATION ANACONDA, the Battalion had been directed to establish blocking positions for a standard hammer-and-anvil sort of operation, which seemed to validate the playbook of infantry tactical tasks. However, would that playbook continue to remain viable?

---

2. I believe only Kelsch, one of our feistiest guys, managed to finagle a spot on the operation. For more on OPERATION ANACONDA, I recommend Sean Naylor's book *Not a Good Day to Die.*

3. Let's also not forget that while the unit had just returned from a deployment, none of the new leadership had. In 2002, recent combat experience was a rare commodity in the Army.

The answer to this question was ultimately, no, and the GWOT eventually forced the entire Army to develop a whole new set of skills. But in 2002, that change hadn't yet occurred. Consequently, we trained on the tasks infantry units had trained on for the past twenty years: clearing trenches, maneuvering through forests, and fixing and flanking enemy rifle squads.

This continued through 2002, and, in January 2003, after having fought Soviet Hind helicopters and armored personnel carriers during a rotation at the Joint Readiness Training Center in Fort Polk, Louisiana, the Battalion prepared to conduct Expert Infantryman's Badge, or EIB, training prior to heading back to war.[4]

The 28-degree-below-zero New York wind howled past my face as I fumbled with the spent shell casings to pick them up so we could fire again. I was miserably cold and the heavy mittens made any sort of dextrous work impossible.

"What the fuck is wrong with these things?"

The range safety NCO, equally miserable, chided me irritably. "Hurry the fuck up, Private. I want to get off this range sometime before my dick freezes off."

I yanked off the heavy mitten, letting it dangle from the string that kept it attached to my parka so I could use my gloved fingers to get at the brass a bit more effectively. But I still couldn't pick it up.[5] After fumbling for a few seconds, I figured out the problem. From my fighting position, I yelled up at the NCO so he could hear me over the wind.

"Hey Sergeant, when the hot brass hit the ice, it melted it for a second, but it all refroze before I stopped shooting, and now it's frozen into the ground. It's not coming up without an ice pick."

The NCO–his face featureless behind snow goggles, a balaklava, a neck gaiter, and a scarf–looked at me for a moment, then keyed his radio to confer with the range tower.

---

4. I was killed five separate times on that JRTC rotation, including once from a lacerated lung thanks to the aforementioned Hind. Each death meant a trip to the Louisiana National Guard's personnel holding area while I waited for the battalion S1 to complete the paperwork that would allow me to respawn. Like pawns in a chess game, privates died a lot at JRTC.

5. Our issued mittens were called 'snotties' because by the end of a road march or field problem, their fur-lined backs had invariably been used for so much nose wiping that they were covered in frozen snot.

In a moment he replied.

"Yeah, that's happening up and down the line. Leave the brass and prepare for the next iteration."

I nodded and turned away from him to face down range. I put my mitten back on, massaging my fingers to get some feeling back into them, and waited for the tower to tell us to start shooting again.

It was January, and we were working through the rifle qualification portion of EIB training. The trouble was, it was Fort Drum in the winter, so the weather wasn't cooperating, and the training was consequently going poorly. For one thing, it was cold. Deadly cold. With the temperature a good way below zero, and the wind chill in the negative twenties, the cold not only made it tough for people to function well; it started to impact machinery, too. Radio and rifle sight batteries died, trucks wouldn't start, and the mechanical pop-up targets on the range didn't function properly. The snow was also a factor.

Fort Drum sits squarely in the lake-effect snowbelt, an area of the Great Lakes region that gets a notoriously large amount of snow. Many years it gets more than 90 inches of the stuff, and this year was no exception. There was more than four feet of snow on the ground while we were conducting EIB, and while shooting rifles in the snow isn't as bad as some other tasks–land navigation for example–the weight of the snow made targets malfunction, and the constantly blowing flakes obscured the more distant targets enough that hitting them required timing our shots between wind gusts so that we could actually see what we were shooting at.

Around the time that I had gotten enough circulation in my shooting hand to use my fingers again, the range control tower's speakers came to life.

"Firers, at this time assume a good prone, unsupported firing position."

I climbed out of the hole and laid down in the snow next to it.

"Lock and load one ten-round magazine."

I picked up my magazine, jammed it into the magazine well of my rifle, and pushed the bolt release to let the bolt slam forward.

"Ready on the right? The right side is ready. Ready on the left? The left side is ready. Firers, at this time switch your selector levers from safe to semi and watch your lanes."

I stuck my face down on the butt stock of my M4, scooching my cheek forward until the tip of my nose pressed against the charging handle. The cold metal burned a little, but I didn't think anything of it as I looked through my rifle's sight.

The section of range in front of me had a dozen small hills, arranged from 25 to 300 meters away. Each hill had a small popup target behind it, and in a prescribed sequence, each target popped up so as to give the firer a three-to-five second opportunity to shoot it.

The 50 meter target popped up.

Shift. Scoot. Squeeze. BANG. Target down.

The 200 meter target came next.

Shift. Scoot. Squeeze. BANG. Nothing. Missed. Fuck, Target down.

I was sure I'd hit the target, but the sequence didn't stop, so I had to keep shooting. I fired at each of the rest of the targets as they popped up, hitting some and missing others, until the tower called.

"Cease fire, cease fire, cease fire."

The cadre ran through the sequence of ensuring our weapons were cleared while the range computer processed the scores. I stood up while I waited. But as soon as I pulled my rifle away from my face, I felt a stinging sensation on my nose. I looked down at the weapon and saw a patch of skin on the charging handle.

Fuck. It was so cold my nose had frozen to the rifle. Ah well, there was nothing to do about it now.

The tower's voice boomed again.

"Congratulations, Privates. Not one of you managed to shoot expert. You all suck. Range detail, bring more magazines to the firing line. All firers will repeat the sequence."

And so we did. Three more times. The wind continued to howl, and targets continued to not fall when they'd apparently been hit. My shooting hand, which had to be pulled out of its mitten to fire, was stiff and numb to the touch. By the end of the third sequence, firers and the cadre alike were visibly frustrated.

On the final 'cease fire' the tower called the range cadre to the center of the range for a conference. A few minutes later, we were all directed to return to the warming tent. The weather had won, and we were rescheduled for the next day. Until then, all we had to do was make camp and conduct patrol base operations until morning.

I walked into the tent, took off my snotties, balaclava, and snow goggles and stretched my fingers next to the blazing potbelly stove to get them working again. The fingers on my right hand were already swollen and pink. "Chilblains probably," I thought. Sobon, who'd been on the range with me, looked at me as he tried to get the feeling back in his own appendages.

"What happened to your face, man?"

"Huh?"

"Your nose, it's bleeding."

"Oh. Yeah. I tore the tip of it off on my charging handle. Froze right to it."

Sobon laughed. "I didn't put my face on that shit for exactly that reason. Now you look like a douche."

I shook my head and laughed. Yeah, I probably did. Nothing to do about it now, though. I returned my focus to getting my fingers working again, massaging them until the numbness was all the way gone and the stinging subsided a bit.

Sobon was doing the same thing. He was one of our new guys, a recent transfer from the 2nd Infantry Division in Korea, and we'd bonded quickly after his arrival. He was a quiet guy who, like me, wasn't in love with the Army.

The two of us stood in the tent, talking and warming up, until Sergeant Stearns leaned in and interrupted the moment of comfort.

"Alright, princesses. Enough time in the warming tent. Time to get out here and enjoy this beautiful winter evening with the rest of us."

"Roger, Sergeant."

Sobon and I stepped back from the stove and bundled up. Our spaces were immediately taken by other guys who'd been shooting with us. We stuffed back on our snotty mittens, pulled down our balaclavas, and pushed back through the tent flap and into the cold. My lungs ached as I sucked in the frigid air, and my nose instantly stiffened as the snot inside froze. Weather this cold made me feel like I needed to breathe to clear my nose, but doing so made my lungs hurt, so I ended up stuck in an awkward middle ground.

Sobon and I walked across the range toward the area where AT platoon was congregating. Like us, nobody else in the platoon had qualified expert, and there was plenty of bitching about the cold, the target lifters, and the Army in general as we picked up our rucks and started walking off toward a section of woodline that the company had designated for our patrol base.

When we got there, Sergeant Stearns, Sobon, and I walked to one corner of the area to set up our guard position. Stearns briefed us on how the evening would go.

"Alright, gentlemen, we're pulling 33% security all night. That means one of the three of us is up at all times."

Sobon and I looked at each other. An NCO who pulled guard with the rest of us?

That was a new and positive trend.

"Roger, Sergeant."

"I'll pull the first shift - 1800-2200. You two bed down, now. Pace, I'll get you at 2200. Sobon, you take the 0200-0600."

"Roger, Sergeant."

This guard schedule put me in a bit of a predicament. In the world of night guard shifts, there is a clear hierarchy of shift quality, and a shift's position in this hierarchy is directly related to the degree to which the shift interferes with the guard's sleep. The best shift is the first one since guard shifts usually start before anyone is actually tired; the guy pulling the first shift will likely have been awake even if he wasn't on guard, so his shift costs him no sleep. The last shift is the next best one, because getting up early is better than waking up, pulling guard for a while, then having to fall asleep again. The middle shift, as you can likely surmise, is the worst, particularly when it's cold.

Knowing this, I pondered what to do. I probably wouldn't be tired until eight or nine, but I had to wake up at ten for guard. Should I try to force myself to sleep now and see if I could get in a few hours, or just push through until ten, pull the shift, and know I'd only end up with four hours of sleep total?

I decided on the first option, pulled out my heavy sleeping bag, and got ready to try to sleep. When it's below zero, this takes some doing. The trick is to keep everything with snow on it outside the bag, otherwise your body heat will melt the snow, soak you, and you'll wake up freezing. Of course, this is easier said than done, because taking off clothing while balancing on something dry and not accidentally putting a foot in the snow takes practice. I succeeded in this, more or less, climbed into my fart sack, and after rolling around for a while, managed to fall asleep.

When I woke up, I was freezing. The sack had been warm enough to make my feet sweat, and as my overall body temperature dropped through the night, my feet had started to freeze. They were aching now, and I was shivering violently. I wrestled my arm up to my face to check my watch. It was 2115. Shit. I had forty-five more minutes until guard. I thrashed around in the sack, trying to get enough blood in my feet to warm them back up, but it didn't work, so I tried my best to ignore the pain in the hope that I'd fall back asleep. But that didn't work either. Finally, after a bit more thrashing, I gave up on sleep and decided to just get up and start my shift early. I figured if I wasn't going to get to sleep, at least someone else could instead.

If there is a worse feeling than unzipping a warm sleeping bag to get dressed for guard in the winter, I've never experienced it. Pulling that zipper down, while knowing that you're about to let frigid arctic air into your tiny cocoon of relative comfort takes an awesome act of will. The achy feet added some incentive, and after a bit of mental fortification, I braced myself and yanked the zipper down.

Agony greeted me, and I responded with a long string of unhelpful profanity.

I sat up and fumbled for my boots. They were zipped into the outer layer of my fart sack to keep them dry, but they were ice cold. Grumbling, I stuffed my frozen feet into them and laced them up. I did the same with my various other pieces of cold weather gear and finally my parka, LCE, and weapon.

I stood up, stamping my feet in the snow to get the blood moving a little, put on my helmet and NVGs, and looked around for Stearns.

"Hey princess, how'd you sleep?"

"Fucking great. It was a real bang-up experience."

"I can imagine. There's nothing to report on my end. Everybody is bedded down. The other squads' guards are up, and the LT already came by to talk to me. You should have a quiet shift."

"Thanks. At least there's no chance of falling asleep on a night like this."

"Nope. It's too cold to sit down, and you'd probably die if you succeeded."

"Ha. Yeah, I guess so. Well, good night, Sergeant."

"Yep, you too, Pace. See you in the morning."

I watched Stearns walk off. He was new to the platoon, but he'd been in the New York National Guard for years. He'd switched to active duty after September 11th, and he'd requested to be stationed at Fort Drum to stay close to home. So far, he'd struck me as a really reasonable and pleasant guy, which were traits I hadn't yet seen in many NCOs, and we got along pretty well. But maybe that had as much to do with how much I'd changed as it did with Stearns' personality.

As I walked around the perimeter of our patrol base, I thought about what pre-Army-me would have done on a night like this.

"Ha. I probably would have died. Or just given up. I sure as hell wouldn't have walked around in the snow for four hours with a rifle and night vision goggles."

There was nobody to hear me, of course, but talking to myself kept me alert.

It was true, though. I wouldn't miss the Army when it was over, but there's no way in hell I would have been able to do any of this two years ago.

Life in the infantry had changed me.

The 10th Mountain Division is a light infantry unit. It shares a higher headquarters with the 101st Airborne and the 82nd Airborne Divisions, both of which are better equipped and better resourced than the 10th. We never had the ammo and training budgets those units did; our equipment was always more beat up; we had no parachutes or organic air assaults; and our personnel had rarely been to as many sexy schools–for example, other than the lieutenant, no one in our platoon had a Ranger tab. Despite all of that, though, we had something they didn't. We had Fort Drum. And Fort Drum made us hard.

The men of the 10th Mountain got used to making do with less in a place where the weather was as hazardous as any enemy unit. From hot chow to transportation assets, to decent lodging or amazing training facilities, our unit never expected much, so we weren't disappointed when we didn't get it. Say what you will, but being the poor kids of the 18th Airborne Corps and living and training in the snow belt had made us a rugged bunch, and this would serve us well a year later when we were patrolling the snowy mountains of Afghanistan.

My shift passed without incident, the hours blurring by in a frosty haze. Fifteen minutes prior to 0200, I walked over to Sobon's cocoon and kicked him awake with my boot.

"What. Huh. Mmf."

"Guard, man. You got guard in fifteen minutes."

"Yeah. Okay."

I stood there for a minute to make sure he didn't fall back asleep, only leaving when I heard the sharp preparatory inhale, the zip of the fart sack being unsealed, and the inevitable stream of profanity that followed. I smiled. Nobody could go back to sleep after that.

I stared off into the night while Sobon got ready, then turned when I heard him crunching toward me in the snow.

"What's up, man?"

"Nothing. Quiet shift. It took about an hour to feel my feet again, but they're good now."

"Ha. Mine are still frozen solid."

"It's what the Army did for me."

"Be all that you can be."

"Ha. Yep. Good luck, man. I'm headed to bed."

"Right. See ya."

I crunched over to my bag and looked down at it resignedly. In terms of sheer misery, getting back into a damp sleeping bag rivaled getting out of a warm one. After much thrashing, though, I succeeded in getting more or less comfortable, and after another half hour of shivering, I heated the damp inner bag enough that it was tolerable.

"Yeah," I thought. "There's no way in hell I would have been able to handle any of this two years ago."

With that, I fell into the light, restless doze of an infantryman in the field, and I spent the rest of the night rolling from one side to the other as some limb or another lost circulation.

At six in the morning, my watch alarm went off, signaling an end to the ordeal and the beginning of a new day. I grabbed the zipper, braced myself for the shock of cold air, and got ready for another day of EIB training. Landmine training was on the docket for the day.

7

March, 2003; Fort Drum, New York

R*IIIIINNNNNGG*

The harsh tones of our phone woke me out of my drunken sleep. Blearily I looked at the clock. 0230 in the morning. My head was killing me. Alycia rolled over and put the pillow over her head.

"Don't answer it"

RIIIIINNNNNGG

I considered her advice as the phone rang again. By the next ring, I'd decided she was right and covered my own head. I *could not wait* for January 2005 to roll around so that I could get out of the stupid Army. Alycia had already been accepted to a PhD program in Madison, Wisconsin. She was going to start in the fall of 2003, and once that magic day in January finally rolled around, I'd follow her there to start grad school myself. No more PT formations, no more stupid Monday morning haircut inspections, and no more middle-of-the-night phone calls!

RIIIIINNNNNGG

RIIIIINNNNNGG

RIIIIINNNNNGG

Silence. I removed the pillow apprehensively and waited. The silence continued, so whoever had called couldn't have wanted to talk to me that badly.

wasted no time digging into the inspection.

"Alright, who are we looking at, here? What's your name, meat rocket?"

"The Hawk, First Sergeant."

"Alright, The Hawk, let's take a look at your uniform."

Uncle Mike stepped on, not around, The Hawk's carefully arranged poncho-gear-display to peer closely at his uniform.

"Jesus fucking Christ. You look like a pile of spilled fuck. Did you sew this rank on yourself?"

"Roger, First Sergeant."

"I can tell. Who the fuck told you that's what it was supposed to look like? Your rank is supposed to be on your upper arm, not your lower one."

"Just tried to copy the picture, First Sergeant."

I heard the rattle and rustle of gear as Uncle Mike turned his attention to The Hawk's gear.

"No-go. This canteen is full of water."

I heard the sound of the cap being unscrewed, then the glug of pouring liquid as Uncle Mike dumped the contents of the canteen onto the poncho.

"That's not water. It's Coca fucking Cola! Holy Jesus, you are a fucking train wreck."

Uncle Mike's tone changed as he addressed our platoon sergeant, who was following behind him annotating deficiencies.

"This kid is a lost fucking cause. He will not be here during the battalion inspection. Lock him and all his shit in a closet until it's done. Do I make myself clear?"

"Roger, First Sergeant."

"The Hawk, secure your shit, and move the fuck out."

"Roger, First Sergeant."

I heard The Hawk start stuffing all of his gear into his duffle bag so he could leave the area. Lucky bastard. He'd been so fucked up he was getting to dip out of this early. Uncle Mike hadn't even had time to say...

"You're Working this Weekend!"

Uncle Mike's words snapped me back to the present. Uncle Mike was now in front of me, kicking over pieces of my gear and picking up the occasional one.

"Dirt in the canteen cup. Ribbon rack is a cunt-hair off center. Haircut looks like it was done with a weed eater. Who's next?"

And that was it. As Uncle Mike continued past me to the next guy, I realized three days of work had just amounted to jack shit. I too was working this weekend.

I remained at attention as Uncle Mike finished the inspection, then fell out with the rest of the platoon to correct our deficiencies and get ready for the battalion inspection later that afternoon. Every person in our platoon had failed the inspection and was working this weekend except The Hawk. Somehow, he'd failed so spectacularly that it had short-circuited Uncle Mike's brain and made him forget to take The Hawk's Saturday away.

"The Hawk's a goddamn genius," said Nevin, a big guy from Jersey, as we packed up our gear.

"Absolutely. Who'd have thought? You heard what we're doing this weekend."

"Yeah. I heard the supply sergeant needs to do a full inventory of the company tents, and that he needs labor for the project. I think we were doomed from the start."

"Fucking great."

"No shit. I was driving home to Jersey to see my old man this weekend, but I guess I get to stay here and work a shift at Uncle Mike's Fun House instead."

"Hooah"

"Hoo-fucking-ah, indeed."

"I can't believe you're working this weekend, again."

One of the best parts about getting bad news at work was that after emotionally working through it myself, I got to bring the news home and experience the disappointment all over again as I shared it with Alycia.

I'd gotten home late–the battalion inspections had taken all day–and I was unpacking my duffle bag of gear back into the entryway closet where it normally lived.

"Right? Luckily, it's not like it's for anything trivial. We're counting tent pegs."

"Oh, good. I was worried you'd be doing something that was a complete waste of life."

The sarcasm ran thick and deep as I continued unpacking and recounting the rest of the day. When I was finished, we sat down on the couch so Alycia could catch me up on the goings on in the neighborhood.

"So, you know Brandy's cousin down the street?"

"Nadine?"

"Yeah, her. Well, the fire department was here today to put out a fire at her house."

"She leave something in the oven?"

"Nope. Her husband is deployed, and her new boyfriend was smoking pot in the house. At some point, he threw the joint in a trash can and walked out."

"And *that* started the fire?"

"Yep. It burned down half the house. The whole neighborhood is talking about it."

We both laughed out loud.

"Wait, it gets better!"

"There's no way it gets better than that."

"It sure does. I've got updates on Brandy."

"Oh, Jesus. Now what?"

"She took the stripping gig to Vegas to make more money. She's been gone for four days."

"How did I not know about this?! Trey's gone; did she take the kids with her?"

"See what you miss when you're working too much? No, at the last minute, she left them with Misty. But she didn't leave enough clothes, diapers, or money, so Misty's freaking out trying to figure out what to do with the kids now. She's not sure if she should call CPS, or what."

"When's Brandy coming home?"

"Nobody knows. She said she was only going for the weekend, but she hasn't come home or called, so Misty has no idea."

"Holy shit. I guess that means we're not playing spades with them tomorrow?"

"Nah, it's still on. You know how it is."

Alycia continued to fill me in on the goings-on of James Street.

A new family we didn't know had left their car in the street before the snowplow came, and now it was so buried in snow that it would be unrecoverable until sometime in May when the colossal ice pile covering it melted. Misty and Tyrone's fighting was getting louder and more visible–they'd been in the streets shouting several times this week already. The neighbors on the other side of the duplex were up in arms because the wife had caught her husband dressing up in her underwear; her chief complaint was that he should have realized he was gay before they'd had two kids together.

The stories (or at least their long-term implications) were almost always tragic, but the details were generally so hilarious that it was hard not to laugh at them. That was life on James Street, though. And how could it have been different? The average resident was an eighteen-to-twenty-five-year-old who'd probably never had a home of his own before, had been married for less than two years, and was living a long way from where he had grown up. Also, Army family housing–particularly junior enlisted housing–lacks the mitigating factors present in most civilian neighborhoods that take the edge off of potentially explosive situations: there are no old, established families on the street to balance out the young new ones, and no deep-rooted cultural norms to normalize things. Even Army-wide norms found in senior enlisted housing are absent, because privates haven't been in the Army long enough to inculcate them. Eventually, soldiers and their families grow up and passions cool. But during those initial years when everyone is thrown together in a half-broke, semi-civilized, hormonally-charged, melting pot of pan-American culture, life is pretty wild.

This being said, Alycia and I had grown to really enjoy James Street. I'm not sure we ever really became a part of the community–we still often felt like strangers in a strange land. But then again, maybe that's how everyone on James Street felt, because there really wasn't a community to be a part of.

Nonetheless, we played cards with the neighbors, had them over for dinner and keg parties, and shared in their tragedies and triumphs when someone landed in jail, deployed to war, got a DUI, or received a promotion. Sure, interpersonal relationships could get petty and gossipy, but there was a straight-forwardness about it that let us all hear about each other's most humiliating secrets one day and share a beer together the next with no awkward feelings.

Without family support or hometown friends, we people of James Street only had each other to rely on when confronted with the uncaring juggernaut of Army life, and, for the most part, in spite of all our differences, we banded together in the face of that adversity.

8

July, 2003: Fort Drum, New York

The deployment order came suddenly, and it caught the platoon off guard. It shouldn't have, of course. We'd been packing equipment, loading stuff on trains, and inventorying weapons for weeks, but despite all of that preparatory work, we were still surprised when the lieutenant told us we were leaving for Afghanistan the next morning.

The surprise, I think, stemmed largely from the skepticism with which our platoon greeted everything we were told. This skepticism was a natural outgrowth of the way the lowest levels of the Army disseminate information. For instance, there were many mornings when I'd leave for work and just not show back up at home again for 48 hours because someone had decided we needed a snap field problem.[1] When every moment of every day was like that, skepticism came naturally.

After receiving the deployment news at PT, we'd been released to get cleaned up, and I'd headed home to pack and (once again) hand Alycia a pile of bad news. I knew how she'd receive it, but I couldn't help but be excited about the deployment. I was still skeptical that any of this was real, but the thought of going back to war tinged the skepticism with the promise of something new and awesome. Surely it wouldn't be as lame as last time, right? War footage from Iraq was common on the evening news, and from what I'd seen, a trip to Afghanistan or Iraq in 2003 looked like everything I

1. In the days before cell phones, this particularly galled Alycia, who often had dinner on the table for a husband who didn't show up to eat it.

imagined Army life would be: driving around, doing exciting stuff, and killing bad guys.

When I got home, I walked into the house to find that Alycia was still in bed.

"Hey babe, sleep well?"

"Okay, I guess." She must have noticed my expression because she followed with. "What's wrong?"

I'm sure I looked sheepish as I replied, "So, it looks like we're going back to Afghanistan."

"What?! When?"

"Tomorrow."

Alycia immediately wilted in despair.

"Shit."

I moved to hug her, wrapping my arms around her before breaking the rest of the now-relevant news.

"And guess what? They're going to keep me on as a .50 cal gunner. Yay."

She pulled back and looked at me.

"You mean like in the movie?! *That* kind of .50 cal gunner?!"

"Yep. Same job. Same trucks even."

"But you're going to die! In *Blackhawk Down* almost all of them died!"

I shook my head and grinned. "Nah, I'll be fine. It's not that big a deal."

"You're an idiot. Of *course* it's a big deal! What if you die? What if I never see you again?!"

I looked at Alycia a little blankly and absorbed what she was saying. It might sound strange, but none of that had really occurred to me on the way home. I'd thought about the excitement of going to war, the actualities of packing my gear, and the sadness of leaving Alycia at home, but I'd never considered I might die.[2]

"But I'm not going to die. Why would I?"

"You think everyone doesn't say that? But people *do* die, dumbass! People are dying in Iraq every day. Why do you think it won't be you?"

I didn't have a great answer to her question, because, of course, there wasn't one.

"I don't know. I just don't think it'll be me. I'll be fine."

Alycia hugged me again, wrapping her arms around me so tightly that I imagined she

2. Even things like packing extra AA batteries for the Gameboy had made my mental predeployment list, but death preparation? Nope.

somehow thought it might keep me from having to leave.

"You're such an idiot. But I hope you're right."

"Of course I'm right. What could possibly go wrong?"

Alycia pulled back and glared at me again, this time with tears in her eyes.

"There are moments when saying something funny isn't appropriate. Don't make light of this."

I scratched my head and looked embarrassed. "Yeah, sorry. I know it's a big deal. I just don't know what else to say during these moments, you know?"

"Then don't say anything! Just shut up and hug me."

So I did. I hugged her as she cried her way through coming to terms with another separation. We didn't know how long I'd be gone, although the word was it would be a six-monther. We also didn't really know where the battalion was going, at least not at my level. Not that it would have mattered to me much. The names of the various provinces in Afghanistan were both unpronounceable and meaningless.

The rest of the day was predictably awkward. Alycia sat on the couch and watched as I packed the rest of my gear, checking items, and stuffing them into my duffle bag.[3]

"So, what do you think? Should I bring the snookie cap and the balaclava, or just the cap? I mean, my face gets cold as hell in the turret, but the cap fits in my cargo pocket better."

"I don't know. The cap sounds fine, I guess."

"What about books? I've got a little space left in my duffle. I think I can fit one or two. Do I go with a classic that I can reread, like *The Lord of the Rings?* Or do I go with something more philosophical I can chew on during boring guard shifts? I still need to finish *The Portable Nietzsche*. Maybe that's too heavy for night guard, though."

I was excited as I thought through what I'd need, and as I packed I talked to Alycia about my back-up battery plans, indigestion medication, reading list, and (of course) the various combinations of snivel gear I'd need to provide some semblance of comfort in a place with blistering summers and freezing cold winters. But Alycia wasn't really interested in what I was bringing or why, so the conversation was one-sided and uncomfortable. When the packing was done, I tried to switch topics and make casual

---

3. Aside from the trunk described by Terry Pratchett in *The Color of Magic*, the Army duffle bag is the least user friendly piece of luggage ever invented. Long and tubular in shape, and equipped with only a small opening at one end, the duffle seems to have been designed to ensure the user has as little access to the bag's contents as possible.

conversation, but that too fell flat.

On the eve of deployment, there was just no way to bridge the gap between us. I couldn't look far enough past the excitement of a combat tour to see how much I'd miss Alycia, and she couldn't look at anything but my being gone and in combat.

By dinnertime, we'd both resigned ourselves to an evening of stilted conversation. After eating, we played a couple of quiet games of *Scrabble*, then got ready for bed, neither of us talking about the elephant in the room.

As we were falling asleep, Alycia's head on my chest, she broached the topic again.

"What if I never see you again?"

I'd learned my lesson earlier, so rather than make a joke or provide a dismissive comment, I just smoothed her hair and stayed quiet. She cried for a bit, thinking about all of the various catastrophes that a tour in Afghanistan could result in, and I lay there thinking about all the stupid things that always get stuck in my head before a long trip. Was the alarm set? What if the power goes out? Did I remember everything? Was there enough gas in the car to get to work in the morning? What could I grab out of the fridge for breakfast in the morning? How long would it be before we got to eat anything once we left? Did I transfer enough cash from savings to checking to cover the phone bill next week?

Physically together, but already miles apart, we drifted off to sleep in each other's arms for the last time that year.

9

July, 2003: Firebase Sh'kin, Afghanistan

It's a little-known fact outside of some military circles, but a single flight in a CH-47 helicopter can expose a passenger to every climate that exists on planet Earth and a few that don't.[1]

The flight from Kandahar took one hour and twenty-eight minutes, and we were currently one hour and eighteen minutes into it.[2] When we'd left Kandahar, it had been one hundred and four degrees outside with zero percent humidity. Now, a few thousand feet in the air, my feet were buried somewhere below an enormous stack of gear, and they were freezing to the point of pain. My head was swimming in a boiling hot miasma of engine heat and a moist cloud of vomit-stink that had been produced by some guy down the row from where I was sitting, and a roaring wind whipped through the interior of the bird as we flew east toward a remote firebase called Sh'kin.[3]

It had been late morning when we'd lined up on the Kandahar airfield to wait for the

---

1. The CH-47 Chinook, or Shit Hook as it is lovingly called, is a dual-rotor helicopter that's been in service since 1962. It vaguely resembles a flying banana and was used heavily throughout the war in Afghanistan because it could move very large things over very high mountains.

2. An interesting fact about Kandahar is that the city was founded by Alexander the Great when he attempted to subdue and conquer Afghanistan in 330 B.C. According to Wikipedia, the name Kandahar itself is a derivative of his name–Iskandar–in Pashto. Alexander didn't subdue the tribes of Afghanistan for very long, and the city remains as a testament to just how long people have been trying to do so unsuccessfully.

3. pronounced shhh-kin'

Chinooks to arrive, and the desert winds had been blowing hard, coating everything we owned with moon dust. Then the birds had landed,which meant we were blasted with even more dust, and we'd boarded. The process of loading a platoon of infantrymen into a Chinook is tricky, particularly when they have a bunch of gear.

A Chinook is shaped like a long metal tube. The interior is open in the center, and the exterior walls are lined with cargo seating that closely resembles the seating in a C-130. The gear gets stacked as neatly as possible in a long row down the center. Then the passengers, each laden with his own individual armor, helmet, weapons, rucksack, duffle bags, and assault bag, smash in along the sides, "nut to butt." Once the passengers can't be smashed any further into the bird, they twist in place as best they can and wedge themselves on the cargo seating, letting their baggage sort of settle on top of them in a heap. The entire process is conducted while the aircraft's engines are running, which means nobody can hear anything over the combined high-pitched whine of the auxiliary power unit and the dull roar of the main engines.[4]

The cargo seating of a Chinook is designed to be light, modular, and easily folded out of the way. It is in no way designed to be comfortable. Some lucky folks ended up sitting between the aluminum bars and had a more-or-less pleasant ride. Others ended up with one of the aluminum bars directly under one of our ass cheeks and spent the flight with the entire weight of our onhand military gear grinding them into a metal pole.

I was in the metal pole club today.

I looked left to where Sobon's head was just visible over our respective piles of gear. He was dead asleep, drooling on the camouflage rucksack in his lap. How the hell had he fallen asleep like that? Clearly there was no pole grinding into his ass.

I looked right and saw Lussi bobbing his head to whatever *Me First and the Gimme Gimmes* song was running through his mind. Eternally upbeat, Lussi was a musical guy who was perpetually singing '60s songs (or their punk remixes) to himself.[5] Even while stuck in the back of this flying Winnebago under 150 pounds of gear, he seemed to be enjoying himself.

---

4. Despite everyone's inability to hear, enraged shouting by all participants is still an important part of the process.

5. He was so musical that he'd bought a full drum set and installed it in his barracks room, much to his roommate Nevin's chagrin. The drums took up about forty percent of the available floor space in their tiny room and used up one-hundred-and-twenty percent of Nevin's available patience.

I shook my head in wonder and looked across the aisle where the snipers from the battalion's scout platoon were sitting. What were their names? Fuller and Thomas maybe? I couldn't remember for sure. The snipers were in our company, but we didn't associate too much. At the moment, they both looked irritated and uncomfortable.

Past them, I could see the world passing through the Chinook's small round windows. From this vantage point, Afghanistan didn't look like much. I could see that it was brown, barren, and mountainous, but I couldn't see much of anything else. I watched the terrain pass for a while, occasionally looking at the bored, uncomfortable faces across from me, until, from the front of the aircraft, the crew chief started waving at us.

"Five minutes," he mouthed, holding up five fingers on one hand and none on the other.

We all blankly stared at him in response.

We'd heard rumors about Sh'kin the whole week we'd been at Kandahar waiting for the battalion to completely deploy from Drum. The days in Kandahar had been hot, long, and mostly boring. We'd zeroed our weapons, test fired our machine guns, and scrounged equipment and rumors from the unit we were relieving from the 82nd Airborne.

"Sh'kin's crazy, man. I heard motherfucking Hajj runs that place. The base is right on the Paki border, so Hajj can rocket your ass all day, and you can't do shit about it."

"Dude, you guys are going to Sh'kin? Good fucking luck. My buddy was there for a few months, and he got fucked up pretty much every day. You won't see Hajj till he starts shooting, and he'll melt away before you can do shit about it."

"You guys are going to the motherfucking Alamo. Enjoy!"

Everything we heard about Sh'kin during our time at Kandahar had sounded bad, and made it seem like we were about to head to war for real this time–whatever that meant.

But even with all that buildup, as I looked around the Chinook, everyone I could see looked bored. Nobody seemed as excited as I imagined they should be, particularly considering where we were headed.

The Army routinely has a way of making wildly exciting things dull and uncomfortable. Almost everything it touches, from jumping out of airplanes, to flying around in helicopters, to throwing grenades and shooting missiles, ends up involving so much waiting around, physical discomfort, and irritating micromanagement that it gets un-exciting.[6] I hated it at the time, but looking back, I think turning the exciting into the mundane reflects unintended brilliance on the military's part.

Motivation is a good thing, and as a leader, you want your people to be enthusiastic about their work. However, when you want predictable, professional performance, it's useful to tone down people's excitement because excitement cuts both ways. In stressful situations, excited people sometimes do awesome things like leap on grenades to save the lives of their comrades or assault machine gun positions. At other times, though, excited people can do less desirable things, like piss their pants and run away, or drop their weapons and tremble uncontrollably. People who are bored, or at least emotionally toned down, do less of all of that. Their responses are more predictable. Routine and memorized responses, which can be absolutely soul-sucking to endlessly repeat in training, help people stay calmer when under stress. In a combat soldier, boredom is preferable to jumpiness.[7]

That response to boredom in the face of rumored danger was keeping our platoon in check today on the Chinook. Nobody had the energy to be jumpy or scared, because we were all too bored and uncomfortable to care about anything other than getting off the helicopter and being able to move our legs again.

The Chinook lurched as we banked into a final turn, causing the mass of cargo to shift and one guy's assault pack to slide off his backpack and vanish into the sea of baggage. The bird set down heavily with a thump, and I looked out of the open ramp to see a wide field of fist-sized, white rocks.

The crew chief gave us the two-arms-waving-toward-the-ramp signal for "get off my helicopter," and we started slowly and clumsily unloading.

The process went about as well as you'd expect, with the platoon shuffling off two-by-two to drop our heaps of gear to the ground and slap feeling back into our legs before returning to the helicopter to help unload the boxes of cargo.

When the last boxes were out, the bird's engines roared to full-throttle and lifted back into the air, blasting us with dirt and rocks. As it vanished into the distance, the roar of noise that had overwhelmed my hearing for the past few hours disappeared.

---

6. As it turns out, combat is often an exception to this rule, particularly when operating at the small unit level.

7. There are exceptions to this rule, of course. I'd rather have nervous soldiers than bored ones on guard duty, for example. Although even then, jumpy guards cause all sorts of problems.

After the long deafening flight, everything seemed unnaturally quiet as my ears recovered, and I looked around. The field of rocks we'd landed on was roughly square shaped and about a hundred yards across. One side of it was open. At the edge of the rocks, the terrain sloped up toward a low mountain that overlooked us. Two of the other sides were bordered by low three-foot walls that looked like they were made of piled up mud. The final side of the airfield butted up against a high, twenty-foot wall made of the same material.

My moment of quiet observation was interrupted by the arrival of two green cargo Humvees from around the eastern corner of the high wall. They rumbled along a small path that ran alongside the airfield and then crunched across its rocky surface to stop near our pile of gear. Two guys in DCUs, green body armor, and helmets got out of the trucks.

"Welcome to Sh'kin, gentlemen. Load up your gear. We need to get off this airfield before we get rocketed. Hajj gets excited anytime birds come in."

We followed these instructions, threw our gear into the beds of the trucks in *Beverly Hillbillies* fashion, and once everything was loaded, the trucks rumbled back the way they'd come, leaving us to follow them on foot.

Lussi was next to me, looking upbeat as always.

"Well, Pace, here we are at the motherfricking Alamo! Look around, man. There's nothing here but us."

I looked around. He was right. Past the low wall, I could see a few scrubby trees that might have been an orchard, but nothing else was visible except distant mountains.

"You're right, man. There doesn't look like much of anything. It looks sort of like the part of West Texas where I grew up."

"I was just thinking it looked a bit like Wyoming."

"I guess they're both big open spaces full of nothing, so there's at least *some* similarity."

"Fact."

SHHEEEEEOOOOO

The weird whistle interrupted our conversation.

Boom

"What was that?"

"Dunno. It sounded strange as fuck, though. Hey, look!"

Lussi pointed to our left, where a plume of smoke was visible on the side of the mountain adjacent to the airfield.

"What's that smoke from?"

SHHEEEEEOOOOO

"There it is again. That weird noise."

It didn't sound like anything else I'd ever heard. It was reminiscent of the sound a vacuum cleaner makes when it's first turned on, but with a quality all its own.

KABOOM

This time there was no mistaking the sound. It was an explosion, and it was closer to us than the first one had been.

"That sounds like something exploding at the range."

"Holy shit, I think somebody's shooting at us."

"Fuck!"

The group of us broke into a jog as we slowly started to process what was going on.

From the wall above, someone yelled, "*Hey dipshits, we're getting rocketed. Get in here, and get under cover.*"

As we ran, I started laughing. "They're actually shooting rockets at us. That's wild. Can you believe it?"

SHHEEEEEOOOOO

KABOOM

I could barely believe it myself. It didn't seem real at all, even when the last explosion was close enough to throw pieces of small rock into the air around us. What the hell was going on? Who was shooting at us and why? I didn't see any bad guys anywhere, so what was I supposed to do about it? Was I in combat?

Lussi seemed amused too.

"Well, boys. It looks like you new guys get your CIBs on Day One. Lucky SOB's."

"Ha. Yeah, I guess we do. Shit, that was easy."[8]

More rockets impacted the ground around us. None of them were closer than a hundred meters or so, though, and by the time we finally made it through the arched gate of Sh'kin and looked around, it looked like none had landed inside the base either.

---

8. The CIB, or Combat Infantryman's Badge, is a decoration earned by serving in direct combat as an infantryman. The badge for the award is a rifle with a laurel around it, and it is worn on the chest of the Army uniform. It is a highly coveted badge and an instantaneous discriminator between a cherry-ass new guy and a seasoned veteran. Lussi had already gotten his CIB during the last trip to Afghanistan during a mortar attack. I had not.

On the other side of the gate, we found ourselves standing in a motorpool. The ground was composed of the same fist-sized gravel that the airfield had been built with, and along the mud wall that separated us from the outside world, a mix of green cargo and gun truck-style Humvees were lined up.

An 82nd guy approached us as we walked through the gate. He was a tall, skinny guy with a scuzzy, half-grown mustache. He was wearing PTs, armor, and a helmet, and his shoulder-slung rifle suggested he wasn't too concerned about anyone attacking the base. We immediately pumped him for information.

"Was that it? Is there going to be an assault on the base or something?"

"Nah, it was just a rocket attack. You'll get 'em most days here."

We nodded. Sure. Daily rocket attacks. Seemed reasonable.

The 82nd guy pointed across the motorpool to the cargo truck that had picked us up.

"Grab your gear and come on. I'm from the AT platoon you're RIPing out with, and I'll show you where your hooch is."

His nonchalance suggested that the rocket attack was over. We headed to the truck, gathered up our gear, and staggered after the 82nd guy as he walked across the motorpool and through a narrow archway. The archway led into a sort-of alley that ran for about thirty feet before opening into a big courtyard. Getting the heavily-encumbered group through the archway took a minute, so I had a chance to look around.

The courtyard was wide–maybe fifty yards across. It was surrounded by low one-story buildings that were built into the exterior walls of the base. Along the tops of these buildings ran a rampart that stretched along the entirety of the twenty-foot high walls of Sh'kin and was accessible by a mud-brick staircase. At one point on the rampart (which I later learned was directly over the base's operations center) was a tall flag pole flying the American flag. The base really did look like the Alamo.[9]

The alley we were standing in had a series of plywood doors set into the walls. The 82nd guy pulled one of them open and led us into a twenty-foot square room. As we walked in, I noticed the door was rigged up with a piece of 550 cord, a few nails, and a plastic bottle. The bottle served as a counterweight that made the door close automatically.

The 82nd guy noticed my observation and commented on it.

"Yeah, the moon dust here is a motherfucker. It gets into everything. It's hell on

---

9. Although the Alamo (of course) flew the Texas flag.

electronics, guns, and trucks, and the shit sits in your sinuses and rots. Keeping the door shut helps keep it out of here for the most part, although when you wake up in the morning you're still going to spend a half hour coughing up gray lung butter."

The room was dim, although a little light came in around the edges of the plywood-and-sandbag plug installed in the one small window. In this light, I could see that two of the walls were lined with plywood bunk beds, each of which had a thin plastic Army mattress on it. The floor was lined with thin, brightly patterned plastic carpets, and in the center of the room sat a card table with a few folding chairs.

"This is where my squad was living, but you guys are a smaller element than us, so I think your whole platoon is in here. Go ahead and drop your stuff and hang out for a minute while I figure out what you're doing next."

The 82nd guy left the room, and we spread out to stake our claims. In these situations, there are two kinds of people: bottom bunk people and top bunk people. To bottom bunk people, the top bunk is a pain to climb into. Additionally, because hanging a poncho off the ceiling is harder than attaching it to the frame of the bunk bed, the top bunk generally has less privacy. Also, the top bunk offers less storage space, and when it's hot outside, the heat tends to rise and make the top bunks hotter. Recognizing this, bottom bunk people invariably try to claim whichever bunk is furthest from the door and immediately construct a sort-of cave-like cube in which they can pretend to have some degree of privacy.

Top bunk people, like myself, see the problem differently. In a twenty-by-twenty room there *is* no privacy, and nobody really has any storage space. We accept that reality and embrace it, trading the pseudo-solutions that bottom bunk life seems to offer for a range of other benefits, including increased air quality. Everyone's boots–even top bunkers'–have to live somewhere, and to avoid cluttering up the limited floor space the place they end up is underneath the bed. Ten pairs of deployment boots reek to high heaven, and while heat tends to rise, bad smells don't, so bottom bunkers spend their days breathing in foot-stink air, which has been proven by science to cause both impotence and cancer.[10] In addition to this olfactory benefit, the elevation of the top bunk means your head isn't at eye level with the ass of every single guy walking around the room. Many a bottom bunker has looked up from the book he was reading to see his

10. Deployment dining and hygiene options being what they are, foot stink is often the most pleasant of the foul odors bottom bunkers have to contend with.

top bunk roommate's hairy, reeking butthole in his face as he changes clothes. Finally, the top bunk provides a lofty perch from which to peer down at the rest of the room and develop increased awareness. For instance, when the LT comes in with a new task from the company headquarters, the top bunkers see him first. When the cook comes in with word that he's found a stash of frozen pizzas, the top bunkers hear the rumor first. What a top bunker trades in privacy, he gains in both air quality and overall awareness about what's going on.

AT platoon had a pretty even distribution of both types, so we secured our bunks with minimal squabbling and relatively little pulling-of-rank.[11] Sobon ended up as my bottom bunker, and, with minimal hassle, we worked out an equitable space sharing agreement regarding our gear.

I pulled my sleeping bag out of my ruck and stretched it out on the plastic mattress, found the small black bag that contained my hygiene items and hung it with my towel on a nail that my bunk's previous occupant had hammered into one of the bed posts, and, finally, pulled out the book I was currently reading, (*Slaughterhouse-Five*) and set it down on the head of my sleeping bag. Unpacking complete.

I looked around the room. The bottom bunkers, like Hogu and Sobon, were already hard at work constructing their poncho-draped hobbit holes. Raimond, Barrett and E-Brake, the newest guys and thus the furthest down the pecking order, were unpacking their bags onto the bunks closest to the door.[12] Lussi, a fellow top bunker, had finished and was climbing down from his bunk.

"Pace, let's go check this place out, man. I gotta see the rest of the Alamo in person."

"Alright, sounds good."

We grabbed our weapons and pulled open the door. The sun immediately blinded me. I hadn't realized how dim our room was until I stepped outside. We turned right, walked down the short alley, and stepped out into Sh'kin's main courtyard.

"Let's climb up those stairs and see what we can see."

---

11. In the Army, when two people are the same rank, whoever held that rank first is senior. This only really matters in a small percentage of Army situations, like deciding which guy has to send the daily LACE report or figuring out who gets priority of choice on the coveted corner bunk.

12. The closest bunk to the door is always the worst bunk. Mission requirements mean soldiers keep erratic hours, and there's no time, day or night, when someone isn't coming, going, or fumbling with something in the dark. The occupants of the closest bunk to the door get the worst of this lifestyle. Nobody else is as likely to get tripped over, stumbled into, or awakened by a slamming door.

We walked across the courtyard, slipping a little on the large pieces of gravel, and climbed the mud brick staircase. When we got to the top, I turned and looked back down at the base. From here, it was apparent just how small it was. Other than the rooms lining the inner walls of the courtyard and the motorpool, there was only a small area with a row of shitters and a howitzer pit. This was it. This was Sh'kin. The whole place couldn't have been a hundred meters across.

I turned back around, walked over to the shoulder-high wall, and looked south and east across Afghanistan. In both directions, all I could see was a long expanse of brown wasteland. Sh'kin itself sat on a small hill. But, past that, the terrain looked flat and featureless until it reached a distant range of mountains. Much of it was a light tannish color, although this was interspersed with rectangles of darker soil that were obviously plowed fields of some kind. Most looked empty, although the ones close to our base appeared to have apple trees. There were a few buildings scattered across this expanse that looked like tiny mud brick shacks. In the distance, near the mountains, I could see a group of larger buildings that might have been a small town of some sort.

"Man, Lussi. This place really does look like West Texas."

"Yeah, and it's flat as hell. I figured Afghanistan would be more mountainous, but this place sure isn't."[13]

We stood for a moment, looking into the distance.

"So, what do you think, man? Are we at war for real this time? Are we going to do something other than pull fucking guard?"

"I sure hope so, man. All the RUMINT suggests it's pretty hairy out here. I think we'll see some shit."

"I hope so, too. That last deployment sucked. Hopefully this one'll be a bit more interesting."

"Yeah, for sure."

A yell came from below us.

"Hey! You two. Get the fuck off the wall without your helmets on! You trying to catch a bullet?"

We walked back to the staircase and made our way down to the courtyard. The guy who'd been yelling at us was an 82nd staff sergeant whom neither of us knew. As we

---

13. As it turns out, it wasn't flat at all. The entire area around Sh'kin was actually a rolling area carved by valleys, hills, and sudden drop-offs. You couldn't see much of it until you were down in it though.

stood at parade rest, he eyeballed us critically, looking at our 10th Mountain combat patches and Lussi's CIB.

"You two should already know this shit. Especially you, *Sergeant.* There're snipers around here that'll take your head off. Nobody walks the walls without their kit on."

"Roger, Sergeant."

The sergeant glowered at us for another few seconds before grunting, "Carry on," and walking away.

Lussi and I walked back to our hooch to find our gear. On the way we detoured around the courtyard to see what else Sh'kin had to offer.

Along the west wall, we found a covered sidewalk that had been fabricated out of wooden pallets. Walking along it, we passed a screened-in porch area. The floor of the porch was a poured concrete pad, and there were picnic tables lined up along it in rows. The pallet sidewalk continued until it dead-ended in a small room with a couple of tables. On one of the tables were some opened boxes of Nutrigrain bars and a pile of sugar and creamer packets.

"I guess this is the chow hall then? We get the food here and eat it at the picnic tables?"

"Yeah, I guess so."

We turned to leave through the room's other door and continued our exploration.

"I wonder where they do the cooking?"

As we pushed through the door, we found the answer sitting on a stool in the corner of a small room. The man was young and heavily bearded. He wore a long, brown robe, with a short white vest over it. On his feet were flip flops, and on his head was perched a bizarre, mushroom-shaped wool hat. The Afghan–as he clearly was–stood up as we stepped through the doorway and stared at us. We stared back stupidly for a moment.

The Afghan was next to a stove with two huge pots on it. He was clearly cooking, and the room smelled heavily of onions, cooking meat, and the weird body odor-mingled-with-eleven-special-herbs-and-spices smell I came to associate with Afghans.

Eventually my brain processed enough information to direct my body to do something, and Lussi and I pushed back through the door, still staring at the Afghan who was still staring back at us. No words had passed between us.

Lussi and I walked in silence until we made it back to the pallet sidewalk, then the words came out in a flood.

"Who the fuck is that guy?"

"Why's there an Afghan on the base?"

"Do those dudes do the cooking?"

"What if they poison us or something?"

"Did you smell that guy? What the fuck was that smell?"

"Why'd he just stare at us?"

"I think I saw a cot in the corner. Does he live in that room with the stove?"

"What kind of hat was that? I gotta get one."

"Is he the guy calling in the rockets?"

"Why'd he want to call in rockets that might hit him too?"

We continued talking as we walked back to the hooch, trying to process what we'd just seen. When we got back to the room and our eyes had readjusted to the dim light, we saw that the First Section guys were gathered around Stearns and looking at a map on the table. He addressed us as we came in.

"Nice of you two princesses to join us. Looks like we're going to war tonight."

July 15, 2003
Calcium, NY

Dear Daniel,

Yesterday was very hard. So many things triggered sobbing - your water glass and dirty dishes, our chairs pushed together, your side of the bed, your laundry, your flip flops in the entry, sitting down to eat dinner alone, etc. Ultimately, I found that I had to do something, however, and so I set to it. But your waterglass, flip flops, and dirty towel stay where they are. Isn't it odd that I found so much comfort in the disorder yesterday? You know that I typically long for things to be neat and clean, yet here I am leaving your messes intact so it looks more like you're still around.

I thought I was ready to listen to some Louis Armstrong (I've been purposefully avoiding it), but I was wrong. "A Kiss to Build a Dream On" definitely made me cry, so that was all I listened to. I could not help picturing us listening to it together so many times. How will I ever get through these months without you?

Hell, everything makes me cry right now. I'm a mess. I can hold it together in public, but I've been scared to leave the house for more than a few minutes at a time because I don't want to miss another call from you. I was crushed when I checked the answering machine the other day and heard your message. I felt awful for missing you, and I cried about it much of the evening and night. I was exhausted, but I was too upset to sleep until after 1:00 am. I really need to get a cell phone, or I'm never even going to go to the commissary to get groceries! At least when I go to a friend's house I can use call forwarding. Have you ever done that? It rings on their phone even when you dial our number. But I still worry when I'm driving there and back, because you know that's probably when you'd call. Murphy's Law.

I just had a funny confrontation with Brandy. I noticed major 4-wheeler tracks right through our yard, within two feet of the porch and then up the hill behind the house. Finally, I saw Brandy going up that way - I guess they bought an ATV. I went out and politely asked her to go up in between our yards like everyone else with an ATV does. She rolled her eyes and kept asking why it mattered. I said that driving through our yard leaves a trail. She got really bitchy and kept saying it was too steep the other way. I nicely repeated that I would prefer she go up the path everyone else manages to use. As I was going back in the house, she said something about having to protect my precious grass. I'm sure I'm going to be in for some passive-aggressive retaliation now. Here come the trash and broken bottles. Oh well, I'll be leaving for Wisconsin soon enough.

I spent the rest of the afternoon reading Cosmo *and trying to ignore the incessant 4-wheeler trips up and down the hill by Brandy and Trey. Since they never go farther than the top of the hill, I can always hear the damn thing. They also love to rev the engine-it's so freaking loud it makes the cats jump every time. Sometimes I miss air conditioning – at least then I could close the windows. I HATE those people.*

Speaking of our least favorite couple, Trey must be leaving too. I saw him in DCUs today. Apparently, he's attached to the 1-87, so he'll be with you guys soon. Then, for reasons I don't understand, he's going to Iraq. They're not slated to return until October 2004 - it could be worse.

I miss you so much. I hope I get to hear your voice soon.

Much love,
Alycia

Sergeant Stearns seemed irritated as he stood next to our passenger side door and addressed the section.

"Alright, I'm going to count down from five this time."

"Five"

"Four"

"Three"

"Two"

"One"

"Mark. Sobon, run a comms check."

"All Killer elements, this is Killer One Delta, respond in sequence, over."

"Killer One Delta, this is Killer One Alpha, over."

Killer Six should have been next, but there was silence. From across the motorpool, we heard The Hawk shout, "You guys hearing me?"

We weren't.

As Sergeant Stearns quietly ground his teeth, I jumped out of the turret and walked over to The Hawk's truck. He was the LT's driver, and he was in charge of getting the truck ready for our upcoming mission.

"Hey man, what's up?"

"Not sure. I think my time's off."

"What's it showing?"

The Hawk punched a few buttons on the PRC 119 radio.

"The time's showing twenty. It should be right."

"What's the date show?"

He punched a few more buttons.

"Ah shit. That's it. I'm a day off. I should be ready now."

I walked back to the truck and told Sergeant Stearns we were good to run the comms check again. He shook his head in frustration and started the process again.

We'd been working on getting our radios to talk to one another for twenty minutes or so. Like all military units, we used an encryption system to keep our radio transmissions secret. The encryption system had two components. The first used a small box, called a CYZ-10, to install a cryptological key. This key changed monthly, and, when it did, one of the platoon's drivers had to go from truck to truck and physically hook up the device to install the key. The key also had to be reinstalled whenever one of the radios' batteries died. The second component was a setting on the radios called "frequency hop," which caused the radios to adjust their frequency multiple times a second. The radios' internal clocks ensured that as long as every radio was on exactly the same time, the frequency adjustment happened across the entire network simultaneously. This ensured that even if the enemy somehow recovered one of our keys (from a lost CYZ-10 for example) they couldn't listen to our transmissions unless their radios were in sync with ours.

When all of this worked, it was great, and our communications were absolutely secure. Unfortunately, the system wasn't terribly reliable. Radio batteries died frequently, particularly when it was cold outside, causing them to lose their keys. Moreover, radio clocks mysteriously drifted out of sync with one another, which meant to sync every truck's radio to exactly the same time required all the drivers to be sitting in their trucks and to physically push a button at exactly the same moment. This could be a difficult thing to coordinate since the tool we normally used to talk to one another–the radio–was exactly the tool we were trying to calibrate.

All of this meant that a good portion of our platoon's average day was spent trying to gain, maintain, or repair our trucks' communications. Even in training, this was a frequent source of frustration, and, on this particular night, as we prepared for our first combat mission, tensions were high.

And the radios weren't the only thing causing problems.

The radio troubleshooting was part of an overall system the Army uses to get ready for missions: Pre-Combat Checks, or PCCs. PCCs list everything a unit should do to verify that it is in good working order before conducting a combat operation, like 'update MEDEVAC plan,' and 'verify all magazines are full.' PCCs are designed to ensure nobody forgets anything. They are only really effective, though, when units have been using them for a while and have a refined understanding of what can go wrong in combat. A fresh unit like ours was using a list that was developed in training, and while it was better than nothing, it still had a lot of holes.

As we struggled to make our designated start time for our mission, we realized just how many things our list missed.

"Where are the smoke rounds for the 203?"

"Do we need to grab a thermite grenade per vehicle? Do we unpack it, or leave it in the can? Where do we put it, then?"

"Shit. The batteries in my M68 died. Anyone got a replacement?"

"Did you zero the laser on the PEQ-2?"

"Who's got the keys to the truck?"

"Where's the aid bag?"

"Where the fuck is The Hawk?"

"I think he had to take a last minute shit."

The list of problems seemed endless. Every guy had his own guesses about what he'd need for combat, and, as we later learned, most of us brought too much of the wrong gear and not nearly enough of the right gear.

Eventually though, after much wailing and gnashing of teeth, First Section was lined up, trucks running, at the main gate of Sh'kin and ready to roll out on our first combat patrol. Our truck, Killer One, was in the lead. The sun had gone down about half an hour ago. We were all wearing our NVGs, our comms were good, and our weapons were all locked and loaded. When the LT called up to HQ to announce that we were leaving the wire, it was about thirty minutes after our scheduled departure time.

Sergeant Stearns looked over at Sobon and me. "Well, princesses, you ready for combat?"

The radio squawk box in our truck crackled to life with the LT's voice. "Killer One, this is Killer Six. Roll out."

Sergeant Stearns grabbed the hand mike and keyed it. "Killer Six, this is Killer One, roger. Moving out, time now."

Sobon shifted the truck into gear, and we rolled out, pausing to give the gate guard time to open the huge metal door for us.

We drove through the gate and turned left on the road that curved around the outside of Sh'kin. As we made the turn, I pivoted the Ma Deuce to keep it pointed out across the Afghan wilderness. I looked back to confirm that the rest of the trucks were behind us, and relayed that information down to Sergeant Stearns before turning back to face the most likely direction of enemy activity.

I couldn't see much of anything. Were my NVG's broken or something? I reached up to adjust them when I heard the high pitched screech of metal on metal and felt the truck shudder.

Stearns must have felt it too, because I heard him yell at Sobon, "What's up, Sobon, did we hit something?"

"I didn't see anything, Sergeant, but it feels like we did. The truck is pulling to the right."

"Did you see anything, Pace?"

"Nope." I didn't mention that it seemed like my NVGs were broken, and I *couldn't* see anything.

The truck rolled forward, squealing, for another few yards before it jerked to a halt.

"What's up, Sobon?"

"She won't roll forward anymore, Sergeant. Something's fucked with the passenger side wheel."

"Alright. Wait here. I'll check it out."

Stearns picked up the hand mike. "Killer Six, this is Killer One. Something's wrong with our truck. I'm going to check it out."

He climbed out of the truck and walked to the front wheel, pausing for a second as his leg got caught in something.

"Pace. Come here for a second. I think I'm caught on a strand of wire or something, but I can't see what it is. Bring a flashlight."

I climbed out of the turret and hopped down off the hood. I pulled out my flashlight and clicked it on.

The flashlight revealed why Sergeant Stearns was stuck. He'd walked into a strand of concertina wire, and his pants were caught on the razors embedded in it. I followed the strand down from his leg and found that it stretched under the truck. It looked like we'd run over a long coiled strand of concertina wire that had been stretched across the road, invisible in the darkness. As Stearns carefully unhooked himself from the wire, I squatted down under the front of the truck to see where the wire led.

In the light, I could see that the concertina wire was so tightly wrapped around the axle that it had stopped the truck. There was no way we were going anywhere until we cut out all the wire. Moreover, as I investigated further, it was apparent that the razors embedded into the wire had shredded the front passenger-side tire and pieces of the undercarriage, so those would need to be replaced. In short, the Humvee was toast for the next twenty-four hours, so we'd have to tow it back to Sh'kin to get a replacement to continue the mission. On the one hand, since we were only a hundred meters outside of the gate, this was relatively easy. On the other hand, since we'd managed to lose a vehicle on our first patrol before we'd even left the wire surrounding the base, it was pretty embarrassing.

Sergeant Stearns never swore, which was unusual for an infantryman, but as he silently stared at the deadlined truck, it was clear he wasn't happy with the situation.

"Pace, get back in the turret and pull security. Sobon, get out here and hold the flashlight for me while I cut this wire out of the wheel."

As he and Sobon got to work, I climbed back onto the hood of the truck and into the turret. Lussi and the LT walked up from the other two trucks to see what was going on.

"Hey Pace, what's the verdict?"

"Looks like the truck's deadlined. Sergeant Stearns is cutting the wire out of the axle, then we'll have to swap the truck out to continue the mission."

The LT shook his head in irritation.

"This shit is fucked up. How did we hit the goddamn wire across the road in front of our own base?"

I replied, "It's dark as hell out here, sir. Nobody could see the wire until we were on top of it."

"That's bullshit. You guys need to tighten that shit up."

The LT turned on his heel and walked back to his truck to report the situation to the company commander, who was manning the radio at the base operations center.

I called after him, "Hey, Sir, it might be quicker to skip the radio and just walk back to HQ to tell the commander the situation directly. We're only about a hundred yards away."

The LT ignored me and continued back to his truck. He seemed irritated, but fuck him. His criticism had irritated me, too. He could take point himself if he thought he could see in the dark better than we could.

As Sergeant Stearns and Sobon continued to struggle with the wire, Lussi and I looked out across what little of the landscape we could see. It really was dark as hell; even with my PVS-14, I couldn't see more than a dozen yards. I twisted the focus lenses on the night vision device to try and improve the situation, but that didn't help. I fiddled with the brightness knob to try and get a bit more clarity, but maxing out the brightness only transformed the dark fuzzy landscape into a bright green, pixilated landscape. The contrast was still too rotten to make out any details.

"Man, I've never had this much trouble with my NVGs. I can't see shit."

"Me neither, Pace. Never seen anything like it. True zero illum kind of night."[14]

I flipped my goggles off and looked around. He was right. There was zero ambient light out here. Since we'd left the wire, either the moon had set, or the last bits of sunlight had vanished, and now I couldn't see my hand in front of my face.

"I guess it's never really this dark back home because there's always some light coming from somewhere, even when it seems like you're in the middle of nowhere."

14. Illum, pronounced 'ill-oom' is the Army's slang for illumination. Every mission brief includes a section on the illumination percentage for the night.

"Yeah, even out in the woods, there's usually the glow of a city or streetlights on the horizon. It's enough to make the NVGs function. Here, though? Nothing. It's completely black."

Lussi turned and walked back to his truck to update his guys. I returned to fiddling with my NVGs to see if I could get a little more clarity out of them. Stearns and Sobon continued grunting over the wire.

About an hour later, Stearns stood back up.

"Alright, that's the last of it. We should be clear to get the truck back to the motor pool. Let me update the LT, then I'll ground guide you back up the road so we can cross-load all of the equipment to one of Second Section's trucks."

Sobon climbed into the truck, and after a few minutes, Stearns walked back from the LT's vehicle and slowly limped our injured truck back to the motorpool.

The cross load didn't take too long. Within half-hour, we had re-loaded crypto into the radio, transferred all the ammo and grenades from one truck to the other, and were lined up with the other two trucks to continue our patrol. It was now two hours and fifteen minutes past our scheduled departure time.

We rolled past the concertina wire–now pulled entirely off the road and out of the way–and continued along the road as it wound down the small hill Sh'kin sat on. The road led into a valley, then curved around a single mud brick shack before climbing back up the other side of the valley toward the main road, which was named Route Corvette on our map. From there, Corvette headed down to a large security checkpoint on the Pakistani border, where it then branched into a half-dozen other smaller roads that led into the mountains between the two countries. Our mission tonight was to stop at the checkpoint, let the LT discuss the security situation with the captain of the guards there, then take one of the paths north in order to patrol the border for terrorists.

As we reached the bottom of the first valley, things went awry again. From the turret, I could hear that Sergeant Stearns and Sobon were having a navigation discussion. Both of them were wearing NVGs, and it was still pitch black.

"Sobon! Where are you going? The road curved left."

"No, it didn't. It curves right, just up there. See?" Sobon reached across the radio mount of the truck to point through the Humvee's small windshield.

Stearns turned to face him. The field of vision provided by a set of PVS 14s is tiny, though, so when he swiveled, he completely lost sight of the road and Sobon's pointing hand.

"No. It doesn't curve right. See that house? The road is supposed to curve left around the house."

Sobon stopped the truck and tried to turn his head to see where Stearns was gesturing. His NVGs bounced off the door of the Humvee as he turned, smashing them into the bridge of his nose.

"Oww! Fuck! No. I can't turn my head enough to see the house. I think you're looking at the wrong one, though. The house is up there." Again, Sobon pointed out of the window toward a building ahead of us.

Stearns took a moment to consider it. "I don't think there was another building on the map. Are you sure that's a building?"

"It looks like a building to me!"

"Pace, which one is the building we're supposed to turn at?"

I looked at the area around the truck. There were definitely two buildings. Since the building depicted on the map had been a tiny, featureless black square, it was hard to tell which one was the correct landmark.

"Can't tell, Sergeant."

"Well, which one does the road curve around?"

I looked around again. The road was made of brown dirt, and it ran through an area of brown dirt. During the day, I could more or less make out where the road stopped and the rest of Afghanistan started because the road had tire tracks running along it, but on a dark night–under NVGs–I couldn't tell which was which. It all looked like open ground as far as I could see.

"No idea, Sergeant. I can't see shit. I can't tell if we're even on the road or not. It all just looks flat and open."

Stearns was quiet for a moment. Being in charge of the lead truck for the platoon, he had to make a call on which way to go, but it was clear he had no idea and was trying to figure out what to do. The radio crackled to life.

"Killer One, this is Killer Six, over."

Sergeant Stearns picked up the hand mike. "This is Killer One, over."

"Killer One, what's the hold up, over?"

"This is Killer One. We're having some trouble seeing which way the road runs, over."

"Killer One, this is Killer Six. We're already behind schedule. Get a move on and figure it out. Killer Six, Out."

Sergeant Stearns's irritation was palpable as he hung up the hand mike. He turned to

Sobon and said, "You heard the man. Move out."

"Roger, Sergeant. Which way?"

"Drive straight toward the building ahead of us, then follow the road to the right."

"Roger, Sergeant."

Sobon put the truck back in drive and started rolling forward. A few yards further, the truck jarred violently as we hit something large. We bounced over it, hit something else large, then stopped with a grinding scrape.

"Fuck! I don't think that was the road."

Sergeant Stearns turned to Sobon. "Stop. Let me get out and see what we hit."

He opened the door and walked to the back of the truck to look at the ground behind it. He looked up at me.

"Well, we hit a couple of huge rocks, and now one's stuck under the truck."

He walked over to Sobon's door and pulled it open.

"Alright, Sobon. We're high centered on a rock. See if you can put the truck in four wheel drive and rock us off of it."

"Roger."

Sobon fiddled with the transmission to get the truck into the correct gear, then pushed the gas. The truck roared, the metal undercarriage ground on something large, and we lurched backward over the first rock. The truck rocked to a stop again.

"I think we're clear, Sergeant."

Stearns walked around the truck to verify, then paused at the front right tire.

"Yep. We're clear, alright. We also popped this tire. Tell Sobon to get the jack out and help me get this thing changed."

I relayed the message and Sobon climbed out of the truck. He and Stearns worked together to pull the spare tire off the rear hatch (where we held in in place with a ratchet strap) and then Stearns rolled the tire around to the front of the truck while Sobon dug around to find the small scissor jack that was standard issue equipment for our Humvees.

As they got the jack into position, the LT walked up.

"What's the status, Killer One?"

Stearns continued his work as he replied. "Flat tire, Sir. We hit a sharp rock and have to change it."

"Well, that's just great. How long will it take to fix?"

"Should be pretty quick. Maybe fifteen minutes."

"Make it quick. We're still running late. I'm going to call this up to higher."

The LT turned to walk back to his truck. Stearns started breaking the lugs loose–our Humvee's wheels had eight of them–while Sobon started cranking on the scissor jack.

I felt the truck start lifting, then I heard a screech and it thumped back down.

"What was that? Why'd you lower the jack, Sobon?"

"I didn't. I think it just stripped out."

To his credit, Stearns took the news well. "Alright. At least it failed before we had all the lugs off. Throw the jack in the trunk and walk back to Lussi's truck and grab theirs."

"Roger, Sergeant."

As Sobon walked back to the third truck in our convoy, the LT walked back up to our truck again.

"Now what? Where's he going?"

"The jack stripped out, Sir. He's going to get Sergeant Lussi's jack so we can get this tire changed."

"Holy fuck, Sergeant. Are you serious right now? Did we do PCCs at all tonight?"

"Roger, Sir. We've had these jacks for years, and they've always worked fine."

"Well it seems pretty ate up to me, Sergeant. We look like total shitbags right now."

"Roger, Sir. We'll get the tire changed."

The LT turned to leave again, passing Sobon as he arrived with Lussi's jack. Sobon set the jack in position and started cranking again. It held this time, and he and Sergeant Stearns muscled the new tire into place, tightened the lugs, and lowered the jack. They strapped the flat tire back onto the rear hatch and brushed themselves off.

Stearns turned to Sobon. "Alright, Sobon. Let's look around on foot to find the road."

The two walked around the truck, found what they were pretty sure was the road, and walked down it for a hundred yards to confirm. When they were satisfied, they returned to the truck, and the platoon continued the mission.

It was now three and a half hours past our scheduled departure time. We were about three hundred yards outside the gate of Sh'kin. I could still see the base on the hill behind us. Even in the dark, it's bulky, brown walls were easy to pick out against the black sky.

I looked around as our truck rolled along the road. We passed the small house and started winding our way back out of the valley toward the main road. The maintenance problems made it easy to forget that we were, in fact, in a war zone. By all accounts, Sh'kin was even a particularly nasty spot. This building, or any one of the ridgelines around us, could conceal Taliban or Al Qaeda fighters lying in ambush. I fiddled with the NVGs again, staring at one of the ridges as I struggled to see. As dark as it was, Genghis Khan's Mongol Horde could be sitting on that ridge, and I wouldn't be able to see them because I couldn't see jack shit.[15]

My scanning was interrupted by another tremendous thump. The truck smashed over something large and lurched to a stop. The sudden movement caught me off guard, and I staggered off balance, smashing my NVGs into the butt of the Ma Deuce.[16] The pain of the NVGs digging into my eye socket was terrific, and it blinded me for a second.

"Fuck, Sobon! What the fuck was that, man!?"

"Sorry, Pace. We hit something. Another rock, I think."

The truck continued to roll, but now its gait was punctuated by the obvious 'thump' of another flat tire.

I couldn't hear him grind his teeth, but I could *feel* Stearns' tension.

"Pull over, Sobon. Let's get this tire changed." Then, into the radio hand mike, "Killer Six, this is Killer One, we've got another flat tire, over."

The pause before the LT responded felt long. We all knew what he was probably thinking right now. "Killer One, this is Killer Six. Roger, over."

This tire change went smoothly. Lussi brought up his spare tire–the only other one in the convoy–and Stearns and Sobon changed it out quickly. When they were finished and putting away the equipment, the LT walked up.

"Alright, Sergeant, I think we gotta call this patrol here and return to base. We've gone less than a mile, we've blown two tires, and we don't have any more tires in the convoy. Our meeting is shot, and, at this rate, we probably won't even make it to the checkpoint. The only thing more humiliating than the way this patrol has gone so far is how humiliating it would be if we had to spin up the QRF to tow us back to base. Let's get turned around and get back to base."[17]

---

15. Much later in the tour, it occurred to me that if I couldn't see, the Taliban probably couldn't either. Unlike us, however, when it was too dark (or too cold, too snowy, or too whatever else) to do anything, they mostly just stayed at home instead of fumbling around in the dark.

16. This happens pretty frequently to Humvee gunners, and because of the way NVGs sit on your face, invariably, either the eye cup smashes into your eye socket, or the main housing of the NVGs cracks into the bridge of your nose.

17. Because the 10th Mountain had just arrived at Sh'kin, the QRF that would have to respond if we got into trouble would be the 82nd platoon that was waiting for their helicopter home. In the words of Inigo Montoya, calling them would result in "Humiliations Galore."

"Roger, Sir."

We loaded back up, turned the trucks around, and slowly limped our way back to Sh'kin.

We made it back to the motor pool about five hours after we'd left. In total, we'd traveled about two miles, blown two tires, found zero enemy, and accomplished none of our objectives. As soon as we parked, the LT got out, grabbed Sergeant Stearns and Sergeant Lussi, and walked off for the debrief.

After we finished unloading our gear, Hogu, The Hawk, and E-Brake walked over to our truck. Hogu lit up a smoke, and we all started talking about the mission.

"Holy shit, what was that?"

"I know, right? Could you see shit? I thought my NVGs were broken."

"Nope, I couldn't see a damned thing."

"What happened to the jack?"

"I think the extra weight of the ammo, the tires, the water and fuel cans, and all the other shit we brought stripped it out. We'll need a heavier jack or it'll probably happen again."

"Fuck. That's crazy."

"Man, the LT was pissed."

"Yeah, he sure was. He was bitching at us like crazy for getting lost and running off the road."

"Yeah, well, fuck him. I didn't see him taking point."

"True."

"So how the hell are we ever going to patrol at night? That shit was impossible."

"I dunno, man. Maybe the 82nd guys had better NVGs than us or something."

"At least we didn't run across any Taliban dickheads. I don't know what the hell that would have looked like."

"How did your gun setup work? The turret handle on my truck kept getting caught on my kit, and it was pissing me off all night."

"Yeah, same here. I think I'm going to re-rig my kit so the first aid pouch is on the right side instead of the left so that doesn't happen."

"Good idea. I'll try that shit, too. You find a magic fix to keep the strap from digging into your ass? I feel like that shit is giving me hemorrhoids."[18]

"That's not the strap's fault. You gotta quit taking it in the ass every night, bitch."

"Fuck you."

"The radio hand mike is shit where it currently hangs. Every time Stearns was out of the truck I had to reach all the way across the radio mount to root around for the damned thing."

"Yeah, we rigged a string up over our radio mount so it'd hang in the middle. That way anyone in the truck can get to it."

The LT and the leadership had their debrief, and we had ours. Tonight, just like every other night of the deployment, the privates sat around the trucks, smoking and bullshitting about what had gone well, what hadn't, and what changes we needed to make so things ran more smoothly. Some of the ideas we came up with were pretty good, some were super impractical Rube Goldberg solutions that consumed hours of time and added zero value, and some were so obvious it seemed amazing nobody had thought of them previously. All of them, however, were the product of a bunch of guys forced to cobble together solutions from the grab-bag of material and creativity on hand. Eventually, we did get pretty good at what we did, and in the years afterward I saw many of the solutions we pieced together in those early years result in fundamental changes to both equipment and tactics as our crazy stretch plays became the blocking and tackling of the mid-war years.

That night, though, we were just a pack of newbs who'd failed at our first mission and were trying to figure out if it was us, the environment, the equipment, or something else entirely that had caused that to happen.

July 30, 2003
Calcium, NY

Darling Husband,

18. The gunner's strap is a notoriously uncomfortable device. The turret of a Humvee has no seat, as it's primarily intended for the gunner to stand throughout most missions. For longer missions, though, the turret is rigged with a canvas strap that runs under the gunner's ass for him to lean on. The strap is better than nothing, but it's about as comfortable for extended use as it sounds.

I am planning a major move by myself, and I'm starting grad school in a few short weeks. That alone is enough to stress somebody out, and on top of that, I have the knowledge that my nearest and dearest is getting shot at 8 ½ time zones away.

I had an appointment with the housing office this morning to get ready for the move to Wisconsin. They're being much more lenient than their paperwork indicated they would be. The only sticking point is lawn-mowing and snow removal, which must be done the whole time I'm away. Apparently, they enter the house weekly and really do check for that. Every time the housing maintenance guys have to mow or shovel, $30 is deducted from your paycheck. That sucks. I'm going to try to hire some teenager around here if I can.

I'm also worried because my rent for the flat in Madison is due August 1st, and we don't have enough in savings and checking combined to cover it. I also have to buy a bed and make electrical and phone deposits, among other things. And we are nearly broke. I don't get paid by the school until September 15th, too. Not even my bonus. Oh, well. It'll all work out.

I made buttermilk cornbread, bacon, and beans for dinner. Can you tell I'm trying to empty the fridge before I leave? Plus, I want to wait until payday to go to the commissary. I'm not taking any chances in case you get an opportunity to use your debit card. How sad would it be to have the card declined??

There's been some drama with Misty and Tyrone lately. I was reading in the shade of the garage a few days ago when Misty, wearing a Confederate flag bandanna on her head, stopped by to chat. She pointed at her head and told me that she bought this new do-rag, and when Tyrone saw her in it, he got really pissed off. She was calling it a "graffiti flag" and saying she didn't know why it made him so mad, but now she was going to wear it on purpose to upset him. Apparently, she (a white woman from Upstate New York) has never seen or heard of a Confederate flag and bought the thing completely innocently, but Tyrone, being a Black man from Mississippi, knows exactly what it is. Worse still, later that evening, they were both drinking outside when I heard Misty yell at Tyrone, "Drink that 40 ounce, eat that KFC, [n-word]." Shocked, I peeked out the window in time to see him shove her and start screaming in her face. Things escalated to the point that the MPs came yet again and confined him to the barracks. I don't get it. They're both actually really kind people, and they even seem to sincerely like each other. I guess it's just that they also really like fighting. Oh well. I'm sure Tyrone will be back in a few days, and they'll be all kissy-kissy and sweet again.

Tell Sobon hi for me - he's on my mind quite a bit. I sent him a special present for his 21st birthday, but I heard that all packages are being thoroughly searched, and alcohol is being confiscated. Poor Sobon. I tried.

I love you more than I could ever love anyone or anything ever, my dear. You make my life complete. Please be careful.

With utmost respect and affection,
Alycia

"You were in Korea before this, right?"

"Yeah, I did two years there."

"How was that?"

"Pretty cool. Lots of field time. Field time and hookers, that was Korea, mostly."

"What kind of hookers? Like, on-the-street hookers? Flagging your car down and shit?"

"Nah, I didn't even have a car. In Korea, they call them *soju* girls. You go to the bar with your friends, buy the girl you want a *soju*, and then she heads upstairs with you. The system is all pretty aboveboard."

"I gotta think that catches dudes off guard the first time it happens."

"Oh yeah. That's part of the 'welcome to Korea' experience. The dudes take you out, convince you some girl is interested in you, and then stick you with the bill when you find out she's a hooker. Since most guys are only there for a year or two, that gag gets played out pretty regularly, but it's hilarious every single time."

"You ever been to Texas, Sobon?"

"Nope."

"Well, if you ever go to a bar in Texas, just know that that's not how the system works at all. FYI."

Sobon laughed.

"It's funny, but that's the kind of shit I don't know that much about. I grew up in a pretty weird family. My folks were first generation Polish immigrants, and I was homeschooled. I joined the Army the second I turned eighteen and immediately moved to Korea. I don't really know much about normal America, I guess."

"Whoa, you were homeschooled? Were your folks in a cult or something?"[19]

"Nah, they were just socially uncomfortable immigrants. Most of my childhood, it was just me, my parents, and my sister."

At this point, Sergeant Stearns jumped into our conversation from the other side of the Humvee.

"Is your sister hot?"

"What the fuck kind of question is that? It's *my sister.* How am I supposed to answer that?"

"Honestly, of course."

Sobon thought for a minute, turning the wheel back and forth to avoid rocks as we rolled along Route Corvette toward the Pakistani border checkpoint.

"I don't know. Maybe? I guess she is?"

"Can I get her number?"

"Absolutely not."

Stearns laughed. "You're a real Blue Falcon, Sobon. And here I thought we were friends."

We lapsed into silence for a bit, and I looked around. I was in the turret, behind the Ma Deuce as always, and we were rolling east from Sh'kin on a patrol. It was mid-morning, and the sun was shining. The last chill of the morning had lifted, and it was getting warm. In another few hours, it would be brutally hot–desert hot. But for now, with the wind blowing on my face as we drove, it was pretty comfortable, except for the dust.

The 82nd guy had been right. Even with my goggles down and a scarf wrapped around my nose and mouth, the dust got everywhere. Our platoon had been on several patrols since that first ill-fated night mission, and on the first of them, I'd learned that the dust scarf and goggles we'd been issued in Kandahar weren't just for show. I'd arrived back at Sh'kin that evening with my face, hands, and lungs coated in the omnipresent beige dirt that seemed to comprise the entirety of the Afghan landscape. The scarf and goggles helped with that a bit, although they didn't keep the dust out of the Ma Deuce, the turret bearings, my pistol, my magazines, the radio connections, or anything else. All of these things had to be meticulously cleaned between missions, or the dust would keep them from working.

19. My kids are all homeschooled. We're not in a cult.

There was nothing I could do about the dust now, though. Right now, I had another problem: target discrimination. We were driving east toward the border checkpoint. The border checkpoint sat on a large dirt road that ran southeast from Ghazni and led into the tribal areas of western Pakistan. For a road in eastern Afghanistan, it saw a good bit of traffic. Moreover, that traffic came in shapes and sizes I'd never imagined existed.

There were camel convoys–straight up rows of camels, each tethered to the one in front by a ring in its nose–walking in a file behind a group of guys who looked like they'd just stepped out of *The Arabian Nights*.[20] We passed enormous flocks of sheep and goats–most of which had been spray painted pink–herded along by swarthy men dressed in the same vest-and-hat-combo the cooks at Sh'kin favored. Motorcycles piled high with three, four, or sometimes five or more people, zipped past our trucks recklessly, and Toyota Hiluxes packed so full of people that I couldn't even count them bounced down the road toward us. Since I was the machine gunner in the lead truck, each of these required me to decide: Do I kill it, or do I not kill it?

Was that enormous Bongo Truck driven by a terrorist?[21] Was it going to smash into us and detonate in a fiery explosion? Did its bed conceal stockpiles of rockets and machine guns just waiting to be distributed to our enemies?

What about the driver? Does he look suspicious? Is he nervous? Is he about to jerk the wheel? Can I see both of his hands, or is one of them holding something? There's his AK-47. Am I sure it's an AK, or is that a longer barrel? Is it a PKM? What *is* it? Is he slowing down, or is he speeding up? In short, do I shoot, or do I not shoot?

That decision had to be made a hundred times on the drive to the checkpoint. We'd been briefed that the Taliban and Al Qaeda forces dressed and acted exactly like all of the other Afghans. We'd also been briefed that every Afghan was allowed to own one AK-47 assault rifle, but no heavier weapons. This meant that each time a car, animal caravan, or pedestrian approached our trucks, I had about fifteen seconds to figure out

---

20. Interestingly, the women always walked in a group *behind* the camels. Whatever else the Afghans may or may not be, they definitely aren't egalitarian.

21. Bongo Trucks, or Jingle Trucks as they were sometimes called, were the tractor trailers of Afghanistan. They were about the size and shape of dump trucks, and they hauled enormous loads of cargo over the country's precarious mountain roads. Their most interesting feature, though, was the way they were decorated. Each one was covered, bumper to bumper, with bangles, charms, metal disks, tassels, and brightly colored paint, and no two were the same. I'm unsure whether the decor was purely cosmetic or served some sort of supernatural, protective purpose.

whether or not the assault-rifle-carrying men inside posed a threat.

And they *all looked like threats*. Every guy I saw looked exactly like one of the Taliban fighters I'd seen pictures of on TV or on slides in our rare intelligence briefings. He had the same huge beard, wore the same suspicious, unwavering expression, and carried the same weapons. Often, as we passed a group of people on foot, the men would squat down in a group and stare at us as we drove past, looking like nothing less than a group of terrorists plotting an attack.

"I got a guy off to our three o'clock with a donkey, Sergeant."

"Yeah, I see him. He got a weapon."

"Yeah, roger. It looks like an AK though."

"What's on the donkey?"

I strained my eyes as we bumped along through the dust.

"Umm. I *think* it's just a big pile of sticks under a blanket."

"Any of those sticks look like RPG rounds?"

Did they? I wiped the lenses of my goggles with my scarf and stared again. By this point, we were passing the guy. Shit. *Were* they RPG rounds?

"I can't tell for sure."

"Fuck it then, we're late for the meeting. If he's a terrorist, then he gets away with his terrorism today."

"Roger."

I stared behind me at the guy for another few seconds, watching him stare back at me with unwavering expressionlessness. As the truck behind me added to the dust cloud our truck had made, I lost sight of the guy, his donkey, and his potential stash of RPG rounds, so I turned back around to face forward.

We were near the end of Route Corvette now, entering the small village of *Angoor Ada*. We passed the small pharmacy, with its gaggle of colorful dusty shelves outside, a mud hut that was under construction by some sort of chain gang, and a handful of staring locals who paused to watch us drive past. As we crossed through the village's single intersection and drove up the hill to the actual checkpoint, I scanned a field to the north where a mass of donkeys, sheep and goats waited to file through the security station on their way to or from Pakistan. I watched as a couple of bored looking Afghan guards gave them a cursory glance. It made me think of the scene from *The Odyssey* when Odysseus and his men hid themselves under the bellies of a bunch of sheep to evade a searching cyclops. Dozens of these herds must have passed through this checkpoint every day. How many weapons were hidden among them?

"I don't think those guards are doing much searching, Sergeant."

Stearns regarded them for a few seconds. "Nah, it doesn't look like it. They're probably bored as hell out here, and they probably see the same fifty herders every day."

"You think there's anything getting smuggled across the border?"

"Yeah, probably."

Stearns didn't seem properly interested.

"Well, aren't we going to do anything about it?"

"What would you like to do, Pace? You want to stand here and search all those goats? They probably file through this checkpoint for twelve hours a day. Are you going to check them all? Hell, even if you did, don't you think the bad guys would just move their smuggling route north or south to one of the checkpoints you're not standing at? I can't imagine it costs too much to bribe one of these Afghan guards."

I stewed on his reply as we turned into the checkpoint, which consisted of a small compound with a poorly repaired six foot wall around it, a parking lot, and a couple of small buildings. One of them, an outhouse, sat on a small rise. Three walls of the outhouse had been destroyed by some sort of explosive. So taking a shit meant climbing onto a small hill, dropping your pants in clear view of every guard and transient person on either side of the border, and relieving yourself into the hole in the ground that served as a toilet.

We parked the truck, and Lussi's and the LT's trucks parked next to us. The LT and Sergeant Stearns got out to walk into the main building for their meeting. I looked around. There were Afghans in the guard towers at the corners of the checkpoint.

"Hey Sobon, do you think we can trust the Afghan guards enough that I can get out of the turret and stop pulling security? Think they'll shoot us?"

Sobon looked around. "Nah. They don't look like they care enough to do that. I think we're good."

I looked around again. He was right. The guards did seem totally uninterested in us. Maybe that's what they wanted us to think, though? Were they trying to lull us into complacency? I balanced this possibility against the reality of having to stand in the turret for the duration of the meeting if I wanted to maintain security myself. Nope. In the end, laziness got the better of me, and I decided if I was going to die, then I'd rather die comfortably. I climbed out of the turret and jumped to the ground. Next to me, Hogu and Raimond jumped out as well, and we congregated around the front of the trucks to kill time while our bosses negotiated with the Afghan commander.

Raimond was a Hispanic guy who'd grown up in and around the Latino gangs of Southern California.[22] He'd quickly established himself as someone who made astute and often hilarious off-color observations, and he opened up the conversation today with: "So I know I'm the cherry fuck and all, and so of course I don't understand these things, but what the fuck are we doing here, exactly?"

My understanding of the overall security situation in the area was fuzzy at best, but as a guy who had two more years in the Army than Raimond, I felt obliged to take a swing at answering his question to maintain the prestige of my seniority.

"Basically, the guards on both sides of the border don't have any real control over the mountains north and south of the checkpoint. That means terrorists come and go across the border as they like, smuggling shit and shooting rockets at us. The hope is that if we meet with these guys here, we can figure out where the holes in the system are and plug them up.[23]

Raimond looked around at the checkpoint's poorly repaired walls and lax security. He regarded the row of dirty pickups, several of which looked like they'd been cannibalized for parts, and the piles of garbage heaped into the corners.

"These guys look like they're wired pretty fucking tight. I'm sure they've got this shit on lock down."

"Yeah, no doubt. I think the best we can hope for is that they aren't actively supporting the bad guys."

"Shit, you think they aren't? It was the same back home with the *cholos* and the cops. Why would I tell some cop from across town something when I'd known the gangbanger he was asking me questions about for my whole life? So what if he's running drugs. At

---

22. Raimond had us all watch *American Me* when it was his turn to pick a DVD on movie night so we'd get a taste of his experience growing up. For a guy like me, whose upbringing was better depicted by *Clerks* or *Half Baked*, it was pretty eye opening.

23. Twenty years later, my understanding (while still imperfect) is somewhat better. The tribes who live along both sides of the Pakistan/Afghanistan border don't recognize the border as applying to them, so they move across it freely. Moreover, in 2003, the central government was still functionally AWOL in this part of Pakistan, so it served as a staging area for the Taliban after we ousted them in 2001, separatists who hated the government of Pakistan, and for smuggling networks. This problem was bad enough that Pakistan had to invade this portion of itself several times through the early 2000's, and over the course of the U.S. war in Afghanistan, the tribal area across the Pakistani border never stopped being a problem.

least he was a dude I knew was a part of the community. What was the cop doing for me?"

Raimond pointed at one of the guards standing across the courtyard. The man was wearing an unbuttoned U.S. Army camouflage uniform shirt over his standard Afghan tribal wear, complete with flip flops. He was heavily bearded and carrying a standard Afghan AK.

"That motherfucker probably has more in common with the terrorists than us. I'm like some *gringo* cop that's come to the hood to arrest his family. Shit, aside from that, just look at him. If he was walking around in the mountains, I wouldn't know if he was on our side or theirs."

"I guess you just gotta wait until he shoots at you. Then you'll know, right?"

Raimond laughed. "Yeah, I guess so. That's not so different from back home either."

We jawjacked for another half hour or so until the LT and Stearns came back out to the trucks. We huddled around them as the LT pulled out his map and spread it across the hood of his truck to give us an update.

"Alright, so here's the scoop. The commander here laid out how the terrorists are getting across the border. Basically, there are Pakistani military checkpoints here and here." He pointed to a spot north of where we were and another to the southwest. "Those checkpoints are both being paid off by the terrorists to stay out of their business, so the bad guys can move and operate around them with no interference."

We all nodded. Damned Pakistanis. Of course they were on the terrorists' side.

Hogu said what we were all thinking, "Are we going to go fuck them up then, Sir?"

The LT shook his head. "Nope. We can't do anything in Pakistan, and we can't shoot at the Pakistanis unless they shoot at us."

"But they're letting people shoot at us. Doesn't that count?"

"Nope. Not good enough."

"That's some bullshit."

"Yes it is. That doesn't matter though. We still can't do it. Instead, we're going to patrol along the border roads that pass through the terrain the terrorists have to move through to try and interdict them."

The Hawk chimed in now. "What's interdict mean?"

"We're going to keep them from crossing the border?"

"Like, by killing them? If they're dead are they interdicted?"

"Yeah, by killing them. Once they're dead, they can't be terrorists, right?"

We all nodded again. We'd interdict the shit out of them.

"Alright. So, today we're going to drive up Route Saturn. That's the road north of here that winds through the mountains along the border. Our mission is to locate and destroy any terrorists we find. There aren't any real landmarks along the route, so we'll just drive for a few hours, then circle back around and come home." He paused and looked around at all of us before continuing. "Along the way, we're going to pass one of the Pakistani Army checkpoints. They are our allies. *Do not fire on the checkpoint unless fired upon.* Am I clear?"

We all nodded.

"Are there any questions?"

Raimond this time. "I have one, Sir. How do we know if the people we see are terrorists?"

The LT thought for a minute before replying. "That's a good question, Private. Basically, if they shoot at us, if they act suspiciously, or if they're carrying any obviously illegal weapons, they're terrorists."

Raimond considered the LT's answer. "What's acting suspicious look like, Sir? Like, if they run away, is that suspicious? What if they stand off in the distance and watch us through binoculars or wave to someone or something? Can we shoot them?"

The LT looked a bit hesitant. "You can shoot if you feel threatened. Also, if the enemy is maneuvering to get into a better position, you're cleared to fire then. You can't just shoot anyone for just standing there though."

"So, if they're walking up a hillside, is that maneuvering to get into a better position?"

"It *could* be."

"One more question, Sir. We're going to be driving on the border, right?"

"Yeah, that's right."

"And we can't shoot at people in Pakistan, right?"

"Yeah."

"How do we know if the guys we're looking at are on the Afghan side of the border? Is there a sign?"

Now the LT was fidgeting. "Just call it up before you shoot, and we'll figure it out."

Raimond didn't look satisfied with the answer, but it seemed that was all he was going to get.

"Any more questions?"

Nobody had any more, so we all climbed into the trucks, got comms up, and rolled out to start looking for terrorists on Route Saturn.

Our truck took point, as we usually did, with the LT's truck in the middle of the convoy and Lussi's bringing up the rear. We drove back through the small village around the checkpoint and turned north at the town's only intersection. We passed a few small shops, then quickly left the village behind and started the long drive up Route Saturn.

This road was worse than Corvette, so the truck rumbled slowly along it, zigzagging around to avoid large rocks and washed out sections. There was no other traffic, which made sense to me, since on the map it looked like this road didn't really go anywhere anyone would be going.

The terrain around the checkpoint was relatively flat, but I could see that we were approaching a much hillier area ahead, and when we reached it, the mountains on either side of the road quickly grew steep. On top of the tallest one, I could see a goat herder with his goats. On a whim, I tested the Ma Deuce to see if the way the turret was shaped would allow me to point it at the guy. I couldn't.

"Hey Sergeant, these mountains are too high and close to the truck for me to engage targets on top of them. If that guy up there is a shit head, there's not much I can do about it."

Stearns looked out of his window to consider the situation.

"Yeah, there's not much we can do about that today. If we wanted to secure our passage, I guess we could put a team up there so they could overwatch us. Looking at that mountain, though, it would take us half a day to get a team in place, and they'd only be able to cover us for a klick or so."

He thought for another minute as we continued along the road.

"I think it's a risk we've just got to accept. There's no practical way to secure the high ground with the guys and time we have available."

"So we're just fucked, then?"

"Nah, he's probably not a bad guy anyway, right?"

"Yeah, I guess probably not."

KABOOM

The explosion rocked me forward in the turret. It had come from behind us. I whirled around to look back at the LT's truck. A huge smoking crater had appeared between us.

I shouted down into the turret hole, "*Contact, six o'clock. Explosion behind us, I think.*"

The radio blared to life.

"Killer One, this is Killer Six, Contact front. What the fuck was that, over?"

Stearns turned to Sobon. "Push through the kill zone, then stop us a few hundred yards up." As Sobon floored it, Stearns picked up the radio hand mike and shouted over the engine's roar, "Killer Six, this is Killer One, unknown. Pushing through so we can figure it out."

"You see anything, Pace?"

I scanned the area around us, looking for terrorists as the truck bounced heavily down the unimproved road. The only guy I could see was the one with his goats on the mountaintop, but he didn't seem to have reacted to the explosion. Was that a good thing or a bad thing? I had no idea.

"Just that goat herder, Sergeant."

"Stop here, Sobon."

The truck pulled over, and the two trucks behind us pulled off on either side of the road. I looked back at the smoking hole in the road behind us. It was a ways away now, but the crater and the lingering cloud of white smoke were still obvious. *Holy shit, somebody had just tried to kill me.*

We sat quietly waiting to see what would happen next. Would there be an ambush? Would there be rockets? Were there more mines?

For a good ten seconds, nobody seemed to know what to do as we held our breaths, listening to the Humvee's rumbling engine and processing what had just happened. Stearns broke the silence by opening his door and getting out of the truck.

He peered up at the mountains around us. "Well, ladies, any ideas what that was?"

Sobon spoke up. "It sounded like an anti-tank mine going off to me, Sergeant."

Stearns nodded. "Yeah, maybe so. Those are set off by a tilt rod though.[24] If that thing had been tilt rod activated, we'd all be dead right now."

"Maybe whoever set the mine rigged it up with something so it'd blow up another way?"

"Maybe, yeah. Like what, though?"

"No idea. Maybe it was manually detonated like a claymore?"

"If so, there'd be a wire. Yeah. That's a good idea. Let's go check it out."

After calling the LT to tell him what he was about to do, Stearns walked back to

24. A tilt rod is a foot-long antennae that is attached to an anti-tank mine when it's armed. The antenna sticks straight up, and the mine's fuse is designed so that if the antenna is bent, when smashed by a passing tank for example, the mine detonates.

Lussi's truck, and, together, they walked back to the crater to assess the first IED any of us had ever encountered.

Stearns' and Lussi's assessment was crude. They walked back to the crater, looked around it a bit, and walked back to the trucks. Stearns grabbed the hand mike and pushed an update to the LT.

"Killer Six, this is Killer One. No wires found leading off the crater. I've got no idea what that was or how it detonated, over."

"Killer One, this is Killer Six. Roger. Try to keep a better eye out for those in the future. That last one hit pretty close, over."

Stearns paused for a second, no doubt marshalling his substantial reserves of professionalism, before continuing. "Roger, Sir, we'll keep an eye out for them. Continuing mission, over."

He put down the hand mike and turned toward us. "Well, ladies, you heard the man. Keep a better eye out for landmines in the future. You could have gotten us killed."

Sobon responded. "Roger, Sergeant. We'll try harder to spot the buried mines in the dirt road as we drive along it. Any hot tips on how to do that better?"

"Nope."

I chimed in. "Try squinting your eyes a little. Or maybe just close them altogether and use the fucking force, Sobon. It worked for Luke Skywalker."

"Zen driving. I like it."

Over the next twenty years, IEDs, or improvised explosive devices, were such a hallmark of the Global War on Terrorism that IED became a household word.[25] In 2003, though, nobody in my platoon had even heard the term IED, and we certainly hadn't mastered any of the techniques to deal with them that most of us would eventually acquire.

This early in the war, IEDs weren't that sophisticated–the enemy wasn't any better at using them than we were at dealing with them. But these early attacks started the cat-and-mouse game the American military played with insurgents in Afghanistan and Iraq throughout the GWOT. Either by trial and error or training received from wealthy sponsors in Iran or Saudi Arabia, enemy forces learned to set off larger payloads with greater accuracy and harder-to-detect triggers. Similarly, driven by the need to survive increasingly devastating attacks, U.S. ground forces developed the tactics and technology

25. https://crsreports.congress.gov/product/pdf/IF/IF10899

to detect and defeat command wires, pressure plates, and eventually cell-phone and infrared-activated IEDs.

However, despite the plethora of agencies, task forces, and contracted solutions that were eventually devoted to the IED problem, it always felt like we were one step behind, and few platoons hit the ground with the knowledge and experience they needed to mitigate the threat. Instead, most platoons got smarter slowly and painfully through trial and error. The cost of this ignorance was tragically high, and over the course of the war, IEDs accounted for more U.S. combat deaths than any other enemy action.[26]

Back in Sh'kin that night, I had trouble falling asleep as I thought about the IED. *Somebody had tried to kill me.* And not in an impersonal, shooting-rockets-from-over-the-horizon sort of way. They'd tried to kill me *personally* by burying a landmine in the road and detonating it as we drove past. The idea seemed ludicrous. Why would they bother? We hadn't done anything but talk to a few Afghans, and we certainly hadn't killed anyone, right? And yet, the more I thought about it, the more local hostility did start to make sense.

We were here, in the Afghans' country, driving around like we owned it. The 82nd guys had certainly killed a bunch of locals. Did the Afghans even know we weren't them? What would I do if a bunch of Afghans occupied Texas? I'd probably shoot at them, too.

As I drifted to sleep that night, the first notions of what war actually was and what we were doing here started to penetrate the video-game-and-'80s-action movie-derived image I'd had in my head since childhood. We'd experienced rocket attacks, but for whatever reason, almost getting blown up by an IED was the first time in my life my brain had grappled with the idea that this was serious business, and that people *could actually die.* This realization created a sense of exhilaration that bordered on giddiness, and I had trouble not giggling a little as I thought about it. It was all so surreal! Of course, my views would continue to evolve as I saw more of war's many facets. But the sheer adrenal rush combat induced was exciting, and I liked it.

The rest of the guys must have been having similar thoughts that evening, because over the next few days it became a frequent topic of conversation. Interestingly, though, we didn't all reach the same conclusions, and everyone didn't find it exhilarating. Those differences would become more pronounced as the months wore on, and eventually

26. https://reliefweb.int/report/world/decade-global-ied-harm-reviewed

they would lead each of us to draw our own inferences about war and what we wanted to do with our lives.

For now though, we'd all just taken our first steps into a wild new world, and nobody was quite sure where those steps would lead.

10

August, 2003: Firebase Sh'kin, Afghanistan

"You're dead fucking wrong. *The Last Boy Scout* is clearly the better movie."

Lussi replied heatedly, spitting his response out as best he could without also spitting out the peanut butter and crackers he was chewing on. "Bullcrap. It's *Die Hard*, all day. The broken glass. The Twinkie scene with that guy from *Family Matters*. The 'yippie ki-yay, motherf-er.' Hands down, it's the better movie."

"False. Does *Die Hard* have Damon Waynes in it? Nope. The comic genius behind Homey D. Clown was the perfect co-star for Bruce Willis. The dialogue between those two–like the argument at the end about how action heroes have to be clever nowadays–there's nothing that good in *Die Hard*. And him telling his wife to 'buy a dog' when he catches her fucking someone else and she says she's lonely? Priceless."

"Two words: Hans Gruber."

I thought about that response. Hans Gruber was a pretty great villain. Frankly, he was much better than the villain from *The Last Boy Scout*, but I couldn't just let Lussi win.

"He's alright, I guess, but I'm sticking to my guns–just like that running back stuck to his in that badass opening scene. What do you think, Sobon? Break the tie. Which movie's better?"

Sobon looked thoughtful as he chewed on whatever he was scooping off the cardboard plate in front of him.

"Well–"

SHHEEEEEOOOOO...SHHEEEEEOOOOO...SHHEEEEEOOOOO

KABOOM KABOOM KABOOM

The scream of incoming rockets interrupted Sobon's reply.[1]

"Fuck." I picked up my paper cafeteria plate, mopped up a bit of stew with a piece of footbread, and stuffed it into my mouth. I stood up from the table and threw the plate in the trash can as I ran toward our barracks room. Second Section was on patrol today, so First Section was on QRF. We had to get out of the wire ASAP to try to respond to the rocket attack.

When I got to the room, Hogu and The Hawk were already pushing past me to get to the trucks. Lussi, Sobon, and I grabbed our kit and followed them to the motorpool.

As we did so, we heard (and felt) the enormous report of the 105mm howitzers from the other side of the base add to the explosions of the incoming rockets. Sh'kin was equipped with a Q36 radar, and standard practice was that once the radar generated a launch grid for where the indirect attack had come from, the gun bunnies would light the area up with 105mm high explosive rounds.[2] Response times were usually pretty good, and sometimes (like now) they were returning fire before the rockets even stopped landing.

Over the last month, our ability to get out of the wire quickly and smoothly had improved. I only had to run back to the hooch once for a forgotten item (the headspace and timing gauge for the Ma Deuce), and Sobon and the other drivers only had to refill one radio as they ran through the pre-mission radio checks. In the five minutes it took Stearns to get the POO from the operations center, we were lined up, running, and generally ready to roll.[3]

As Stearns climbed in the truck, he caught us up on the situation. "Alright, ladies. It looks like the radar's got us something to kill, today. Let's get out there and find the bastards who've been firing rockets at us."

Sobon and I were genuinely enthusiastic–it was about time we got to actually find

---

1. Although he did eventually agree with me– *The Last Boy Scout* is the better movie.

2. Wikipedia describes the Q36 as "a 'weapon-locating radar,' designed to detect and track incoming mortar, artillery and rocket fire https://en.wikipedia.org/wiki/Counter-battery_radar ." My experience suggests this wasn't as precise a science as the brochure would have us believe, though, and Wikipedia neglects to mention that the Q36 was responsible for more wild goose chases by infantrymen during the GWOT than pretty much any other source, with the possible exception of the tip line.

3. A POO, or Point of Origin, is the point from which an indirect fire attack originated. This could be the location of the mortar tube that fired mortar rounds at your platoon, or it could be the location of the rock on which the 107mm rocket was leaning when it was fired at your base.

the motherfuckers who'd been shooting at us and do something about it.

"Fuck, yeah!"

Stearns got on the radio to push out guidance to the rest of the section.

"Killer One Alpha, this is Killer One, over."

"Killer One Alpha, over."

"I'm working out the grid to the POO on my map now, but it looks like it's generally southeast of here, toward the border–break–I'll pinpoint it as we drive that way, and we'll stop short to plan our attack when we get closer–break–The operations center says Second Section is already on their way there, so we'll either support them if they are in contact, or work out a plan with them if they aren't. How copy, over?"[4]

"This is One-Alpha, I copy all, over."

We sped down the road from the base and took the road south that led toward the border. This was more direct than taking Corvette, although the terrain was worse, so we couldn't drive as fast. Our route took us through the village of Sh'kin proper, a small cluster of shops that sat a kilometer from our base and served as one of the only two commercial districts in the area, along with the border checkpoint village.

The term 'commercial district' might be a bit misleading. The village really only had four shops: a butcher whose shack was adorned with the corpses of a half-dozen fly-covered, skinned goats; a baker whose store-front consisted of a set of shelves stacked high with footbread; a table with a tent over it where an old woman sat with a pile of vegetables; and a bedazzled bodega that had a UNICEF flag flying above it and looked like it sold everything else.

As we passed through the small cemetery that was just outside the village, the radio crackled to life.

"Killer One, this is Killer Seven, over."

Killer Seven was our platoon sergeant's call sign. The platoon sergeant position had been gapped since Sergeant First Class Wolf left the platoon before we deployed, and Staff Sergeant Collazo, the senior NCO in the platoon, was filling the role. He was out

---

4. The word 'break' is used when passing long messages over the radio. When a user says 'break' he also releases the key on the hand mike and pauses for a few seconds to clear the net. This ensures if critical traffic has to get pushed by another user, that user has the chance to do so. Some people are better at using 'break' effectively (and in a non-irritating fashion) than others. My personal favorite *faux pas* was when people said 'break' but didn't actually release the hand mike, thus defeating the purpose of the word entirely.

with Second Section today, and he sounded stressed out.

"Killer Seven, this is Killer One, go ahead, over."

"Killer One, this is Killer Seven. Current location is on Lozano's Ridge investigating the rocket POO. My vehicle has been hit, and is disabled–break–we are in contact and taking enemy fire, and my other vehicle is suppressing–break–request immediate support, over."

"This is Killer One, I copy. First Section is en route to your location. ETA, seven mikes, over."

The radio hand mike sat silently in Stearns' hand. There was no response to his transmission.

He turned to Sobon. "You heard him. Floor it."

Sobon did, and the Humvee's engine roared loudly as we turned into the village at full throttle. I shouted and gesticulated wildly at the small clot of people gathered around the bakery to get out of the way, and they did, although not before we startled a heavily burdened donkey enough that it bolted away from its master. The owner immediately chased after it in frustration.

"That guy probably won't think well of us," I thought. "Ah well, fuck him. He's probably helping the terrorists anyway."

We were through town in a few seconds, winding our way across the high plateau that led toward the border, which Lozano's Ridge overlooked. The ridge, named after an 82nd soldier who had been killed in a fire fight there in 2002, was a long, open piece of high ground sandwiched between two valleys, and its commanding position made it an easy place from which to observe the surrounding area.

The radio was dead quiet–not a good sign. But as we started down into the valley that bounded Lozano's Ridge we could hear the sounds of a fire fight to the south.

"They're still kicking, up there. Move it, Sobon! We gotta get to 'em."

"I can't go any faster, Sergeant, or we'll veer off the road."

The Humvee's tortured suspension supported Sobon's claims. We were already bumping down the steep, narrow, rock-filled mountain road faster than we reasonably should have been. One particularly large lurch sent the Ma Deuce bouncing almost out of its mount. I dove forward to put my weight on the gun to keep it in place. My effort succeeded, but as I caught the gun, I smashed my forearms into the sharp metal lip of the turret.

"You just smashed my fucking arms and almost bounced the gun loose. Are you *trying* to hit every rock in the road?"

"Sorry, Pace. Doing what I can, man."

We made it to the bottom of the valley and neared the fork in the road that would take us up the ridge.

"Left or right, Sergeant."

Stearns was still working on his map. He absentmindedly said, "Right."

We turned right, but I noticed the truck behind us turned left.

"Hey, guys, Lussi went left. We going the right way?"

Stearns snapped back into the present. "What? Fuck. No, we should have gone left. Back up and follow them, Sobon."

"Roger, Sergeant"

Sobon slammed the Humvee in reverse, guiding us back down the road at top speed. As we reached the fork, he jerked the wheel right to turn us around so we could take the correct turn. As he did so, though, I felt the truck suddenly jump sideways.

I fell to the right side of the turret as the left side of the truck lurched into the air."

Sobon had miscalculated and we were rolling over.

The world went into slow motion.

As the truck rolled onto its side, I could hear myself shouting "*FUCK FUCK FUCK*" as I was thrown out of the turret from the waist up, and my arms hit the ground. I watched as the Ma Deuce spun sideways and its barrel dug into the dirt. Inside the truck I could feel Sobon grabbing my legs, and I could hear him shout, "*PACE!*" The Humvee continued its roll, and I felt my back and head smash into the ground as the top of the truck neared the ground.

"Fuck," I thought. "I'm dead."

Then, everything stopped. My head and neck were wrenched against the ground, but there was space between the truck and the dirt, and I could move. I pushed up a little and twisted to try to find space, when suddenly the truck lurched back a little the other direction and settled onto its side.

I was lying on my side, half in the truck and half out of it. In a stupor, I climbed out of the turret and walked to the front of the truck. Inside, I could see Sobon and Stearns piled together in a struggling heap in the passenger side seat.

Still dazed. I looked around. The truck was lying on its side, and it didn't look like it was at risk of rolling further. What had stopped us from going all the way over?

I walked around the truck again and stopped at the gun. The Ma Deuce barrel was full of dirt, and the ammo can mount on the side was wrenched out of position. Mindlessly, like any good gunner should, I kicked the mount back into place and started picking dirt out of the barrel. I was still doing so a few minutes later when Sobon and Stearns, who'd managed to climb out of the turret hole themselves, were standing next to me.

"Pace! Are you alright, man?"

Stearns' voice snapped me out of my daze.

"Huh? Oh, yeah. I guess so." I pointed at the Ma Deuce. "I think the gun somehow got stuck between the truck and the ground while we were rolling and kept us from flipping all the way over."

"Well you better give her a kiss then. She saved your ass."

I stared at the Ma Deuce for a second. "Yeah, I guess she did."

We heard more shots from up the ridge. M4 fire this time.

"Enough lollygagging, ladies. Let's get up there."

Sobon and I nodded, and the three of us started pushing as hard as we could on the top of the Humvee. After a few rocks back and forth, it crashed heavily back onto its wheels. We climbed back into our respective seats and quickly got all of the gear that was out of position back into some semblance of order.

Stearns turned to us. "You two ready? Everything working?"

I racked the Ma Deuce. The action felt solid, so I squeezed the butterfly and fired a couple of rounds into the dirt ahead of us. "Gun's good."

Sobon cranked the truck back up. "Engine sounds good, too."

"Then let's go kill some f'ing Taliban."

The engine roared to life as the truck pulled onto the left fork and we climbed up to the top of Lozano's Ridge. As we crested it, the world opened. To my right (south), I could see the wide valley that divided Afghanistan from Pakistan. On the far side of that valley–about a kilometer from us–I saw a Pakistani Army checkpoint. To my left was all the area that stretched between Sh'kin and the border checkpoint at the end of Corvette.[5] Ahead of me, though, was the object that most drew my attention–Second Section's ruined Humvee.

Lussi's truck had pulled up alongside it, and I could see Dickey and Collazo standing between the two trucks, pulling gear out of the damaged vehicle and loading it into

5. Our base's tall radio antenna stood out clearly against the sky, marking Sh'kin's location and providing an easy aiming point for amateur rocketeers.

one of Second Section's other two vehicles, a beat up old Humvee pick-up truck that had been carrying a bunch of guys from Charlie Company whom I could see bounding quickly down into the valley toward Pakistan on foot to see if they could catch the enemy.

Dickey and Collazo both looked dazed.

I looked past the Charlie Company guys and tried to figure out where the fire fight was, but I didn't see or hear any more shooting. Hogu was scanning from the turret of Lussi's vehicle, but he wasn't firing at anything either.

Stearns spoke up. "Stop here, Sobon. Pace, pull security."

"Roger Sergeant."

Our truck stopped short, and Stearns got out to help with the recovery of the damaged truck. He relieved Dickey and helped Collazo pull equipment. Dickey came over to our truck and climbed into the back seat.

Sobon and I immediately plied him for information.

"Dickey! What the hell happened, man?"

Dickey was a Utah native with a quiet, easygoing demeanor that was almost comically out of place in the infantry. This might have caused him problems, but he was also the fastest runner in the battalion, which more than offset his lack of hate and anger.[6] Today, though, his normally casual expression looked pained. His face was blackened and sweaty, and he was rubbing his neck like it hurt.

"We hit a mine, I think. We were driving along, doing our thing, when we started taking fire from across the valley in Pakistan. The dudes couldn't have been more than a few hundred meters from that army checkpoint. Anyway, we started pulling up so I could get a better angle to fire back and cover the Charlie Company guys while they assaulted the bad guys, when all of the sudden there was this huge explosion and we got launched into the air like rockets."

"You and Collazo?"

"The whole truck, man. It launched straight up. All of it."

"Holy shit!"

---

6. In the infantry, being a fast runner was one of the surest signs of being squared away. Conversely, being a slow runner who fell out of formation runs was the kiss of death. In the weird world of all-male eighteen to twenty-two year old peer dynamics, this one attribute could offset pretty much any other personality trait. An alcoholic lunatic with seven girlfriends and a rash of personal problems would get promoted ahead of a reliable, smart, fat, non-runner every time.

"Yeah. Whatever we hit was *big.* The gun launched out of the mount and flew off onto the ground, and I launched all the way out of the turret."

I looked back at the truck in front of us, trying to picture what Dickey was describing.

The truck looked like it had been cut in half. Black smoke poured out of what remained of the engine compartment, and it looked like the front wheels were missing.

"Look." I pointed at a hunk of something black next to us. It was one of the front wheels–or what was left of it anyway. "Those things must weigh a hundred pounds, and that blast threw it a hundred yards from the truck.[7] Holy shit."

Dickey continued. "Yeah, I believe it. It threw my ass about twenty feet in the air, too. I bounced off the roof and fell back into the turret. My neck is killing me."

"Well, take it easy, man. The cavalry's here. Did you guys kill any Taliban?"

"I don't know. One second I was shooting over the infantry guys' heads. Then, the next, we exploded. I'm not sure if I hit anyone or not. The guys firing at us were far enough away that it was hard to see them very clearly, and at three hundred yards, their clothes blend in with the dirt really well."

He stopped to think for a second, staring out the window. "Looking back on it, what I really should have done was to put a hundred rounds into that checkpoint. There's no way those Pakistanis didn't see the guys firing at us."

I looked across the valley at the military outpost. It looked like an unremarkable, walled compound with the usual radio antennae poking out of a building in the middle. From here, it still looked quiet, and I couldn't see anyone moving along the walls. That seemed weird. Even if they weren't helping the enemy, I figured the huge explosion and the bullets impacting outside their base would have provoked some sort of response. Why didn't it?

"Yeah, you're probably right. There's no way those fuckers aren't helping the enemy." My brain jumped to another related problem. "Hey, Sobon. Did Stearns ever finish plotting that POO from the rocket attack on his map."

"One sec, I'll check."

Sobon leaned over and pulled the map out of Stearns' seat. He'd left it there when he got out to help recover the gear from Collazo's truck. Sobon looked at the map, then looked out of the truck to translate what was on the paper to the terrain around us.

"Yeah, it looks like he got it marked. The attack came from a spot right at the top of

7. Actually, a Humvee wheel weighs one-hundred and sixty-eight pounds.

that ridge on the other side of the valley."

"Let me see that shit."

I held up the map and tried to imagine where Stearns' dot was. The dot had the rockets coming from a spot about two hundred yards from the checkpoint, but given the margin of error on our estimation, the rockets could easily have been fired from the base itself.

"What the fuck? The POO is practically right on top of that base."

Our conversation was interrupted by Stearns opening his door.

"Well, ladies. It looks like we're going to put our air assault school skills to work today. We've got a bird coming to sling load our downed truck. Once the dismounts are back from their little stroll down the mountain, we'll get her out of here. Sobon, you're air assault qualified, right?"

"I have completed the hardest ten days in the Army, yes."[8]

"Great. Get your gear on, and I hope you remember your link counts. Let's get to work. Pace, keep an eye out for terrorists so we don't get shot."

"Roger, Sergeant."

Sobon climbed out of the truck, and the two of them walked around to the trunk to grab a set of sling load chains. Together, they lugged the heavy can over to the disabled truck, climbed onto it, and started rigging the Humvee for its ride home.

To those unfamiliar with the process, slinging a truck involves connecting heavy chains to the four corners of the vehicle, then securing those chains to a flying helicopter so that the vehicle can be lifted into the air and flown somewhere.

In a perfect world, a rigged Humvee looks something like this:

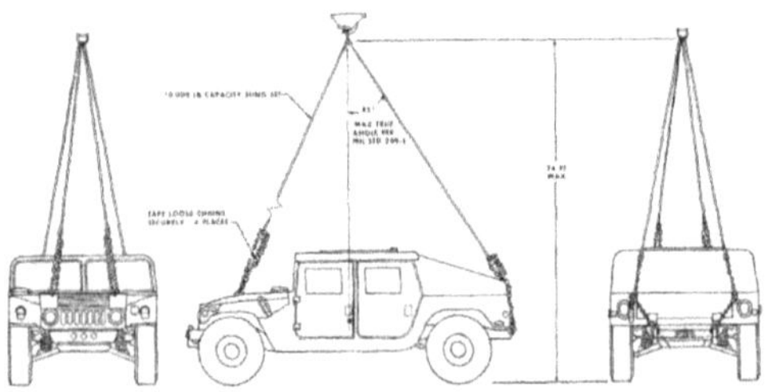

Image from Army Technical Manual (TM 55-2320-280-14, dated 25 AUG 1988)

8. Air Assault School is an Army course focused on helicopter operations, climbing ropes, and walking around with weighty backpacks. It's designed for entry-level soldiers and ROTC cadets, and describing it as the hardest ten days in the Army is wildly hyperbolic.

Unfortunately, our platoon did not inhabit a perfect world. A Humvee weighs about four tons, so if it shifts in flight, the motion can create a pendulum effect that can cause the helicopter carrying it to crash. Consequently, slinging a Humvee requires that the chains be attached properly to keep the truck's weight from shifting under the helicopter as it flies. Our damaged Humvee had been hit by a mine, though, and was missing a good number of its parts and pieces. Were there enough of the right pieces left to sling it at all?

As I scanned for terrorists, Sobon and Stearns climbed around on the destroyed truck to determine just that. After fifteen minutes, and lots of arguing, they concluded that it could indeed be slung.

The dismounts returned, the helicopter was called, and thirty minutes later, we heard the dull roar of an approaching bird. It was a Chinook, and as it settled over the damaged truck the propwash created by its rotors created a hurricane like wind that blasted us with dirt and rocks. Braving the mini-cyclone, Stearns stood on top of the truck to hook the pintle of the sling chains onto the hook that hung below the Chinook while Sobon held a long stick with a metal tip on it–called a static discharge wand–against the helicopter to keep it from electrocuting them both.[9]

Or at least we were pretty sure that this was how it was supposed to work. None of us had ever actually slung a vehicle from a helicopter before, and we didn't have a manual with us, so if we were being honest, we were somewhat guessing. Also, since these were the same sling legs the platoon had been inventorying in the motorpool for as long as anyone could remember–remember way back in Part One?–we weren't totally sure they were even serviceable.

When the hookup was complete, Stearns and Sobon both jumped off the damaged truck and ran back over to ours. They climbed in, and we watched as the Chinook's slow lift-off took the slack out of the sling load chains.

Was this going to work, or were we about to eat a magnesium and aluminum

9. The rumor was that if you touched the Chinook with the sling chains before someone else touched the bird with the static discharge wand, the resulting electrical explosion would kill everyone in the area. I have no idea if this is true or not, although my Google research suggests the discharge from a Chinook can be up to 20,000 volts. I'm not an electrician, but this sounds like it would either kill you or, if you lived in a comic book, would grant you some definite superpower.

sandwich when the bird smashed into us?[10]

As the Humvee's remaining tires–two were smoldering lumps–lifted off the ground, the moment of truth arrived. I held my breath, straining my eyes to see through the torrent of wind, and watched as the Chinook swayed a little, then slowly turned and took off toward Sh'kin. As it passed overhead, pieces of rubber and metal from the damaged Second Section truck rained down onto the ground around us.

"Like manna from heaven."

"What?"

"Nothing, Sergeant."

With the helicopter gone, the world seemed unnaturally silent, and I blinked and looked around. The magic moment had passed, and we were back on some ridge in Afghanistan where someone had just tried to kill a friend of ours.

Again, Stearns snapped me out of my head.

"Pace!"

"Sergeant?"

"You see anyone take a shot at that bird while it was on station?"

"No, Sergeant."

While that was technically true, if I was being honest I hadn't really been pulling security. I had been watching the Chinook. With all of the background noise, I'm not sure I would have noticed anything less than an artillery shell hitting the ground right next to me, but I *do* think I'd have noticed bad guys firing at the helicopter I was staring at.

Stearns grabbed the hand mike. "All Killer elements, this is Killer One, you ready to move?"

Lussi and the Second Section trucks replied that they were, and with the cargo truck full of dismounts between our two vehicles, we rolled back to base without incident.

When we arrived at base, Sobon and I started the usual post-mission truck reset. Clean out the trash. Clear, clean, and lube the Ma Deuce. Cover the gun. Clean, lube and cover the ammo. Turn off all the radios. Secure all the grenades and sensitive items.

I looked down as I worked and saw that Dickey had climbed out of the truck and was standing next to Sergeant Collazo. The two of them looked dazed, and without a truck

10. Fun fact: Since Chinooks are in large part made of magnesium, they explode into flames and burn to the ground when they crash. I learned this almost twenty years later when I watched it happen elsewhere in Afghanistan. It's probably for the best that I didn't know this back in 2003.

to reset, a little purposeless.

I called down to them, "Dickey!"

He and Collazo looked up at me.

"Look who gets to sham for the rest of the day! You dirty shit bag. I'll bet you got your truck blown up on purpose so you wouldn't have to clean that shit any more."

Dickey smiled and nodded his head. Collazo just stared at me. He and I had never gotten along all that well.

They walked off to the aid station to get checked out by the medics while Sobon and I finished up the truck reset. When we were done, we walked to the other end of the motor pool where the Chinook had dropped the wounded Humvee.

The truck–one of the original four we'd brought from Drum–squatted there in a heap. It was an unarmored truck with the doors taken off, so it was in no way designed to handle an anti-tank mine. The blast had destroyed the front suspension, so what was left of the front wheels sagged, leaving the bumper to drag the ground, and giving the Humvee a despondent look.

"Man, Sobon. Just think. This truck was sweating her ass off in Panama twenty years ago. Then she was in New York training with us. She's been to JRTC a bunch of times. She went to Uzbekistan with us two years ago. And now..."

"You remember how hard it was to get her going when it was cold outside?"

"Ha. Yeah. I remember you standing in the snow with your beanie cap over the exhaust pipe while I tried to crank the starter to get her bitch-ass running. She was a moody truck."

"But she never left us stranded, you know? Even when we buried her up to the axles in mud and Brock had to dig her rear wheels out with a shovel while the rest of us pushed. She always came through."

"Yeah, that's true. She never did. And now..."

Sobon finished my sentence. "Now she's dead as fuck here in this shit hole."

We continued to stare at the remains of the truck for a long while, Sobon smoking and me thinking.

Eventually he finished the cigarette, flicked it onto the ground, and turned to leave.

"I'm going to check on Dickey."

"Why, what's up? He seemed okay."

"I dunno. Yeah, he probably is. I just have a bad feeling about the whole thing."

I shrugged and turned to follow him.

As we left the motorpool, I turned to look back at the fallen truck one last time. She'd done everything her country asked of her for twenty-something years, and now her service was over. Tomorrow we'd pull everything useful out of her and transfer it to another truck, the mechanics would cannibalize what was left for parts, and then the hulk would be left here to rot.

"That's a hell of a way for her to end up."

Shaking my head and thinking dark thoughts about what the truck's fate said about how the rest of us would fare, I walked back to the hooch, took off my boots, and climbed into my bunk.

Was Dickey alright? That truck had been almost cut in half. What if that mine had hit a foot further back? What would Alycia think about our own crash today? Should I tell her, or would it just worry her? When was our platoon's next turn with the phone, anyway? That rollover had been close! Another few degrees, and I'd have been smashed flat. What was chow going to look like tonight? Should I head to the gym while I have a sec? Fuck, my head hurts. Had we even killed any bad guys today? Or was all that shit we just did a waste of time?

I fished around in my assault pack for my Discman, put on my headphones, and let Fat Mike's dissonant, eighteen-minute rendition of "The Decline" drown out the noise in my head for a while.

August 16, 2003
Madison, WI

My darling,

This move is cursed. Where on earth to begin?

We had to take two days to drive from New York to Wisconsin because our U-Haul topped out at 63 miles per hour, and we got a late start to begin with. Camille (my roommate) and I drove our own cars (she is moving from Albany and spent the night before we left with me), and my mom drove the U-Haul. Things started off poorly with Mickey peeing in the cat carrier before we even got to Syracuse, so I stopped to soak up what I could and give him a tranquilizer. His constant yowling was making Pandora literally foam at the mouth so much that it was starting to cover the bottom of her carrier and soaking her fur, but the meds mercifully put a stop to that.

Believe it or not, on the 14th (our first day of travel), power went out in the entire Northeast – NYC, Toronto, Ottawa, Detroit, Cleveland, Buffalo, and pretty much everywhere else. Unfortunately for us, that meant that working gas pumps, food, and open hotels in cities without curfews were difficult to find. We had to listen to the radio constantly to find little islands that still had power so we could pump gas! It was downright eerie driving the loop around Cleveland at twilight – I've never seen a city dark like that. We ended up finding a place to sleep in Norwalk, Ohio.

The second day's drive was uneventful, and we got to Madison as it was getting dark. We had planned to unload our few things and spend the night in the apartment, and we were relieved to be nearing the end of our journey. Our apartment was supposed to be cleaned out and unlocked with the keys on the counter. I found the house first – our place is the third floor of what was built in the 1890s as a three-story house - and I was horrified to see gang sign-type graffiti spray-painted on the siding and utility box. There was trash, not even in bags, all over the curb in front. The interior hallway up to the third floor was perilously narrow and dark and winding, complete with broken steps, shattered plaster, and lightbulbs removed from the very necessary hall lights. AND the door was fucking locked. To make a long story short, we spent the night in a hotel.

This morning we finally got into the apartment. The guy who lived there before us was gone, but HE NEVER BOTHERED TO MOVE OUT!!! The place was filthy and full of clothes, old food (fridge included), bongs, and tons of other miscellaneous crap. Horrified, we demanded a cleaning crew ASAP. We got one, and they carted all the shit away and cleaned the place so we could finally get our stuff in. This took hours, of course. This place was like your dorm room at its zenith.

The apartment is in dreadful condition, and the decades-old stove has a pilot light that makes it too hot to touch. That means we essentially have a heater in our un-air-conditioned, third-floor apartment on a 90-degree day. To make matters worse, we're surrounded by very noisy, party-hearty undergrads. I'm in absolute hell, and I want to go right home now.

Now, the icing on the cake... something I carefully prearranged and the ONE thing I really wanted to work – the phone – was NEVER set up! And I can't even talk to someone about it until Monday! I might not even get the same number - the one they told me would be in operation by the 15th. Meaning, of course, that I will probably miss your call, and if the army needs me, their calls too. I don't even have the Internet because the jackass here before me didn't bother to cancel his cable, so although I've been trying for weeks to set it up, I still can't. Fuck it, I'm getting a cell phone.

It feels like the whole universe has been telling me not to come here. I'm waiting for a tornado or fire next.

I'm exhausted, love. I'm going to try to get some sleep – hopefully Mickey won't howl all night again.

Yours always & forever,
Alycia

"*No good, Mistah! No good*!"

The camp laborer's cries were lost on me as I painfully emptied my bowels into the hole that served as the toilet seat of our plywood latrine. God, I felt awful. My head was fevered, and my guts were twisted with pain.

The laborer shouted again, "*No good, Mistah! No good*!"

Fuck this guy. *"Shut up! Go the fuck away and let me shit in peace!"*

"No good, Mistah!"

My stomach rumbled again. Uh oh, it's coming up this time! I stood quickly, trying not to dribble shit onto my pants, braced my arms on the seat, and vomited copiously into the toilet. As I wretched, I felt something run down my leg. Fuck!

I reached one hand out to grab the roll of toilet paper, dabbed the drip off my leg, and puked again, holding the roll of toilet paper in place to keep my ass in line. I wretched for half a minute or so before my system calmed down enough to provide a moment of respite. I wiped the phlegm off my mouth and nose with my sleeve and looked down into the hole.

Normally there was a metal bucket on the floor under the latrine, but today the results of my indigestion were splattered all over the concrete pad that served as an underfloor for the outhouse. Oops. I guess that's why the guy was so worked up.

"*No good, Mistah! No good*!

I shouted back through the door, "Yeah, yeah. Sorry, man. You're right. It's no good."

I heard his flip flopped feet crunch through the gravel as he walked closer to the latrine. As I sweated and shivered inside the wooden shack, trying to regain control of my digestive system, he quietly circled it to observe the damage I'd caused to his outhouse. His crunching steps stopped in front of the door. This time his reply was a little quieter, and a little sadder sounding. "No good, Mistah."

I regarded my phlegm-covered sleeve and the fouled roll of toilet paper in my hand as my guts twisted in pain yet again. I was sad, too. We were on opposite sides of the wooden shitter door. Hell, we were on opposite sides of pretty much everything else in the world, too, from women's rights to our feelings about bacon. On the issue of sadness, though, the Afghan tribesman and I were united. Despite our divergent upbringings and experiences, we could at least share this moment.

My head was still spinning as I sat back down on the seat and sprayed more burning liquid onto the pile of intermingled shit and whatever else that was below me. "No. It's really not good at all," I replied. The Afghan walked off, muttering something irritated and unintelligible, and I returned to my intestines' battle with foreign bacteria. Our fleeting moment of shared sadness had passed.

The latrines we used on Sh'kin were simply designed. They looked like Old-West style wooden outhouses, but instead of emptying into a hole in the ground, each outhouse emptied into the bottom half of a cut apart fifty-five gallon drum. This made the process of resetting the latrine a fairly simple but labor-intensive process. Once the drum was full, someone had to drag it off base, fill it with gasoline, and set the whole mixture on fire so the waste could burn off.

This duty, appropriately labeled 'shit burning detail,' was a particularly unloved task. So, whenever possible, the camp commander tried to wrangle the Afghans who were contracted to provide logistic support to the base to do it for us.[11]

However, my priority today wasn't to worry over whether or not I'd eventually have to burn the shit I was producing. That was tomorrow's problem. Today, I just had to figure out how to stop producing it quite so frequently.

As I was mentally wrestling with this problem a fly landed on my lip. My brain sounded a claxon of disgust as I contemplated where that fly had been just a few moments previously, and I lashed out at it, missing the fly entirely and triggering a fresh wave of nausea and cramping.

I collapsed on the toilet seat in despair, swatting ineffectively at the fly yet again as I retched on the floor and sprayed the concrete with more liquid. As I convulsed there

11. Which they did (more or less) without complaint. Except when the commander grew suspicious that they were feeding information to the enemy (which they probably were) and threw them all off base for a few weeks. Then shit burning was left to the privates until negotiations with the new cleaning crew was complete.

helplessly, I felt the fly–or one of its innumerable friends or associates–land on my neck.

Fuck this.

This drama continued to unfold for another half hour or so, until I felt that my system was sufficiently empty that I could safely relocate back to my bunk. I cleaned myself up as best I could, hitched up my pants, and pushed open the door. The dusty sweltering air that hit my face was refreshing, and I immediately felt a bit better.[12]

I grabbed my rifle and stepped out of the latrine. The walk back to my room was only around fifty yards, but I'd need to take it slowly or I'd probably crap my pants. Hunching slightly to relieve pressure on my abdomen and keeping my steps small, I started the trip. As I did so, I looked left and saw the Afghan who'd expressed his concern about my latrine etiquette standing with his two fellow workmen. Our eyes met briefly, and although his face remained completely neutral, I suspected he was marking me for special attention from his Taliban friends who aimed their daily rocket attacks at our base.

"Sorry bro," I muttered. What else was there to say?

I shuffled across the main courtyard, passing the operations center and the chow hall and receiving helpful commentary from my fellow soldiers as I went.

One Charlie Company guy: "Ha! Got the bubble guts, Pace? You gotta stop going down on the workers. It'll give you The Shits."

A mortarman: "Aw man, Pace. You got The Shits? Fuck. I hope that's not going around."

A sniper: "Man, you look like a bag of ass, Pace."

I nodded in response to them all, continuing my slow walk and trying not to vomit until I made it back to the room. I climbed into my bunk, took a couple of Imodium, pulled a watch cap over my head, and tried to make the world stop spinning long enough to go to sleep.

Sobon inquired after my health from across the room, "You good, Pace? You look like smashed ass."

---

12. The relative merits of burn shitters versus Porta-potties was a hotly debated topic in the infantry community. While the Porta-potty requires less maintenance (from the private's point of view) and offers better reading material (the walls are easier to write on), the burn shitter is much better ventilated, and therefore more comfortable when used in hot weather. Which of the two has the better aroma is a true toss-up: do you prefer other people's feces and urine to be served up to your nose *au naturel*, or with a healthy topping of blue chemical biocide?

"Yep. I'm good. Just my turn to have The Shits today, it seems."

The Shits were a part of life at Sh'kin. The CIA guys who ran the base had contracted locals to procure food and do the cooking, and though the Army usually provided its own chow, Sh'kin's remote location meant we didn't have regular access to food from the major bases at Kandahar and Bagram. Consequently, we ate what the locals ate.

Usually this was a good thing. The footbread was fresh, the beans and rice were good, and whatever meat was in the stew (goat? cow? camel? dog?) always tasted great. About once a week, though, the local food sanitation situation caught up with us and we got The Shits.[13]

If you were lucky, like I was today, you got The Shits on a day when you didn't have a mission. If you were less lucky, then you and the people around you got to have an extra pleasant day as you spent the entire patrol wrestling with your various sphincters in and around fire fights.[14]

In either case, the treatment was always the same. Go see the medics for your bottle of Imodium: "Take two tablets now, and take another after every loose stool. Come see me if you run out." Every soldier knew the mantra by heart.

Since most guys' digestive systems never completely stabilized, even between outbreaks of The Shits, plenty of soldiers were on Imodium for the entirety of the deployment: one pill after every trip to the latrine. Since trying to kick the habit risked crapping one's pants during the middle of a four-day patrol, an Imodium addiction seemed like the best bad option.

As I lay in my bunk, trying to go to sleep, I thought ahead. We had a night ambush planned for the next day, and I hoped I'd be over this before then, or it would be a very long mission indeed.

13. As a former holder of a Texas Food Handler's License, I can assure you that the food handling policies in Afghanistan were a bit different from what we were accustomed to back home. For one thing, the bread was called footbread because it was kneaded by the baker's feet. For another, I never saw any meat stored at or below forty-five degrees, unless it happened to be cold outside. Instead, the meat was generally hung from a hook in front of the butcher's shop, where flies congregated on it until someone bought it. Finally, the standard method of cleaning one's ass in rural Afghanistan is to use one's left hand.

14. Many a pair of pants were lost because of combat diarrhea.

August 30, 2003
Madison, WI

Dearest Daniel,

Well, my first week as a CMB graduate student is over. Yesterday dragged by in a blur of twenty faculty talks - we've had nearly a hundred this week, each ten minutes long - so we can hear a little about each lab in the program. We have to rotate through our top three choices, one month each, to see which one is the best fit. Then we will pick a permanent lab to do our research in for the duration of grad school. I also managed to buy the materials necessary for my microbial genetics class, as well as sneaking in two scoops of Babcock's delicious ice cream.

Despite our dislike for our apartment, Camille and I are starting to enjoy Madison. Last Saturday after breakfast, Camille and I walked down to the farmer's market, which is enormous. I saw vegetables I've never heard of and others I'd heard of but never seen. You can buy beautiful, fresh-cut flowers for a couple of bucks, clothes, jewelry, pastries, breads, and more. It's like a carnival, almost. We've also run a few times on beautiful trails along the shores of Lakes Monona and Mendota.

Right after class was over today, Camille and I had to pick up some paperwork from our apartment headquarters. Then we headed to Madhatter's for happy hour with three other CMB students. This bar has $2 pitchers on Friday from 4-9, and they play good music. Plus, it's only about ¼ mile from our house. I had my fair share of beer, and about an hour later we all headed home. So I guess you could say I'm settling in.

I've enclosed two articles for you – one I cut out of the Fort Drum Blizzard *and another one I found online at AfghanistanNews.net.*

The Blizzard *one was about a chaplain who visited Sh'kin, which he found quite personally challenging, saying, "This was a place rumored to be the most evil place on the face of the earth because of all the stuff that goes on." Why the hell do they put things like that in the Fort Drum newspaper? Are they trying to freak people out? Also, isn't it kind of lame for him to complain about being there a few days?*

The other article reads like something from The Onion, *but it isn't meant to be satirical. Apparently, there are crazy rumors in northern Afghanistan that our military has been dropping bloodthirsty killer cats out of airplanes in order to terrorize the population as a more effective alternative to enacting a curfew. Several villagers claim to know someone who was killed by these cats or to have bravely fought off these fearsome felines themselves with little more than a pocketknife. Let me know if you see one. I probably shouldn't hold my breath.*[15]

I think about you all the time. How are you doing? How are the other soldiers handling the stress out there? I am very curious about their reactions. How do the guys who have been there some time feel? And how about most of you guys who just got there? Are you really getting goat stew? If so, what is it like? Any other cool local foods?

Be safe, honey. I'm actually very proud of you for taking out some of the bastards responsible for September 11, 2001...It pleases me to know that you are there to kick their asses, believe me.

Missing you,
Alycia

Sobon had been right about Dickey. There was more wrong with him than it had at first seemed. After the blast, he'd been alright the rest of the day, but the next morning he had woken up unable to move his head or hear anything, and he'd had to return to the aid station. The medics found out that the blast had ruptured his eardrums and messed up his neck and back badly enough that he had to be medically evacuated to Landstuhl. Collazo had been similarly injured, and he was pulled out of country as well.[16]

That left us down two people, including our acting platoon sergeant. But since we were the only mounted element stationed at Sh'kin, the patrol schedule continued unabated.

Thus, with a skeleton crew cobbled together from both First and Second Sections, we prepared to roll out of Sh'kin's front gate to conduct a night ambush mission. Since it was a full-platoon mission, the LT was going to brief us on the plan from a map spread out on the hood of his truck before we left the wire. He didn't usually brief the platoon, but with the platoon sergeant down, he had to, and he seemed a bit nervous.

"Alright men, we're going to set up our ambush on Route Camaro. We've had reports of terrorists moving across the border at night, and we think this will be a good place to

15. https://archive.seattletimes.com/archive/20030811/tigercat11/afghanis-blame-the-us-for-violent-cats

16. The body's ability to use adrenaline to ignore pain still fascinates me. It's even an old joke in the paratrooper community. "How'd you feel after that shit landing? No, wait. Don't bother. Tell me in the morning."

intercept them."

Barret interrupted. "Which one is Camaro again?"

The LT looked irritated. "Haven't you gotten this straight yet? Mark it on your map. Camaro parallels Saturn, but it's one route further from the border."

"Roger, Sir."

"We're leaving the wire as soon as this brief is over. From here, we'll head to the border checkpoint where I'll meet with the Afghan commander and we'll wait until dark. Once it's close to midnight, we'll leave the checkpoint and head east."

He traced our route along his map with a metal pointer.[17]

"First Section, you'll set up here, by the intersection of Saturn and Camaro. From there, you'll be able to see the ridgeline between the routes and have eyes on any vehicles or dismounts coming that way. I'll push the ITAS truck to you so it can help with PID.[18] Any questions on that part?"[19]

Stearns spoke up. "Sir, what's our ROE for any vehicles or dismounts we see?"[20]

"For dismounts, you're looking for unauthorized weapons. It should be pretty easy to pick them out with the ITAS, and if you see anyone carrying anything illegal, he's fair game. For vehicles, stop them and search them. If you assess them to be terrorists, you're authorized to PUC them."[21]

"Do we have a terp for the mission? Can we talk to the people we find so we can tell what they're doing?"[22]

"Nope."

---

17. Always with a pointer. Never with his finger. If there's one thing every infantry officer has drilled into his head, it's that he must never point at a map with his finger. Since a person's finger covers up a thousand square meters on a normal Army-issued map, this is a reasonable practice to reinforce.

18. PID stands for positive identification.

19. The ITAS, or Improved Target Acquisition System, is a large thermal sight. Originally designed to fire anti-tank missiles, the ITAS was the best thermal sight available to most infantry elements, so it was repurposed in Afghanistan to scan areas for insurgents.

20. Rules of Engagement. The list of conditions that described when you could and could not shoot someone.

21. To PUC someone is to detain him. The term is derived from the acronym PUC, which stands for Person Under Custody.

22. Terp is short for interpreter.

"So we just assess whether they're terrorists by looking at what they've got in the car?

"Use your best judgment."

Stearns thought for a second. "Roger, Sir. My best judgment."

The LT continued. "Second Section, you'll be set up here, with me. We'll be covering the west side of Camaro and looking out over the open fields here and here. Anyone that gets past First Section will have to come this way, and we'll have a clear view of them."

The LT paused and started folding up his map. "Any questions?"

There were none, so we spread out to our trucks to get ready to roll. In a few minutes, the trucks were lined up at the gate and ready to leave the wire. Our trucks wound down the road outside of Sh'kin and passed the houses near the fork in the road that had perplexed us so much a month prior. Two little girls ran out of the house and waved at us as we drove by. Eyes sparkling and huge smiles plastered across their faces, the girls leapt in delight as Hogu and I tossed them pieces of candy. They scooped them up cheerfully and hugged one another playfully, comparing their respective hauls.

As alien as the Afghans could sometimes seem, every Afghan kid I met acted exactly like any other kid I've ever seen anywhere else. Aside from these two girls' dresses–full-length colorful robes with head scarves–they wouldn't have been out of place in any Texas yard back home, playing in a sprinkler, throwing a Frisbee, or running around in Halloween costumes.[23]

We took the fork, drove east along Route Corvette, and pulled into the border checkpoint. As usual, the checkpoint had the general feel of a deserted small town gas station. The expressionless gate guard, wearing a mixture of '90s U.S. Army gear and traditional Afghan clothing, waved limply as we rolled past his shack and parked along the western wall.

We had two or three hours until dark, so while the LT and Stearns walked over to the commander's office, the rest of us settled in for some quality bullshitting. After a few minutes, the jaw-jackers settled into two rough groups, each of which was immersed in a different topic of conversation. On one side of the truck line, Lussi and The Hawk were trying to develop a version of "In the Still of the Night" that incorporated Sh'kin, rockets, and getting shot at. Lussi was on lead vocals with The Hawk providing the doo-wop background sounds. Hogu was explaining some of the background details

23. I don't like to think about where those two little girls might be now. It's a thought that haunts me regularly. They're probably nearing thirty in a part of the world that's extremely hard on young women.

about 50 Cent's life and the merits of his music to the other group. Neither subject appealed to me, so I stood in the turret and stared mindlessly across the border. The Pakistani soldiers there were searching a herd of goats, and I half-heartedly wondered if the guards would find anything, explode, or do something else worth watching.

They didn't, at least not during the two hours I watched them work, and as it got dark I pulled out my NVGs to check the batteries, tweak the focus, and clean the lens. As I stood there doing that, I noticed the checkpoint was getting more crowded. Afghan guards started materializing from nowhere, moving in small groups toward the main building where they usually ate chow.

As one of the groups walked past our truck, they gestured toward Hogu and started talking to him.

"Come! Come! Good!"

We looked at one another and laughed.

Hogu was curious. "What do you think they want?"

"No idea, man. Go check it out. It's gotta be better than standing out here."

"Yeah, true. It's boring as hell, and we've still got a couple of hours. I'll see what's up and tell you if it's cool. Maybe they got a hot party cooking in there or something, or at least some good chow."

"Bring me some footbread if they've got it, man!"

Hogu nodded and waved at the Afghan, gesturing that he'd come with him. Hogu grabbed his SAW and followed the group of soldiers up toward the main building.

Lussi and The Hawk were still perfecting their song, so Sobon and I were left alone at the truck to speculate on what might be going on.

"Do you figure they're eating dinner?"

"Dunno, man. I haven't seen the PL come out of his meeting, so I figure the commander is still busy in the other building. I imagine it's just the soldiers doing whatever they normally do in there."

We sat in silence for a while as more Afghans drifted toward the building.

"Man, there must be fifty of them in there."

"No kidding. It's got to be crowded as hell."

"I bet it smells like complete ass in that room. Poor Hogu."

"What do Afghans even do when they get together? Is there a TV in there or something?"

"Right? And there're no chicks. Total sausage party."

"Maybe they'll bring in some goats instead."

We both laughed. Several guys in the platoon had reportedly seen farmers fucking sheep or goats during our various night security missions. The ITAS could pick out such activity from more than a kilometer away, and there usually wasn't a whole lot else to look at.

As even more Afghans piled into the room, we could now hear music playing from somewhere inside. One of them came out with a stack of footbread for the platoon and handed it out.

"Come! Come!"

I shook my head. "No thanks, friend." Then pointing at the bread. "Good! Thank you!"

The Afghan nodded and continued his rounds. I dug into my bread with gusto. Even if it did cause explosive diarrhea, it was hot and delicious. I could always find more pants, right?

The music got louder over the next hour, and by 2200 it sounded like a full-on party. Suddenly, Hogu popped out of the building and started walking back toward the trucks. His posture was a bit deflated and he was shaking his head as he approached us.

"What's up, Hogu? What the fuck was going on in there? Sounds pretty wild."

"Man. I don't know. That shit was weird."

"What do you mean? Were they busting out the hookah?"

"Nah, I'm good with hookah. I thought maybe I could catch a little contact high, you know? But that shit in there got too fucked up for me. A few minutes ago, they brought in some weirdo from Pakistan. It was, like, a boy dressed up as a girl or something, maybe? Then this she-male started dancing, and all the soldiers were clapping and shit. They loved it."

Hogu was clearly disturbed, and his tone was uncertain as he tried to process what he'd just seen. I was briefly stunned into silence.

"What the fuck are you talking about? Like, they brought in a dude stripper or something?"

"Nah, not a dude. Or at least not a grown dude. I think it was a kid."

"That. Is. Fucked."

"Dude, so very fucked. What the fuck is wrong with these people?"

Hogu had inadvertently stumbled upon 'Man Love Thursday,' as one part of the Afghan custom of Bacha Bazi was called by U.S. forces in Afghanistan. I'm still fuzzy on the nuances of the Bacha Bazi, but, basically, young boys are sometimes purchased or borrowed for sexual pleasure by older Afghan men. At other times, the youngest boy–called the chai boy–in an Afghan military unit serves in this role instead. For whatever reason, Bacha Bazi parties often happen on Thursday nights.

"No idea, man. That shit is fucking gross."

"It is absolutely disgusting. I feel like I need to take a shower or something."

Hogu walked back to his truck, shaking his head, and climbed into the turret. He pulled out his cleaning kit and started to oil down the MK-19 for the upcoming mission. His posture expressed what I felt. Confusion. Disgust. Horror. Who the fuck were we working with over here? Why would we be working with a bunch of creepy pedophiles? This seemed about as bad as anything I'd heard the Taliban did–were *these* supposed to be the good guys?[24]

Not long after Hogu's return, the LT and Stearns finished their meeting and returned to the trucks. Stearns climbed into ours as Sobon cranked the engine.

"Alright, ladies. You ready to kill some Taliban tonight?"

I sure was. I needed to get out of this weird ass checkpoint and its pedophile garrison and focus my brain on something else.

"Yeah, for sure. Think we're going to find any?"

"Of course we are, Pace. Of course we are."

"Did the commander say anything interesting?"

"Nah, not really. We didn't have a terp, and his English was shit, so we just sort of back and forthed a bit on what a good job everyone was doing. Then he brought in some food and we ate on the floor. I fucked up when I walked in. There's no furniture in the room, but apparently, some of the rugs are for walking and sitting on and some of the rugs are for eating on. I guess I walked on the table rug. Dude gave me a weird look."

"Did he bring in a little boy to strip for you, too?"

Stearns either didn't hear me or chose not to reply. Instead, as we rumbled into line and out of the checkpoint he talked through the night's plan.

"Once we get to the intersection, we'll orient our trucks to cover south and east with the ITAS truck in the center. Once we get set up, we'll be on sixty-six percent security

---

24. U.S. forces, on the whole, never got used to Bacha Bazi, and it caused friction over the course of the war. For further reading on the friction between U.S. forces and their Afghan partners caused by this custom, see *Grey Zone Ethics*.

for the night. That means six guys from the section are up at all times. One guy on the ITAS. Two guys in turrets with the heavy weapons. Three guys in the prone, scanning with their NVGs. The other three guys get some rack, but they sleep in their kit, ready to roll."

Sobon and I listened and rogered.

"How long are we in position for?"

"Until daybreak. It's going to be a long night. Nobody had better fall asleep, or it's your butt. If some Taliban bastard gets the jump on us and kills me on your shift, I'll haunt you from hell."

"Roger, Sergeant." We were in combat–who the hell would fall asleep?

The drive to the designated intersection was short, and within a half hour we were in position. We killed our engines, and while Stearns got out to prepare the section for the night, Sobon and I started getting our own gear ready. Turn the radio volume down so the squawk box couldn't be heard across the entire area? Check. Prep the range card on the Ma Deuce so the next guy in the turret would have it for reference? Check. Test the PEQ-2 laser? Check. Pull out some snivel gear cause it was going to get cold lying in the prone? Check.

By the time we were done, the rest of the section was ready as well. Stearns and Lussi were finishing the guard roster, and all that was left for the evening was to stay awake and look for terrorists.

A few minutes later, Stearns briefed us on the plan.

"Alright, it's 0100. We're in position until 0700. Pace, you've got first shift. You're up until 0300, down until 0500, then back up until we're done. Sobon, you've got second and third shifts. You're down until 0300, then up straight through the rest. Lussi and I are going to pull the whole thing and sleep tomorrow.

"While you're up, tag out with the other guys who are up so nobody falls asleep. And don't jaw-jack. Just scan your sectors and we'll see what we see. When you're on the ITAS, only have it running for fifteen out of every thirty minutes so you don't hammer the battery."

Sobon and I nodded, and I shifted my weight to lean back and stare out across my sector while the group broke up so that everyone could get to work. I heard Sobon below me trying to fidget around and get comfortable enough in his seat to grab a few hours sleep. It was going to be a long night.

The first two hours passed uneventfully. *Not one thing happened.* The lum wasn't terrible, so I filled the time staring off into the distance at the endless sea of green, looking for movement, trying to pick out constellations, and listening to equipment creak, people sniffle and fart, and all the other sounds of a platoon trying to be quiet. Half-way through the shift, I tagged out with Horn, a Louisiana native from Second Section who lived down the street from us.[25] He took over on the Ma Deuce, and I took his spot on the ITAS.

"You see anything yet?" I whispered.

"Fuck no, bro. Not one goddamn thing," he replied.

I nodded and climbed into the other turret, where I flipped on the ITAS and pushed my eye up to its rubber reticle.

The world instantly changed. Under night vision, as long as the lum is decent, you can see people and objects pretty well–particularly if they aren't too far away. They don't pop out at you though, because NVGs make everything green and monochromatic, so it takes a bit of practice to get used to them. In contrast, thermals aren't like that at all. Viewed through the ITAS, anything that emits heat stands out as sharp white against a dark blue background. Even the best camouflaged vehicle in the world looks screamingly obvious under thermals if its engine has been running, as do people and animals.

So, all of the sudden, instead of seeing nothing but an endless green expanse, I could see *things.* I slewed the ITAS across the terrain in front of me.

At one end of my field of view sat a small dirtbrick house. The walls, like everything else around them, were cold and blue, but there were white areas visible where the metal windows glowed from the heat inside the building. Nearby, a glowing white donkey ambled around the yard, munching something indiscernible that was on the ground. Curious, I clicked the ITAS's built-in laser range finder. Twelve hundred meters. Way too far away to shoot at if I saw something bad. But *was* there anything bad to see? I scanned a bit more. Old car. Random boxes. The remains of a freshly burned pile of garbage. Nothing that screamed 'terrorists live here' that I could see.

I pivoted the ITAS back across to the other end of my sector, where I could see both the road that led from the checkpoint and Route Saturn, the other road that ran along the border. There were no cars on either road. But past them both, I could see a Pakistani checkpoint glowing white-hot in the distance. I zoomed in on it and spent a

25. Horn actually held the keys to his hometown of Lafayette for the Purple Heart he received as part of OPERATION ANACONDA in 2001. His wife made a mean lasagna.

few minutes trying to tell whether one of the guards had an AK-47 or some other kind of gun. Suddenly, in my peripheral vision, I saw a flash of white light.

I pulled my head out of the ITAS. In the distance, coming toward us from the checkpoint, was a car with its headlights on. It was still several hundred meters away, so I slewed the ITAS toward it to see if I could make out who was inside.

As I fiddled with the focus settings, Stearns walked up next to the truck.

"You looking at that vehicle, Pace?"

"Roger, Sergeant."

"It looks like...a Toyota Hilux."[26]

"Got it, Pace. How many people inside?"

"Driver. Maybe two more in the cab. I think I see a guy in the bed, too. I'm not totally sure."

"Weapons?"

I scanned a bit more closely. The car was moving slowly over the bumpy unimproved road, so it was still at least three hundred meters away. We could hear its diesel engine rumbling in the quiet nighttime air.

"None that I can see."

"Keep looking."

Around me, as Stearns called the truck up to the LT, soldiers across the platoon were repositioning to aim their weapons at the Hilux. We were set up for an ambush, and if we got the order to fire, everyone wanted a piece of the action.

The tension mounted as the truck neared our position. Two hundred meters. One hundred and fifty. Lasers played across the vehicle as people tested their aim points. Cheeks squelched down onto stocks and elbows were repositioned for good recoil management.

"Any weapons?"

"Nope. Still not seeing anything."

Stearns echoed the information to the LT again.

One hundred meters. The suspense built further. Thumbs depressed selector levers

---

26. Across the U.S. military, the Toyota Hilux is known unofficially as the 'car of evil' because it's used extensively by all manner of bad guys in Africa, the Middle East, and Central Asia. This speaks well of the Hilux, as the bad guys use it because it's rugged, reliable, and easy to maintain, although I'm not sure the brand loyalty by organizations like Al-Shabaab and the Taliban easily translates into good marketing material for Toyota.

to switch rifles from 'safe' to 'semi.' Everyone wanted to blast the hell out of this vehicle. Wiping out a truck full of terrorists would feel *so* good. It would validate everything. The long nights. The sickness. The explosions. The injuries. The endless guard shifts. The shitty phone. The family separation. The deployment. The war. Our decisions to join the Army. *Everything.* We just needed to see *one* illegal weapon.

Fifty meters. The truck stopped. The door opened and after a few seconds, the driver got out. The platoon's vision was washed out by his headlights, so nobody could tell if he was armed. Everyone tensed in preparation for something to happen.

"What's he doing? Why'd he get out?"

The headlights didn't affect the ITAS, so I was the only one who could see the guy clearly. "He's just standing there. He's not armed."

"He's just standing there?"

"Yeah. He's just leaning on the door and looking around like he's waiting for something."

Stearns, who'd been leaning across the hood aiming at the vehicle as we talked, stood up.

"Well, crap."

He reached for the hand mike in the Humvee.

"Killer Six, this is Killer One, over."

"This is Six, over."

"This guy is unarmed. He's just standing there. What do you want us to do, over?"

The tension eased as uncertainty replaced enthusiasm. Guys picked their heads up from shoulder stocks. Weapon barrels drooped toward the ground. Lasers flipped off.

The radio was silent for a solid minute. Stearns tried again.

"Killer Six, this is Killer One, did you monitor my last transmission, over?"

"This is Killer Six. I monitored. Umm...Break. Just...Break. What do you think the guy wants, over?"

Stearns stared at the hand mike for a moment. "Killer Six, this is Killer One. No idea. Might be getting ready to kill us. Might just be trying to drive past us. If he was going to shoot, though, I think he would have already done it. Break. You want us to search the car, over?"

"This is Killer Six, roger, search the car, over."

"This is Killer One. Roger, over."

Stearns turned to Sobon. "Come on, Sobon. Let's grab Lussi and figure out what this guy is up to."

As the rest of us covered them, the trio walked up to the car. Stearns approached the driver while Sobon and Lussi circled the Hilux from the other side to look in the windows. Stearns motioned for the guy to turn off his truck, which he did, so it was quiet enough for us to hear the situation unfold.

"Sir. Please step out of the way so we can search your vehicle."

Blank stare.

"You" hand gestures.

"Go there" more hand gestures.

"Me search" even more gestures.

The driver started talking. He was a young guy–probably in his mid twenties– and he didn't sound terribly excited. He spoke for fifteen seconds or so, accentuating his words with hand gestures. Not one of us understood a single word. Nor could we glean whatever concept he was trying to impart from his gestures.

Stearns tried again. "You. Go. There."

More words. More lack of comprehension.

Frustrated, Stearns grabbed the guy by his shoulder and physically pushed him to the spot where he wanted him to stand. The guy didn't resist, but when he got there, he said more words and walked back to the car.

Stearns looked across the car at Lussi. "What's this guy's deal?"

Peering into the truck, Lussi replied, "Looks like the cab has a bunch of people in it. Bed has a couple more. Maybe he's trying to get them out. I don't see any weapons."

"Yeah, maybe so."

Stearns turned back to the guy. "All of them. Out. There." He circled his arm in the general direction of the car for emphasis.

The guy now appeared to understand what Stearns wanted. He opened the rear driver's side door and talked to the people in the cab, and they started piling out, one after the other, until seven people–the driver, two women, another man, and three kids–were standing alongside the pickup. The two young men in the bed climbed out as well.

"Holy shit," I said to nobody in particular. "How'd they fit that many people in that cab?"

Lussi and Sobon started searching the vehicle, while Stearns talked to the assembled group.

"Where are you going?"

No response.

"Do you have any illegal weapons?"

Nothing.

"Do you understand anything I'm saying to you?"

Silence.

Stearns shook his head in frustration. "Lussi, you find anything?"

Lussi, who was digging around in the rear of the cab, popped his head up. "Nope. Nothing in here but some blankets and a sack of, maybe, onions? It does smell like absolute ass, though."

"Sobon, you got anything?"

"Nope. Nothing here."

Stearns looked back at the group of people. They were all wearing long flowing garments of one cut or another and flip flops. The men had beards and mushroom hats. The women's faces were invisible behind their head scarves.

Stearns looked like he was weighing his options. Should he search these people? Let them go? Ask the LT for guidance? The Afghans might have been concealing something under their robes, but it couldn't have been a machine gun or an RPG. Were they terrorists? Did he have any way of determining that? It was impossible to tell.

"Go. Leave. Drive wherever you're going. Get out of here."

The group stood, staring at Stearns. They appeared confused.

"Go!" He waved his hand at them. "Go! Go!"

The driver slowly moved. When Stearns didn't kill him, the rest of the Afghans cautiously started to walk back to the Hilux. They piled back in, one after another, and the driver started the engine, pausing to look at Stearns again.

Stearns again waved his hand. "Go!"

The Afghan ground the truck into gear, then started slowly rolling along the road, passing between our trucks and continuing down the road. In a few minutes, only the headlights and taillights were visible and we could no longer hear the diesel engine as the Hilux picked its way down the unimproved road.

Stearns walked back to our truck in silence. Lussi and Sobon followed. Nobody said anything for several seconds until the squawk box came to life.

"Killer One, this is Killer Six. What happened? Why did the truck leave? Give me an update on the situation, over."

Stearns stared at the hand mike for a moment before picking it up and responding.

"Killer Six, I'll come to your location, over."

Stearns walked off to express whatever frustration he had with either the situation or the LT in person. The rest of us resumed our security positions. We knew we would never hear whatever words passed between Stearns and the LT, so we did what privates have always done when presented with the unknown–we speculated.

"Hey, Lussi."

"Yeah?"

"What the fuck just happened, man?"

"Well, I guess they weren't terrorists, so we let them go."

"Go? Go where? Where does a Hilux stuffed full of Afghans go at two in the morning?"

"Dunno. That way, I guess. Their house? The grocery store? Disney World? Who friggin knows?"

"And where the hell was the LT that whole time?"

"His truck, I imagine."

"Kind of left Stearns in the lurch, didn't he?"

"I guess he could have probably been up here providing a little more guidance, yeah."

"Seems like a pattern."

"Maybe so, yeah. He's definitely not the most engaged leader I've seen."

We were quiet for a while, thinking.

"It's a bit silly for us to remain in this ambush site at this point, isn't it? I mean, isn't that Afghan going to get to his house and call every person he knows to tell them we're here?"

"Yeah, probably so. Having a truck drive through your ambush site pretty much compromises it."

"So, we're breaking down the position then?"

"I think you know the answer to that question."

I nodded in the darkness. I did indeed.

Lussi walked off to check on the other guard positions, and I continued to scan until 0300, when I woke Sobon to take over for me.

"Did I miss anything?"

"Not really. A car full of Afghans came through. Did you wake up for that?"

"Yeah, I did. I searched the vehicle, remember? Anything else?"

"Not one goddamn thing."

"Hooah."

Sobon flipped down his NVGs and climbed into the turret to take over on the ITAS. I grabbed my rifle, laid down in the prone next to the truck, and let my head fall forward to try and catch a little sleep. It was impossible, of course. I was still spun up from the near-ambush, and every time I closed my eyes, I imagined a bearded Taliban guy with an AK creeping up on the truck and killing us. Still, the opportunity to turn off my NVGs for a while and relax my eyes was welcome enough that I quietly dozed, waiting to see if anything exciting would happen.

Which it eventually did.

I had just checked my watch–0440, almost time to go back on shift–when the night behind me was lit up by the distinctive flash and *THUMPTHUMPTHUMP* of a MK-19 grenade launcher firing. The thumps were quickly–too quickly–followed by a series of loud explosions, which meant whatever the gunner was shooting at was close.[27] I flipped my NVGs back on and turned to look at where the fire was coming from. It was the LT's truck!

The radio suddenly came to life. The volume was turned down so low that I could barely make out the call.

"Contact. Dismounts maneuvering in the open. Three hundred meters."

The firing was behind me, but I couldn't ditch the area I was covering. My own sector was infuriatingly devoid of terrorists.

"Sobon!"

"Yeah?"

"What's up, man? Can you see anything?"

"No. And there's no other shooting, either from the enemy or our guys."

The radio remained quiet, and there was no more firing from anyone. I could hear the crunch of people–probably Stearns and Lussi–making their way over to the LT's truck in the darkness, but otherwise, everything was still. Long minutes passed. What was up? Were we under attack? I continued to scan the high ground in front of me.

More minutes passed, until Sobon whispered down to me, "Stearns is coming back."

Then, I heard, "Alright, ladies. Pack it up and get ready to roll back to base. This ambush is over."

---

27. MK-19 grenades fly very slowly. You can actually see them as they fly through the air. This means that in most cases, the time between when you hear the *thump* of the MK-19 firing and the explosion of the grenade landing is several seconds.

"Huh? We're leaving? What happened, Sergeant?"

"We'll talk about it later, Pace."

Irritated with the lack of information, but happy to be doing something other than lying in the prone, I brushed myself off and climbed back into the turret while Sobon prepped the truck to move. In moments, the trucks were noisily rolling back to Sh'kin.

"Alright, Sergeant. Spill it. What the fuck happened?"

"Somebody got jumpy and fired at what they thought were enemy forces."

"Somebody?"

"Yeah. Somebody."

"Who"

"I told you. Somebody. Anyway, by the time we got over there, it was obvious there wasn't a squad of terrorists maneuvering on the platoon. In fact, we couldn't see evidence there was anyone out there at all. It was probably a dog or something. Still, between the Hilux and the grenades, it was pretty obvious that everyone in the surrounding area knew we were here, so the LT decided we'd break down for the night. Mission accomplished."

Mission accomplished indeed. Well, at least we weren't lying in the dirt anymore.

Nobody spoke the rest of the way home. I think we all shared a general sense of disappointment and embarrassment about our performance, and nobody was interested in discussing it much.

We made it back to Sh'kin without incident and started our usual post-mission reset. Once the leadership left to put together their debrief, Sobon and I compared perspectives on the mission as we worked.

"Well that was dumb."

"Yeah, it was. I get the sense that nobody in charge really knew what the hell to do out there."

"Stearns seemed good enough."

"Yeah, but he's not the guy calling the shots. With Collazo gone, it's really just the LT, and whatever 'it' is, the LT just doesn't seem to have it."

"Yeah, true. I don't know if he's scared, or just unconfident, or what, but he sure as hell isn't taking charge."

"Ah well, at least we're not doing anything important, right? I mean it's not like it's life or death."

I laughed, figuring Sobon would laugh, too, but he didn't. He just looked at me for a few seconds then shook his head.

"I dunno, man. I guess I don't think that's funny anymore. Since Dickey got hit and we had that rollover, I'm failing to find much humor in near death experiences."

"Ah, lighten up. We'll be fine."

"Yeah. Maybe. Or maybe we won't."

I finished oiling down the Ma Deuce, and Sobon finished powering down the radios and locking up the grenades. We gathered our gear and walked back to the hooch.

"So, who do you think fired those MK-19 rounds?"

"No idea. I'm sure we'll hear the details at some point, though."

"Yeah, probably so."

We got back to the room, dropped our gear, and got ready to grab a few hours of sleep. As for the details on what actually happened, we never did find out which 'somebody' fired the MK-19, although we *did* find out pretty quickly what he'd fired at.

As it turned out, the mysterious figures had been neither dogs nor terrorists, but they also hadn't been figments of the gunner's imagination. Instead, they'd been a farmer and his son walking through their fields in the early morning hours. Later that morning, the farmer had arrived at Sh'kin to demand compensation for the bloody wounds his boy received from the grenades. The base commander–Captain Lopez–paid the man the required indemnity, and the farmer left without further complaint.[28]

The farmer's family had survived the incident, but the LT's career as a combat leader did not. The failed ambush convinced the battalion commander that the LT's judgment couldn't be trusted, and he'd been pulled him from the position.

Nobody in the platoon was sad about the decision, but now we were even more short-handed, and we were entirely bereft of senior leadership. Between the loss of most of our leaders and the injury of Dickey and Collazo, the platoon was starting to fray a bit around the edges, and we needed someone competent to take charge and straighten things out before they got worse.

---

28. The system of condolence payments seemed strange to me at the time, but over the course of the GWOT I got used to the idea that direct fiscal compensation was a legitimate way to reimburse someone for the injury or death of a family member. It sort of nested with the Pashtun system of justice, and it worked about as well as anything else we tried.

II

September, 2003: Firebase Sh'kin, Afghanistan

I woke up shivering and sweating so heavily I thought I might have pissed myself. It was dark in the hooch, but I could see flashlights moving around me. Lussi was staring up at me, blinding me with his headlamp.

"Pace. Wake up, we gotta roll."

"Lussi. What's up, man? Something happen? We on QRF?"

"Yeah, we gotta go. Scouts got hit or something, I'm not sure yet. Second Section is rolling out, but we're going with them since they're short on dudes. You feeling alright? You look like hell."

My head swam. "I feel like ass. Dunno what's up."

He looked at me for another few seconds. "On second thought, stay here. I'll pull Hogu instead."

I nodded, only half comprehending what he was saying before rolling over to go back to sleep.

When I woke up again, the light coming in through the cracks around the door and window indicated it was late morning, and most of the platoon was asleep around me. I felt sweaty, but otherwise better, so I climbed out of my bunk to go find some chow. I slipped on my flip flops quietly to avoid waking anyone else up and stepped outside the hooch. After recovering from the blinding light, I walked over to the kitchen, grabbed a granola bar and a Gatorade, and made my way to the screened-in chow area. The Hawk was there eating, so I sat down to join him. He looked terrible. His eyes were red, and he was staring off into the distance, seemingly lost in thought.

"What's up, man? Did you go on the op last night?"

My question seemed to snap him back to the moment, and he looked at me for a second before replying. "Yeah, man. Shit was fucked up. Two guys got killed."

"*Killed?!* Fuck! Who?"

"Fuller and Thomas. From the scout platoon."

"How?"

"They were in a 24 hour hide position watching for enemy movement. I don't know the details, but at some point this morning, something happened, and somebody crept up on them and fucking shot them. The other two scouts were in another position nearby and started shooting at the enemy forces, but it was too late. The QRF mission was us responding to the radio call and recovering the bodies."

I listened to The Hawk in stunned silence. Holy shit. Dudes were dead. I didn't know Fuller or Thomas particularly well–they were just guys I'd seen around the company area–but it still seemed unbelievable that they'd been killed.

"Did they kill the fuckers who shot them?"

"No. They got away."

"That seems real fucking weird. How'd that happen? I mean, did nobody see them? Weren't they scanning their sectors?"

"Dunno, man. Nobody's sure what happened out there, and none of the rumors make much sense."

I looked at The Hawk. It was obvious he'd been crying and was having trouble dealing with the situation. "You alright, man? I know you were buddies with them. They partied at your house a few times, didn't they?"

The Hawk nodded. He was a social guy and was on close terms with almost everyone in the company. He was also surprisingly sensitive, once you got past the thorny exterior that we all had to maintain in order to avoid getting ridiculed to death.[1]

"I don't know what his wife's going to do. They haven't been married long."

"Man, I didn't even think about that aspect. Alycia would melt down if I got killed over here. What's that even look like?"

"Don't know, Pace. I guess she'll have to leave housing and move somewhere else or something."

"Really? The Army just throws her out?"

"That's what I heard. Soldier's dead? Out goes the family. Paycheck stops too. Figure

1. Sometimes literally. It's a brutal community.

it the fuck out."

"Man. Fuck. That'd be a lot to deal with."

"Exactly. What's she supposed to do? Move back in with Mom? Marry someone else real quick before she can't make rent?"

"I guess she gets the life insurance, though, right? 400K?"

"Yeah, maybe. I wonder how long it takes that shit to pay out."

"No idea. Probably a long-ass time. You'd think the Army would have figured this out by now."

"Man, fuck the Army. The Army don't care about our families."

"The big green machine keeps rolling along, right? There're always more dudes to fill in the gaps."

The Hawk only nodded in response. He was usually pretty talkative, but he didn't seem to have much conversation left in him this morning.

It was hard to tell how much of what we thought the Army would do if we got killed was fact and how much was fiction. There were all sorts of rumors floating around about what your family would or wouldn't get if you died, how long the surviving wife would be able to stay in housing before she had to leave, and what they would do with your remains. We'd all received briefs before. But without context, none of the information meant much, and most of us had slept through half of them anyway. Of course, our uncertainty was compounded by the fact that this early in the war, the leaders who were briefing us usually didn't know all that much about the processes either.

Over the next twenty years, 'casualty notification' and 'casualty assistance'–which is what the Army called the processes employed to help surviving family members get on with their lives after soldiers were killed–became battle drills that units rehearsed just as often as they rehearsed attacking enemy positions or dropping bombs.

Black cars drove dress-uniform-clad officers around to give the 'I regret to inform you' speech to tearful widows. Status blocks on slides were changed from red to green. Memorials were conducted. Posthumous medals were announced. Wives were informed of their rights, benefits, and the support that was available. Gear was inventoried and turned in. Families moved out of housing. New families moved into housing. And so it went. And so it goes.

However, in 2003, combat deaths were still fresh and confusing, and nobody had used the systems often enough to fully understand them, much less be able to communicate the pertinent details to the guys on the ground. And so, we guessed our way through things and imagined how we thought they would turn out. In the end, as with so much else, most of what we thought we knew turned out to be nothing more than apocryphal bullshit. At least speculation provided us the illusion of certainty, which brought us a degree of comfort. And comfort, in that time and in that place, was a sorely needed commodity.

September 1, 2003
Madison, WI

My love,

In many ways, yesterday was pretty hard on me... I saw on the Afghanistan news site that two soldiers were killed near Shkin in a 90-minute battle. That was a terrible feeling, I assure you. I felt like frozen dog shit, if that makes any sense —both numb and horrible. I tried to be normal, so I went with some girls downtown to the food festival (Taste of Madison), but I literally was having trouble breathing because I was so upset. Pretending everything was ordinary was just too hard. Finally, Nelly Collazo called to tell me that the KIAs were indeed in your company, but they were scouts. She didn't know the names yet. I started crying then, so I walked home alone. This brings home very strongly the fact that it could be you or Sobon at any time. You have no idea – and probably never will – what it's like to hear that your spouse is possibly dead or wounded, but to have to simply wait. And wait. It's unbelievably stressful even though I try to be normal on the exterior. I just felt alternately ill and pumped with nasty stress hormones. It's been very taxing.

I need you, honey. You're the foundation of my entire life, and right now, that foundation is undergoing a small earthquake. I don't want to live without you, and I don't think I would. People really can die of grief. Basically, I'm telling you that you have to come home.

Love always,
Alycia

In *Dune,* Frank Herbert describes Gurney Halleck as "an ugly lump of a man." He further describes Gurney as loyal, fearsome, and possessing a good sense of humor that easily endeared him to his guys.

Our new platoon sergeant, SFC Gilbert, exactly matched that description. He'd shown up fresh off the bird from Kandahar just as we were getting ready to leave on the day's mission, and he'd made an immediate impression.

I had been oiling up the Ma Deuce when he walked up. Seeing his rank, but not knowing who he was, I had stiffened to parade rest.

"What's up there, killer? At ease. Keep doing what you're doing."

"Just getting ready for the mission, Sergeant."

"You taking good care of that gun? She ready to kill some motherfucking Taliban today?"

"Roger, Sergeant, she's good to go."

"That's good to hear, 'cause I'm coming with you today, and if we get hit, you better light them motherfuckers up."

"Roger, Sergeant."

"I'm your new platoon sergeant, by the way. Good to meet you."

"Good to meet you too, Sergeant."

After that, Gilbert walked off to talk to the rest of the platoon. He had obviously been checking up on us to see if we knew our business, but his raspy voice and gruff-but-good-natured demeanor made him come off like he was just bullshitting with the boys.

When we'd left the wire for a patrol up Route Camaro, Gilbert and one of the Charlie Company platoon sergeants, SFC Philpot, had climbed into our truck for the mission.

Fast forward another hour. We were now rolling north past the spot where we'd set the night ambush the month before, and Gilbert and Stearns were deep in a conversation about current events.

"I don't get it, Stearns. What do you mean you haven't been able to kill these sons of bitches? Why not? You being aggressive enough?"

"Roger, Sergeant. We're shooting plenty–when we have something to shoot at. It's not that clear most of the time, though. Haj just looks like Haj, and they've all got friggin AKs, so we can't tell who's bad until they shoot at us."

"Hmph. Well, that shit is going to change. We're going to find those bastards and kill the fuck out of them." He smacked me on the leg. "Right, Pace?"

"Roger, Sergeant!" I hollered down from the turret. I still had no idea how to kill the right people, but it seemed appropriate to reply enthusiastically.

"Fucking hooah!" He replied.

The day was dusty, dry, and hot as hell as we bumped along Route Camaro for another half hour. To our left–west–the terrain opened to the horizon. To our right was the ridgeline that separated Route Camaro from Route Saturn, and past that was Pakistan. I looked behind us, where I could see Hogu sucking down the endless cloud of dust from our vehicle. He had a dust scarf wrapped around his mouth and nose, but I knew he'd still be coughing up brown crap for the rest of the day. Such is the gunner's life. At least I was in the lead vehicle today where dust was much more manageable.

I was bored, and it was hot, and before long I was zoning out a bit thinking about whether or not Carr–a buddy of mine and the only U.S. cook on Sh'kin–would be able to scam us some frozen pizzas tonight since we were going to miss dinner.[2]

Suddenly, my idle speculation was interrupted by what sounded like sleet hitting the truck.

pingpingtinkpingtinktinkping

I looked around confused for a second. We were in the middle of the desert. Why did it sound like it was sleeting? Then I heard the *crack* of something zip by my head. Shit. We were getting shot at. I looked left and saw faint puffs of smoke on the mountainside a few hundred meters to our right.

"*Contact right!*"

I wheeled the Ma Deuce to the right and started blasting the mountainside.

BOOM BOOM BOOM BOOM

Hogu's MK-19 quickly joined in.

BOOM BOOM THUMP BOOM THUMP BOOM THUMP

As Sobon accelerated out of the kill zone, the Humvee's burst of speed sent the gun bouncing, which meant most of my fire went wild. Still, the sheer volume of fire caused the mountainside to be shrouded in clouds of smoke as the heavy machine gun bullets and high-explosive grenades struck it repeatedly.

2. Preoccupation with chow was a constant during my time in the infantry. When's the next meal? What's the next meal? How can I scam more or better food? In a world that was so often dusty, hot and miserable, the next meal was something positive one could focus on when the rest of the world went to hell. Carr understood this. On the days when we had a double patrol or got ambushed, he always found a way to get us something good to eat, and we loved him for it.

Gilbert started shouting. "Bound! Bound! Hold what we've got here, Sobon. Pace! Suppress those fuckers while the other truck pushes past us."

The Humvee jerked to a halt as Sobon hit the brakes. I was facing sideways, so the sudden change in velocity caused me to smash my forearm into the front of the turret and jerk the gun sideways.

I recovered my footing and got down behind the gun to improve my accuracy. Bullets were still *tinking* and *plinking* off the sides of the truck around me, but we were far enough away from the AK fire that it wasn't very effective.

BOOM BOOM BOOM

BOOM BOOM BOOM

Now I could really zero in on the enemy's puffs of smoke, and after a few bursts from the Ma Deuce whoever was shooting at us stopped.

I heard Hogu start firing again from somewhere ahead of us which meant his truck must have bounded into position.

Gilbert shouted again, "Move! Move! Bound past them now–a hundred meters!"

I braced myself this time, and held the gun so that we wouldn't flag Hogu's truck as we roared past them. I could hear grenades exploding in the distance, but by the time we stopped, I saw no more enemy fire. I hollered down the turret hole, "They stopped shooting–I think they're either dead or gone."

"Well, we won't know until we clear through the objective, will we?"

I looked across the valley toward the ridgeline we'd taken fire from. The base of the ridge was easily three hundred meters away. Also, the shooters had been a good way up the side of the mountain–probably two hundred meters up or so. That meant clearing through the objective would entail walking across open terrain toward an unknown enemy force in an elevated position, and then climbing up to meet them. We had five total dismounts in the convoy.

Stearns beat me to the observation. "Sergeant, I'm not sure we've got the men to do that. Even if we set the trucks up to provide supporting fire, we'll need a bigger element to get up there and clear the objective."

Gilbert thought about it, analyzing the situation. "Yeah. I guess so. We gotta get Charlie Company out here then."

"Yeah, we'll need their guys for sure. It'll take them an hour to get out here, though. We gotta swing around the ridge and cut off the Taliban. Otherwise, they're going to break contact into Pakistan."

"Well let's get the fuck over there then."

Sobon turned the truck around, and we drove back south on Route Camaro to get around the ridgeline. While we bumped along, Gilbert and Stearns coordinated the Charlie Company support over the radio. They also requested close air support to cover our movement.

I was excited. This had morphed from a long, boring presence patrol to an effort to destroy actual Taliban forces. We were even getting air support.

The drive took around forty-five minutes. Even trying to hurry, the roads were bad enough that we could only drive so fast without battering the trucks to pieces or blowing a tire. By the time we were in position on Route Saturn and scanning the backside of the ridgeline, I was pretty sure whatever enemy forces had been up there were long gone. Sill, maybe the aircraft would see something.

Charlie Company was still ten minutes out when the aircraft–two Harriers–arrived. Our truck didn't have a radio on their net, so all I could do was speculate about what they were going to do.

I watched as they first flew over the targeted ridgeline, then banked around to fly over our position. It looked to my untrained eye as though they were lining up to drop bombs on the area where we'd seen the Taliban. As they flew overhead, I watched in horror as three large sparkling bits of light dropped out from under one of the planes and started falling toward our truck. Shit! They'd fucked up and dropped on us.

The world slowed down as I stared up at the falling objects. I knew I should yell *'incoming'* to give everyone a heads up, but nothing came out of my mouth. I was totally fucked. What a shit way to die!

Suddenly, each of the balls of light vanished into the air. Confusion replaced panic in my mind. What the hell? Where'd they gone?[3] I didn't have time to think through possible answers, though, as right then, the Harriers executed the live drop, hammering the ridgeline with a pair of cluster bombs that exploded, and then exploded again, in a shower of glorious destruction. The bombs left a huge area of the ridge in smoking

3. Unbeknownst to me at the time, fighter jets commonly drop chaff when the threat of surface to air fire is high. Chaff is a countermeasure that aircraft drop to disrupt incoming missiles. It consists of a series of bundles of aluminum or zinc that fall, sparkling, from the rear of the plane, and it looks roughly (to my untrained eye) like a flare. I learned all of this later and felt like a total newb for being freaked out.

ruin, blasting shrubs, rocks and whatever else was up there to pieces.[4] The panic from a moment ago was forgotten. This was the coolest thing I'd ever seen.

"You see any movement, Pace?"

"No, Sergeant. I don't think anything survived that."

"Yeah, maybe. We'll see when Charlie gets up there."

Charlie didn't get up there for quite a while–several hours in fact. Once their trucks arrived and the infantrymen dismounted, they started the long climb up the steep slope. From our overwatch positions, we covered their advance, ready to blast anyone who started shooting with our heavy weapons. But no enemy materialized, and when Charlie Company reached the top, their radio calls confirmed what we already suspected.

"Killer Seven, this is Cobra One One. Two EKIA up here. Recovering the weapons, then RTB, over."

Gilbert seemed satisfied, but a little disappointed, as he returned the call. "This is Killer Seven. Copy all, over."

"Guess there were only two of the fuckers up there, then. Shame there weren't more."

"Maybe there were. Maybe the rest took off."

"Yeah, maybe. We gotta find a way to respond faster. Can't give those fucks a chance to get away."

I nodded. "Roger, Sergeant. How though?"

"Don't know, Pace."

The truck lapsed into watchful silence as we considered the question. It took a few more hours for Charlie Company to get back to their vehicles and mount up. By then it was dark, and we drove back to Sh'kin without incident.

Once we arrived, Gilbert got out to bullshit with the platoon while we cleaned up the trucks and reset everything.

"You guys did well out there today. Good job getting on that gun and immediately returning fire."

"Thanks, Sergeant."

"You did good, too, Sobon. Handled the truck like a champ."

"Roger, Sergeant."

He left our truck to check on the rest of the platoon, slapping guys on the shoulder

---

4. To be honest, the ridge wasn't in amazing shape before the bombs. After centuries of deforestation, high winds, and dust, one might even say it had already been ruined long before we got there. The smoke, though, was our doing.

and cracking jokes as he went.

Gilbert had been good to work with. He'd been aggressive, but not irritable, and he'd given guidance when we needed it–something the old regime had struggled with. If this was what having a solid platoon sergeant felt like, it seemed we'd be okay.

"I got a bad feeling about this one, Pace."

"You have a bad feeling about all of them, Sobon."

"Well, I'm usually right, aren't I?"

"Yeah, maybe."

The new lieutenant–tonight was one of his first patrols with the platoon–was silent in the passenger seat of the truck.

The drive up Route Saturn was always tense, particularly at night. The route ran directly along the Pakistani border, and it looked exactly like the sort of place to expect an ambush. The road itself was a broken dirt track littered with large sharp rocks, so vehicles had to move along it slowly or risk a flat tire. The terrain around it was steep and mountainous, inaccessible to wheeled vehicles and perfectly positioned to conceal anyone who wanted to take three or four hundred meter shots at passing vehicles with rifles, machine guns, or RPGs. Finally, since the route paralleled the border, enemy forces could hide in Pakistan and attack with impunity along most of the route. We could fire back, but we couldn't pursue, and any aircraft we called in would be limited in what they could do.

There were Pakistani Army checkpoints that nominally secured the route, but the last month-and-a-half had taught us these weren't to be trusted. Theoretically, the Pakistanis were our allies, but we'd been attacked by enemy elements hiding right under the checkpoints' noses too many times to have any faith that they were on our side.

As our truck rumbled north, I could see one of the checkpoints through my NVGs. It sat on a ridgeline a kilometer east of us, and in the glow of its lights, I could see silhouettes of a few people walking around its walls. Other than the checkpoint's lights and the limited starlight, it was a pitch dark, hot, and completely still night.

I scanned, looking left and right at the tops of the ridgelines and ahead at the road in front of us. The night vision made everything look grainy and green, and the odds of seeing an IED or a hidden assailant were low, but trying was better than doing nothing.

I looked back at the other truck–the ITAS truck–and frowned. The ITAS was great for scanning for hidden enemies, but it couldn't be used while the vehicle was moving, and the size of its sight meant that it couldn't be mounted alongside a heavy weapon like an MK-19 or a .50 cal. Instead, the truck had a relatively puny M249 SAW. To make matters worse, the ITAS truck was a light-skinned Humvee. It was the last of the original trucks we'd brought from Drum, but since we didn't have enough up-armored trucks, we had to keep patrolling with it.

Normally, since the ITAS truck was brought along as a third vehicle in our convoys and never took point, none of these deficits posed a problem as the other two trucks provided the heavy weapons and the armor. Tonight, though, the MK-19 truck had broken down at the main border checkpoint, so we'd had to leave it there while continuing the patrol with only two vehicles. This also left us with a total of ten guys, two radios, and one machine gun against whatever Route Saturn might have in store for us.

Sweat ran down my neck. Sobon was right. Something did feel off tonight.

Our trucks continued to creep slowly along the trail. I heard the PL's voice come up through the turret hole. "Pace, you see anything?"

"No, Sir. Nothing but that checkpoint. It's still as hell out here tonight. Nothing's moving."

"Yeah, I agree. I kinda hate it."

"Yeah, me too. It's too quiet."

A few minutes passed. I looked over my shoulder at the ITAS truck to make sure it was still with us. No issues.

"Those ribs were good tonight, right?"

"Yeah, Sir. They were badass. The cooks must have been showing off for you."

"Maybe so. They must have heard what a big deal I am. First Lieutenant is a pretty distinguished rank."

I laughed, "Yeah, that's probably why."

I thought for a second. "On second thought, how would you even know what good ribs taste like, Sir? Aren't you from Massachusetts or something?"

"Connecticut, actually."

"Same difference. Shit, Sir, you're a damn Yankee. You probably wouldn't know good barbecue if it bit you in the ass."

"Quiet there, Southern boy, or I'll have my Union bros come carpetbag your ass again."

I laughed. Carpetbagger jokes–maybe this PL wasn't so bad.

I looked back, checking on the ITAS truck, and saw that it had stopped. Our new Platoon Sergeant, SFC Gilbert, was in it, and we heard his voice come across the radio.

"Killer Six, this is Killer Seven."

"Go ahead Killer Seven."

"We got a runner! Moving up the hill to chase him down."

I jerked my head back. Gilbert and another guy had jumped out of the doorless truck and were sprinting east to climb to the top of one of the nearby ridgelines. From below, I heard the PL on the radio.

"Killer Seven, say again. Do you have visual contact with the enemy?"

Then, off the radio, to us. "Did Gilbert just say he was fucking assaulting the enemy?"

"He's out of the truck now and running up the hill with another guy."

"Fuck!"

Back on the radio. "Killer Seven, this is Killer Six, over."

His driver's voice came across the net. "Uh, Killer Six, this is Killer Seven Delta. Seven Actual left the vehicle."

"Killer Seven Delta, does Seven Actual have comms?"

During the pause that followed, I scanned the ridgeline. I could see Gilbert's small team moving up the steep ridgeline. They were picking their way around the shrubs and boulders. It looked like a hard climb. I swerved the Ma Deuce in their direction to provide covering fire if necessary, but I didn't see any enemy elements. Fuck, it was dark.

The radio came back to life. "Killer Six, Killer Seven has no comms, over."

To us. "Goddamnit, what the fuck do I do with this?"

We assumed the PL was being rhetorical and let him think.

"Alright Sobon, we gotta get up there and cover him." The PL pointed northeast to another section of the ridgeline. "Can you get the truck up there? I think we'll be able to support from there."

Sobon looked to where he was pointing.

"Yeah, I think I can, Sir. We might blow a tire, though."

"I'll take that chance. We can't leave the platoon sergeant exposed. Get us up there."

"Roger. Pace, help me out here. I can't see shit through this tiny windshield."

"Gotcha, man. Mind not rolling us over this time, though?"

"Ha. You're a funny motherfucker, you know that?"

Sobon spun the steering wheel and started maneuvering the truck off the trail. Looking ahead from the turret, I guided him left and right to avoid the biggest rocks I could see, but it was still hard going. The truck lurched alarmingly a few times as the front tires hit things I couldn't see under night vision, and I could feel myself tensing for a potential rollover. After a few close calls, though, we made it to the top of the ridgeline without incident.

As we crested the ridge, the terrain opened up in front of us. Our ridgeline, which marked the approximate border between Afghanistan and Pakistan, overlooked a wide empty valley. On its far side was a higher ridgeline, where I could see the Pakistani Army checkpoint. It was still a good way off and looked like it hadn't noticed us.

When the truck stopped, the PL got out and stood next to his door to scan the ridge for signs of Sergeant First Class Gilbert. Above him, I wheeled the gun to face out over the valley and look for signs of the enemy.

Everything was dead quiet.

After a minute or so of watchful silence, the PL spoke up.

"You see anything, Pace? Where the fuck is he?"

Pop pop pop pop. snap snip whiz snap.

Muzzle flashes from across the valley. Fuck. We were taking fire.

"*Contact right,*" I shouted, flipping on the PEQ-2 laser that was mounted on the gun and wheeling the turret to the right to open fire on whatever was shooting at us.

BOOM BOOM BOOM BOOM BOOM

The Ma Deuce's muzzle flash lit up the night and the stream of tracers sailed into Pakistan toward the unidentified enemy.

I saw more muzzle flashes from a bunch of positions around where I was shooting. Enemy tracers this time. Our muzzle flash had given them something to aim at. Their tracers were flying toward our position and toward something to our right. Sobon and the PL got out of the truck, and their rifle shots joined my machine gun fire.

BOOM BOOM pop pop pop BOOM BOOM BOOM pop pop

I fired several more bursts, hammering the ridgeline, then paused to assess whether anything was still shooting at us. Off to our right, I could see someone else shooting out toward the enemy.

"Sir, I think I see where Gilbert is."

"*What?!*"

I pulled my head out of my sights and looked down. The PL, who'd been standing directly below the barrel of the .50 cal, was holding his ears and had hunched down behind his door. The muzzle blast off an M2 is no joke. On the range, I'd seen the concussion it produced blast apart sandbags and throw small rocks into the air, and none of us in the truck were wearing ear protection. Shit. I'd probably deafened the new PL.

"Shit, Sir. Are you okay?"

I didn't hear his response, as more tracers soared toward us from across the valley.

BOOM BOOM BOOM BOOM BOOM

Out of the corner of my eye, I watched the PL wince as he got rocked by the muzzle blast again. Shit. Sorry, Sir.

BOOM BOOM BOOM BOOM BOOM

The flashes stopped again.

More flashes, this time from another direction.

BOOM BOOM BOOM BOOM BOOM

The PL shook my leg. I looked down. The PL was saying something, but I realized my ears were ringing so badly I couldn't hear anything at all. I hadn't even noticed.

"*What, Sir?!*"

He had the hand mike up to his ear. He was saying something into it, now. Shit. He was probably working with the forward observer to call for fire.

snap snip whiz snap

Tracers zipped past me. I looked up and saw that the gunfire was coming from the original position. But there were also tracers coming from the left, and now it looked like the Pakistani checkpoint had woken up and was shooting too. Hopefully they wouldn't shoot at us.

I returned my focus back to the first enemy position.

BOOM BOOM BOOM BOOM BOOM ***KABOOM***

The last boom was huge, and it didn't come from the Ma Deuce. The sound was accompanied by an explosion on the ridge across from us. A howitzer round had hit.

BOOM BOOM ***KABOOM*** *BOOM BOOM*

Another howitzer round hit on the ridge in front of us. Shit. Those were landing in Pakistan.

BOOM ***KABOOM*** *BOOM BOOM BOOM*

That one hit behind us. The gun must have dropped a short round or something. In the meantime, the muzzle flashes in front of us ceased, so I stopped shooting and hunkered down as the artillery continued.

KABOOM KABOOM KABOOM

Three more booms rocked the truck. The rounds were hitting in front of us again, thank God. My ears were a wreck. They didn't hurt, but they were ringing like mad, and there was a weird quality to the sounds around me, as if I was hearing them through a thick layer of cotton.[5] I looked over at the PL. He was back on the radio and occupied with something I couldn't hear.

I stood back up and looked out across the valley. The Pakistani checkpoint was still shooting at something, but I couldn't tell what. Nothing else was moving, although it was hard to tell because the thick smoke from all of the firing and artillery was hanging in the night air and making it hard to see. There were definitely no more muzzle flashes from the ridgeline, though.

I took a deep breath and wiped my brow with my sleeve. I was sweaty as hell and my eyes were stinging from the smoke in the air. Everything smelled like cordite.

The PL grabbed my leg again. I looked down and saw him trying to tell me something. Shit. It looked like he'd been talking to me for a minute without my noticing.

"What, Sir?!"

"I said, Gilbert's hit. We gotta get a medic over there. Prepare to break contact! Keep scanning, and shoot any motherfuckers you see."

"Roger, Sir!"

Sobon started rolling the truck backward, easing it back down the ridge. Once we were a few feet back, the ridge in front of us blocked my view of the enemy, so I turned around to guide the truck's backward progress. Humvees don't have back-up cameras, and the mirrors don't work well under NVGs, so Sobon was basically driving blind.

"Slow."

"Right."

"Left."

"A little more left."

We made it back down like this, slowly and carefully, to avoid flipping the truck. When we reached the bottom of the hill, Sobon pulled onto the road and spun the truck around, and we sped back to where the ITAS truck was stopped. The PL jumped out of the truck and ran over to where the medic was waiting for him. They ran east, up the

---

5. The best approximation of this feeling I've seen portrayed in a movie was during the D-Day scene of *Saving Private Ryan*.

ridgeline together to find Gilbert's team.

Within a minute, the sounds of their crunching feet were lost in the distance. The valley was once again silent except for the rumble of the two trucks' diesel engines.

I looked around. It was as black as only an Afghanistan night can be. Above us, to the east, loomed the ridge where most of the squad was trying to find Gilbert. Past that was the tribal region of Northwest Pakistan. To our north and south, the road stretched on forever. Nothing remarkable there. To our west was a long sloping hill. Somewhere past that–a long way off it seemed–was Sh'kin. Home. There was no movement in sight.

I looked over at the ITAS truck where the driver, E-Brake, one of our newest guys, was sitting alone. Everyone else in the truck had gone up the ridge. "Shit," I thought. "There's nobody down here right now but me, Sobon, the forward observer, and E-Brake." All of a sudden, the situation looked a little precarious. I looked down through the turret at Sobon.

"Hey, man, if someone attacks us right now, we're pretty fucked."

Sobon thought for a minute. "Yeah, we'd be in some real pretty shit."

"Or what if they get attacked on the ridgeline? What are we going to do about it from here? I can't see shit."

"Yeah. I guess we could try to get the trucks up to them, but that shit's steep and rocky enough that I imagine we'd blow a tire or roll the truck again."

There was silence as I thought for a minute.

"Yeah, we're a bit exposed at this particular moment, for sure."

"It's fucking eerie. I don't like it."

"So, what do you think E-Brake thinks of all this? He's been in the platoon for what, six months? And how old is he? Nineteen or something? He's some dude from Kansas who joined the Army and came straight to Afghanistan. He just heard a firefight, and everyone else in his truck just ran up that hill because his platoon sergeant got shot. Now he's here, sitting in a valley in God-knows-where Afghanistan, alone as fuck. What do you think is running through his brain?"

Sobon thought again. "I guess I'd be pretty worried if I were him. This is turning out to be a pretty goddamn crazy night, and there's a fair shot someone is planning to ambush us right now. We're stretched thin as hell, and even if we can get anyone on the radio from down in this valley, I'm not sure anyone can respond quickly enough to matter. Doesn't seem all that unreasonable to think we'll all get killed. That's some pretty heavy shit to deal with when you've been in the Army less than a year."[6]

"Yeah, that's a lot to take in. Especially for a fucking cherry."

We laughed together.

"Ah well, shit'll buff out I guess. At least if I get whacked tonight, I'll know that I spent my last few minutes on this Earth talking to your dumb ass. That's something."

"Yeah, lucky us. My last memories of life will be talking about dumb shit while surrounded by the smells of diesel fuel and another guy's unwashed balls. E-Brake might be missing this stellar conversation, but at least he only has his own stink to suck down."

"Ball stink is the driver's curse, man. You should have been a gunner. Then you could be up here with the fresh air."

"Ha. Yeah, I'll take living with the stink over death. At least I've got an armored window between me and the AK rounds. We'll see how that fresh air is treating you when you've got a sucking chest wound."

"Fair point." I shifted myself slightly in the turret and let out a blast of rib-induced flatulence. "But I'm happy with where I am. Better to risk death in the fresh air than to live safely in reeking squalor."

Sobon's reaction was a violent punch to the leg. "Goddamnit, man! Your ass smells rancid. Let me face death in peace."

I laughed and continued to peer out across the black night, looking for the tribesmen who were probably maneuvering to fire RPGs at us.

Suddenly, the radio crackled to life with a broken version of the PL's voice.

"KSHHKKKSHKK..ller element this...KSHHKKHSHKK..oving to your loc...KSHHH"

Sobon grabbed the hand mike and keyed it.

"Last calling station, you came in broken. Say again, over."

"KSHHKKHKKKHKK..ing to your location, time nowKKSHHHKKHKKHKK ..DEVAC, over."

Sobon looked up at me. "Dude, can you tell what he's saying?"

"No idea, man. I don't hear any shooting, though, and it sounds like the PL, so he's probably headed this way."

Sobon keyed the mike again. "Last calling station, roger, over."

---

6. Sobon and I, of course, had been in the Army a whopping two years at this point. We were seasoned veterans who had it all figured out–at least in our own estimation.

There was silence following the call. I looked over at E-Brake. I wondered if his radio had better reception than ours. He hadn't moved, so I suspected not.

I continued to scan the silent darkness. By now, the ringing in my ears from the fire fight had subsided, and I could hear normally again. After a few minutes, I started to hear a noise from above us to the east. It was a human voice, talking loudly in the distance. I couldn't tell what it was saying though.

I stared hard in that direction, trying to pick out any movement on the rocky ridgeline. On a pitch black night, NVGs were better than nothing, but it was still hard to discern anything in the wall of green rock and dirt in front of me.

The voice increased in volume as whoever was making it got closer, and for several long minutes I strained my eyes to see where they were.

Then, all at once, I could see people picking their way down the ridgeline.

A squad of Americans was coming down the hill. I couldn't tell who was in front, but whoever was approaching was next to a guy with a SAW who must have been either Raimond or Hogu. Behind them, I could see someone suspended between two other guys, walk-limping down the ridge on one leg with his arms around the other guys' shoulders.

The guy in the middle was Gilbert, and he was the one making the noise, audibly grunting as his bad leg banged off rocks.

As he descended, the PL flashed his NVG's infrared light at me–the commonly used sign to indicate an approaching friendly force. I returned the flash, and in a few minutes the group arrived back at our cluster of trucks. The security guys peeled off for the ITAS vehicle, and the PL and the medic moved to load Gilbert into the back of ours. The forward observer shifted from our vehicle to the other truck to make room.

Sobon hopped out and opened the heavy rear driver's side door for Gilbert. Even under NVGs, Gilbert looked pale and sweaty, and he groaned loudly as the medic helped load him into the Humvee, laying him down across the turret floor to allow his injured leg to stay straight. Under the best of conditions, the backseat of a Humvee is a tight fit for a grown man, and Gilbert was a big guy even without an injured leg. I widened my stance and shifted my legs to give him space. It was tight, and if I had to wheel the turret quickly, I'd have trouble not stepping on him.

Gilbert was clearly in a lot of pain.

"*Gyah, fuck, that hurts.*"

The medic replied soothingly. "Take it easy, man. We've got you settled in and we're getting you out of here."

"Fuck! Goddamn. Yeah, I'm good Doc. Are the boys okay?"

"Yeah. All the boys are fine. They didn't get hit. They're all fine. It's you we gotta take care of."

"Fuck. Two weeks I've been here, and this shit happens. What the fuck. Goddamn my leg hurts. How's it look?"

"Not good, honestly. It's in bad shape, but I can't tell much here. We gotta get you back to somewhere I can see better."

The PL chimed in. "Let me know when you're good to move, Doc. We gotta get out of here. Enemy is probably coming this way."

"Roger, Sir."

With Gilbert stuffed into the back of the Humvee, the medic closed the door. Sobon pushed the truck into gear and we started rumbling down the road. I looked down at Gilbert's sweaty form below me, thrashing as the Humvee bumped over rocks and dips.

Under the NVGs, Gilbert's desert uniform mostly glowed a light green, like the rest of the interior of the truck. I could see the dark green of the bloody leg distinctly though. The extent of the damage was hard to see, but based on the amount of blood everywhere, he'd clearly taken a nasty hit. His pants were cut all the way up that leg, and there was a tourniquet on the exposed thigh. Shreds of dark bloody cloth dangled around it. His face was visibly sweaty, and his eyes were closed and his teeth gritted as he worked through the pain. It was the first time I'd ever seen a guy this badly injured up close, and I was momentarily hypnotized.[7]

Gilbert snapped me out of it.

"Fuck! Sobon, stop hitting bumps, man. That shit is killing me."

"Sorry, Sergeant. I'll try to keep it slow."

The PL: "You're doing fine man, just keep us on the road, we'll get there."

Gilbert: *"Doc! You there, Doc?"*

"Yeah, I'm here. What's up?"

"How's my junk, man? They didn't blow my dick off, did they?"

Doc laughed. "Nope. All the damage looked below the knee. Your junk is fine."

"Fuck. That's good news. I was worried."

The news seemed to relax Gilbert somewhat, and he was quiet for the rest of the

---

7. The image of him lying in the truck, green under NVGS, is still burned into my brain today.

fifteen minute drive to the border checkpoint. During the drive, the PL was on the radio with the company operations center in Sh'kin to line up a MEDEVAC.

"Cobra base, this is Killer Six, nine-line follows, over."

"This is Cobra base. Send it, over."

"Line one: Forty-two sierra, whiskey alpha two tree two, seven zero niner, seven four zero two.

"Line two: This frequency. Call Sign Killer Six.

"Line three: One patient. Urgent.

"Line four: None.

"Line five. One Litter.

"Line six. One platoon. Security established.

"Line seven. IR Buzzsaw.

"Line eight. U.S. service member.

"Line nine. None.

"How copy, over?"

The operations center repeated the instructions back to the PL, and he confirmed they were correct. Afterward, he grabbed the hand mike for the platoon net.

"All Killer elements, this is Killer Six. When we arrive at the border checkpoint, we're going to hold there until the birds are a few mikes out. When we get that notification, we'll move to secure the large field just west of it. That'll be the MEDEVAC site. My truck will push to the far side of the field, and Killer Two will secure the short side. Charlie Company should be pushing a few guys to help out, too. Hold your positions until the bird arrives, and we'll get Killer Seven loaded and in the air."

The ITAS truck, manned now only by Hogu, E-Brake, and the FO who'd bumped over to make room for Gilbert, replied. "Roger, over."

The operations center came back across the company net. "Killer Six, this is Cobra base, over."

"This is Killer Six, over."

"MEDEVAC birds confirmed en route to your location. ETA, forty-five mikes. How copy, over?"

"I copy, forty-five mikes, over."

The PL put down the mike and turned to us. "Gilbert, you're going to be good to go, man. We'll get you out of here and fixed up."

Gilbert was quieter now. Whatever pain meds the medic had given him must have kicked in all the way. His response sounded a little slurred, almost like a drunk guy's. "Roger, Sir."

We eventually arrived at the designated spot for the MEDEVAC. It was a large field near the border checkpoint we'd spent so much time at over the last few months. As directed by the PL, our truck pushed across the field to the far side to ensure that no enemy forces fired at the helicopters as they came in for Gilbert.

Once our truck was in position, Sobon, the PL, and the medic hopped out and moved around to unload Gilbert. Before he flew, the medic wanted Gilbert strapped down on a litter and covered up to keep his leg from bumping around during the long cold flight back to Bagram.

"Open the door."

"Yeah, got it. Grab the litter off the back of the truck."

"On it."

"Alright, now, help me get under his arm."

"Grab his hand."

"Careful of his leg."

"Fuck, he's heavy as hell."

"Can you get him?"

"Yeah, it's just a bad angle. One sec."

Gilbert was a big guy, and in his kit, he was pushing three hundred pounds. The trio eventually got him out of the truck and onto the litter, and the medic wrapped him in a blanket and started strapping him in.

The radio buzzed to life. "Any Killer Element, this is Dustoff One, over."

The PL ran back around the truck and grabbed the hand mike.

"Dustoff, this is Killer Six, over."

"Roger, we are en route to your location. ETA, five mikes."

"Roger, over. I'll begin marking the LZ when I have audible."

"Roger."

The PL grabbed the IR buzzsaw off the radio rack and ran out into the center of the field.[8] The medic and Sobon finished wrapping up Gilbert and waited next to him.

8. An IR buzzsaw is a simple homemade tool for signaling aircraft. The buzzsaw consists of a three-foot length of five-fifty cord tied to an infrared chemlight. Once the chemlight is cracked, the wielder holds the end of the string and spins the chemlight over his head, creating a glowing wheel that's visible from a great distance.

Sobon lit a cigarette, but he didn't have time to finish it. Within a minute, we could hear the distant whir of approaching helicopters.

A hundred yards away, in the center of the field we were securing, I watched the PL start spinning the buzzsaw. From the air, even a place as rural as Afghanistan could have enough light pollution to make it difficult for a pilot to tell which light was marking the MEDEVAC site, so the glowing wheel created by the buzzsaw provided the aircraft with a visible point of reference to mark his landing point.

As the PL spun the light, the buzz of the approaching helicopters turned into a roar, and in seconds, the Blackhawks were overhead. They were flying blacked out, so they were impossible to see in the darkness without NVGs. One of the helicopters remained overhead while the other swooped down over our truck to land. As it flared to touch down, it kicked up a huge cloud of dust that washed out my NVGs and filled my mouth and nose with dirt. Below me, the medic and Sobon tried their best to shield Gilbert from the filth.

Once the bird was on the ground, Sobon and Doc grabbed the stretcher and ran for the bird. One of the crew flipped on an IR light and guided them in, and a team hopped off the bird to take charge of the stretcher and strap it into the helicopter. The whole process took no more than a minute and, once complete, the Blackhawk lifted back off and flew away with its sister ship.

Once again, the night was quiet.

I turned from the departing helicopters and looked back to the dustoff site to watch as the PL, the medic, and Sobon made their way back to our truck. Given the circumstances, they seemed oddly elated, and they were laughing and joking as they arrived.

"Holy shit, when that bird came in, it almost knocked me off my feet!"

"No kidding, and the dust! What the fuck, man?"

"In my head I was like, don't hit the rotors, don't hit the rotors."

"Yeah, that was some wild shit. Alright, let me get on the radio."

The trio climbed back into the truck, and the PL called back to Sh'kin to report that the MEDEVAC had been successfully completed. We drove across the field to pick up the other truck and The Hawk, who'd stayed to keep an eye on it, then made our way to the border checkpoint to hook a sling leg up to the broken vehicle we'd left there before the patrol so that we could tow it back to base.

The drive up Corvette was uneventful, and within an hour the PL was writing his end of mission reports while Sobon and I unloaded the truck and covered up the gun.

Raimond, Hogu, The Hawk, and E-Brake were doing the same thing on the trucks next to ours.

"Hey Raimond, Hogu, were you the two dudes up there with Gilbert?"

Raimond replied. "Fuck yes, we were. We just about got fucking killed up there."

"I bet that shit was wild. What did it look like?"

"Man, I was trying to get as small as I fucking could against those rocks. Over my head, I could see these streams of tracers flying over, and all I could think about was hugging that ground tighter and tighter. That was fucking crazy, man."

"I bet it was. We couldn't see y'all at all from where we were, but there was shooting everywhere."

Hogu chimed in. "Man, now we don't have a platoon sergeant anymore. *Again!* That position is fucking cursed. First, Wolf left. Then we lost Collazo. Now we lost Gilbert after only two patrols. Who's going to take charge now?"

"Maybe they'll put you in charge, man."

We all laughed, but it was a valid point. We'd lost two platoon sergeants in less than a month. How many more platoon sergeants were available?

"Man, we're running out of people in general. Dickey's gone. Collazo's gone. Now Gilbert's gone. Shit, at this rate, in a few weeks we're not going to have a platoon left."

It was meant to be a joke, but it fell flat. We didn't talk about it much, but all of us realized that any of us could easily be the next guy to go. Contestants only got to spin the wheel of fortune so many times before it eventually stopped on 'bankrupt,' and all of us felt like we'd been spinning the wheel a bunch lately.

The adrenaline from the fire fight was long gone, and as I finished oiling the gun and covered her up, I could feel my eyelids drooping. By the time I was done, all I could think about was going to bed.

I skipped a shower and skipped dinner, stripped down to my shorts, and climbed into my bunk. I lay down, pulled my watchcap over my eyes and thought about the evening.

What a night. Where was Gilbert now? What was he thinking about? He'd been on the ground for what, two weeks? He'd been in the Army for twenty years training to do something, and when he finally got on the ground to do it, he'd gotten shot on his second time outside the wire. It was wild. What would happen to the platoon now? I thought Gilbert had been the answer to our problems, but he'd gotten blown up. And what about the other guys? How were Dickey and Collazo? *Which of us would be next?*

My mind jumped around from question to question for a short while, but before long exhaustion kicked in and I drifted off to sleep.

The next few days were sluggish and strange for AT platoon. Between the scouts' deaths and the sense of letdown after the big firefight when Gilbert had been hit, folks were stressed and there was much uncertainty about what was going to happen next. While the new PL seemed competent, Gilbert's loss had left us without enough senior NCOs. Lussi was young–even with his stripes he still very much felt like one of the boys–and Stearns had gotten tarred by the old PL's mistakes and was getting sent to Bagram to take another position. The rest of us still functioned well enough on the surface, since we performed most of our basic tasks on autopilot by this point. But without leadership around to energize things, the day-to-day management of the platoon felt a bit rudderless.

On the personal level, guys also seemed to feel adrift. Some members of the platoon had been friends with Fuller and Thomas, and all of us had been friends with Dickey, which led to some of the guys being infected with "I'm going to be next" syndrome. You could see concern on everyone's faces that hadn't been there before during our daily rocket attacks. This was particularly acute for Sobon, who was due to leave the Army in a few months and starting to get the itch to get out of Afghanistan before he got killed.[9]

Uncertainty about the platoon's future didn't last long. Company and battalion leadership noticed what our platoon was going through. The sudden loss of Gilbert was quickly addressed, particularly since the Sergeant Major was already at Sh'kin for Fuller's and Thomas's memorial, and a new platoon sergeant was already inbound.

The platoon was told to stand down on missions while we waited, so we spent the next few days working on the trucks, cleaning up the base, and watching *Old School* over and over on Sobon's laptop, since it was one of the only movies we had.

The day the head shed arrived, Sobon and I were rebuilding spare tires in the motor pool.

"What happens to the main characters at the end of the movie, Pace? They either live

9. This was before the days of the Army's 'stop loss' program. A few years later, when the Iraq war was burning hot, and the Army was running out of fresh guys, Sobon would have just been formally extended.

happily ever after, or they get fucking killed. There's nothing that makes one of these Army stories better than sprinkling a little tragedy on top. And who usually gets killed? The guy who's about to retire. It's game over, man. Game over![10] I'm fucked."

I thought about his comment as I tightened the ratchet strap around the truck tire to pop out the run flat.[11]

"That's true. You're probably fucked. Look on the bright side though. You also haven't done any preparation to get out of the Army anyway. Instead of applying for college or lining up a job, you've been over here building dramatic tension before your big exit at the end of our war movie. So, there's that, right? At least if you die, you won't have to deal with the consequences of not doing any preparation for real life."

Sobon puffed on his cigarette as he watched me work the ratchet strap. Only one of us could do the strenuous part of tire-building at a time, and it was his turn for a break.

"You know what? You're a real dick, Pace. That shit isn't funny. So I either die over here in Wherever-the-fuck-istan, or I live and end up back on my parents' couch. What the fuck kind of options are those?"

"Well, they're your options." I laughed and popped the run flat out. "Your turn, man."

Sobon bent down and picked up the run flat, dragging it over to the new tire while I popped the strap off the old one. He looped the strap around the new tire and started cranking it down. I could hear helicopters approaching in the distance. It sounded like the ring route birds were here.

"Yeah, I guess those are my options, aren't they? Man, what a shit sandwich. I gotta get on top of my future."

"Assuming you have one. You thought about going to college? Alycia's working on her PhD in Madison, and I'll be heading there in a year once I'm done with the Army to do the same thing. There anything you want to study? College is a great halfway point between the Army and adulthood."

Sobon continued to crank on the strap. He had it about halfway done.

"I want to build robots."

"Really? Robots? Like, you want to be an engineer?"

---

10. Which, of course, was said in the style of Bill Paxton's character, Hudson, from *Aliens.*

11. Humvee tires contain a thick rubber ring called a run flat. This allows the truck to drive–somewhat anyway–on a flat tire.

"Sure. Yeah, whatever it's called. Yes, I want to build robots."

I cracked open a Gatorade and sipped it, thinking for a minute.

"I think that's a pretty strenuous degree, but I'm sure you could study robotics. How hard could it be?"

"It can't be harder than this bullshit, that's for sure."

"And I'm sure the food is better. Well, if you decide to go that way, you can move in with us while you figure stuff out."

Sobon looked at me.

"You mean that?"

"Yeah, sure man. As long as you don't mind cats."

Sobon nodded thoughtfully.

"Yeah, that just might work. I've never done anything but the Army before. Honestly, I'm not sure what real life even looks like. Madison might be a good place to start."

"Well, don't think college is too much like real life, but yeah, I think you'd enjoy it. The chicks are better looking, anyway."

Sobon nodded and bent down to continue his work. He paused though when we heard the gate opening. The Humvees that had been out on the airfield were coming in with whoever the ring route birds had delivered. We watched two cargo Humvees in the center of the convoy continue past us to the operations center while the security Humvees peeled off to park in the motorpool near us.

Lussi, Hogu, and The Hawk got out and walked over toward us.

"Did you guys hear? The new platoon sergeant just got here, and the SMAJ is going to bring him by."

Like any smart private, I'd avoided the sergeant major like the plague since he'd arrived at the battalion, and I wasn't excited about getting forced to meet him. "Great. I heard the new Sergeant Major is a holy terror. What's his name again?"

"Watson. And, yeah, that's what I heard, too. I haven't heard anything about the new platoon sergeant though, except that he was a Marine."

"A fucking *Marine?* A motherfucking crayon-eating jarhead? I bet he's a total rock-farmer dickhead."

"Yeah, maybe. Whatever he is, he's ours now, though." Lussi shrugged and walked off to grab chow. Hogu and The Hawk followed him, leaving Sobon and me to finish up with the tire. We finished installing the new rim, tightened all the bolts that held it together, and rolled the completed wheel over to the truck to be mounted on our

Humvee's rear hatch.[12]

After tightening the ratchet strap that held the tire onto the trunk, I climbed onto the roof and pulled the cover off the Ma Deuce to oil it up for the day. Sobon took off to join the rest of the guys at chow.

I hadn't been working on the gun long, when a pair of guys I didn't recognize walked up. One was tall and rail thin, and the other was shorter and stockier. Both walked like folks who were in charge of things, and when they got close enough that I could read the rank on their uniforms, I saw why.

"Hey, Sergeant Major." I set down the cleaning rag I was holding and locked up at parade rest.

The rail thin man wearing sergeant major rank looked at me. His unsmiling gaze had an unsettling intensity. In the seconds before he spoke, I got the impression that he'd taken my measure, reconciled it with whatever he'd already heard about me or my platoon, and rendered some sort of internal judgment.

"Carry on, there, high-speed. Specialist Pace, right? You tracking who I am?"

I guessed. "The new Battalion Sergeant Major?"

"Roger that. Just down here to check up on you boys and to bring you a new platoon sergeant since the last one didn't last too long." He chuckled at his own joke and turned to the guy standing next to him. "This is Sergeant First Class Romero. He's squared away, and he's going to take over AT Platoon."

Romero nodded and didn't say anything. He had a hawk-like nose, dark piercing eyes, and a serious expression. He looked pretty much like the asshole Marine I was expecting, except that he had too much hair.

The Sergeant Major continued. "So tell me about the night Gilbert got shot. I heard you did pretty well out there. What do you think about what happened?"

"Honestly, I'm not totally sure what to think, Sergeant Major. It was dark as hell, and once we started taking fire, it was hard to tell where everyone was. I'm still not entirely sure whether we killed the guys who were shooting at us, whether they were in Pakistan or Afghanistan, or what exactly happened with Sergeant Gilbert."

12. Humvee hatches weren't built to carry spare tires–the extra weight destroyed the hydraulic struts and wore out the trunk latch–but we popped so many tires on patrol that we needed to carry at least one extra per truck, and there wasn't anywhere else to store them.

The Sergeant Major nodded. "Yeah. Combat is like that."[13]

He lapsed into silence for a few seconds, and I stood there awkwardly, not knowing what to say.

He continued. "Well, the commander and I are proud of how you guys handled yourselves. And I talked to Gilbert, and he's doing just fine. They got him back to Germany, and he'll be alright. They'll get him a new leg, and he'll be good to go in no time."

"How about Dickey and Collazo, Sergeant Major? They good, too?"

"They are. Both of them are fine. They're both injured enough that they won't be coming back out here, though, so I'll be getting your platoon some new soldiers to replace them."

I nodded again, unsure what to say.

"There anything you need from me or the commander, son?"

The list of possible answers ran through my brain. More phones. Better chow. A platoon sergeant who wasn't a dickhead. Hot showers. More comfortable armor. A lucky rabbit's foot that'd make me immune to rocket attacks. There were a million things I could think of that would be great to have. To a private, though, talking to the Battalion Sergeant Major was like walking through a minefield. Best case, I might come out with nothing bad happening. Worst case, I'd find myself on extra duty, transferred to another platoon, or something worse. And my new platoon sergeant was standing next to him. The Sergeant Major might leave on the next bird back to Kandahar, but Romero wasn't going anywhere.[14] I wasn't falling into his trap.

"No, Sergeant Major. I think we're good to go."

Watson looked at me, probably guessing much of what was running through my mind. He smiled a crooked smile and nodded.

"Okay, then. Take care of yourself."

Watson turned to leave, and after staring at me for a moment and no doubt completing his own internal assessment, Romero turned to follow.

Had Romero just given me the 'fuck you' look? Had I said something wrong? Did the guy already hate me?

---

13. And he would know. He was recognized some years later with a silver star for his service in Mogadishu during the operation made famous by *Blackhawk Down*.

14. If I learned one life lesson during my first term in the Army, it was how to shut my stupid mouth.

"Great," I said to myself, retrieving the oily rag and continuing to clean the Ma Deuce. The memorial was in an hour, and I needed to get the gun reset before then.

"Just great."

The morning sun beat down on two pairs of dusty boots standing in the rocky courtyard of Sh'kin. Each pair had a rifle balanced, barrel down, between the left and right boot, and each rifle had a set of dog tags and a helmet balanced on top.

I stood looking at the boots, wondering if they were Fuller's and Thomas's actual gear, footwear the unit had on hand just for these sorts of ceremonies, or a random set of boots from the supply room. I couldn't tell. The boots weren't new, but they weren't in all that bad a shape either. The rifles and helmets were in similarly okay condition. They were dirty enough that they could have been the real deal, but not covered in blood or anything. Had the snipes gotten blood on their boots when they got killed? If so, who'd had to clean that off before the ceremony?

I looked down at my own footwear. One of the boots had a Gilbert blood stain on it, and the other had something brown smeared across the side. Stew maybe? Shit? My desert boots were made of some sort of suede, so if I got killed in them I didn't think anyone would be able to clean them well enough for a ceremony. And what if I got hit like Gilbert and lost my leg before I died? Wouldn't my boot either be blown off somewhere or shredded by the medics when they cut it off like they had Gilbert's? Would they still use my other boot for the ceremony, or would they get an entirely new pair from the supply room?

The memorial was short. The battalion commander and a few other guys–including the scouts' squad leader who'd been on the mission with them– stood up and said few words about Fuller and Thomas. The ceremony then concluded with a roll call, a bagpipe rendition of "Amazing Grace" played on a boom box, and a twenty-one gun salute. It was my first memorial, and while they've grown to become deeply emotional for me–particularly the playing of "Amazing Grace"–at the time, I was more confused than anything else.

I knew that Fuller and Thomas had been infantrymen like me. Just a few days ago, they'd been patrolling, eating chow, calling home, and doing all the other stuff I'd been doing at Firebase Sh'kin. Now they were dead, and they wouldn't do any of those things any more. They wouldn't do anything else any more either. No going home at the end of the deployment. No greeting their families. No living happily ever after. No nothing. Intellectually, I understood that. But somehow, I couldn't grasp the significance of what their deaths really meant. I couldn't grok it. At age 24, death just didn't really seem very real.

At 45, though, their deaths seem far more tragic. When I look back at all of the life I've lived since that day and think about how much water has passed under the bridge since then–the births, the laughs, the first steps, the milestones, the anniversaries–Chad Fuller's and Adam Thomas's deaths feel deeply depressing. Neither of them will ever experience any of those things I just listed.

The world, and all of the people Fuller and Thomas knew and loved, will move on without them, and Fuller and Thomas will become ever older memories for those they meant something to. And the same is true for all seven thousand other fallen service members lost in Iraq and Afghanistan during the war. Eventually, of course, this will be the fate of all of us, to include the memory of the war itself.

All of this perspective was in the future, though. At the time, staring at the two pairs of boots, I only remember feeling that I should have felt something more than a confused sense of loss.

I turned away from the boots to join Lussi in the loose crowd of soldiers who were still lingering after the ceremony.

"Sucks."

"Sure does. They were good dudes."

"Yeah. I didn't really know them."

We stood around awkwardly, nobody really knowing what to say.

"What time are we rolling out?"

"Nineteen hundred. Gotta get out to the trucks in an hour, though. New platoon sergeant wants to check our gear or something."

I looked at my watch.

"Want to grab some chow before that?"

"Yeah, let's do it."

We walked into the kitchen area, where there was still a good amount of leftover breakfast. I scooped out some old rubbery scrambled eggs and grabbed some footbread and a granola bar before sitting down in the screened area.

We ate in silence, watching the crowd thin out around the memorial area. After ten minutes or so, there wasn't any one left, and the boots stood alone and unregarded. After another few minutes, one of the supply clerks or chaplain–I wasn't sure which–started packing up the boots, the rifles, and the other remnants of Fuller's and Thomas's lives with the First Battalion, Eighty-Seventh Infantry. And so it goes.[15]

"You get any mail when the birds came in?"

"Yeah. I got a couple of letters from Alycia. You'll like this one. She read in the Fort Drum paper that some chaplain came out here and called this place the evilest place on earth. Maybe the dude wasn't too far off."[16]

"The evilest place on Earth, huh? Yeah, I can see it. The food is definitely bad enough."

"Why do you have your ammo rigged like that?"

"It's more comfortable this way."

"Yeah, I get that. But when you're executing a magazine change, can you get to it easily?"

"Yeah, I think so."

"You think so? Show me."

I adjusted my slung rifle and popped out the current magazine. I put the magazine back into one of my chest pouches, reached for a fresh one, and stuffed it back into my M4.

"That was too slow, Pace. You're running your reloads all wrong. You need to be doing it more like this."

Sergeant Romero ran through his own reload drill. I was self-honest enough to recognize that his was much better than mine, but it still irritated the shit out of me to have this new guy come in–fresh from the States–and tell us–the guys who'd been at war for *two months*–how to do something different.

"I dunno, Sergeant. I think my way works fine. It's what I've been doing the whole time, and it's worked well enough so far."

---

15. A line made famous by *Slaughterhouse Five.*

16. *Time* magazine later picked up the quote in an article. "Battle in the Evilest Place," Tim McGurk. November, 2003.

"Yeah, got it. Your way isn't good enough, though, so change it. Now."

I tried hard to keep my expression neutral. "Roger, Sergeant."

Romero walked off to the other truck, leaving Sobon and me alone to continue our work while he berated Lussi for the state of his vehicle. I watched the ass chewing as I took off my armor to start rearranging my ammo pouches. Since arriving, Romero had been all over us. Everything we'd been doing, from the way we talked on the radio, to the way we stopped our trucks, to the way we briefed our pre-mission plans, was wrong. Everything had to be redone. Below me, Sobon was emptying the truck to reload it differently so we could get to the jack faster. We'd been doing that wrong, too.

"Man, Sobon, fuck this guy. Who's he think he is, making us rearrange all our shit?"

"I imagine he thinks he's the platoon sergeant. Which is correct."

"Well, yeah. It's still bullshit though."

"I'm not in love with doing work for no reason either, but, I mean, he's right isn't he?"

"Well, maybe. About some of it. But still. You know what? Fuck you, Sobon. You're seeing this all wrong."

Sobon's only response was to laugh at me as he continued to adjust the equipment in the trunk of the Humvee.

My ammunition pouches were attached to my armor using a system of straps called the MOLLE system. The system was great because it was basically impossible for MOLLE gear to fall apart through normal use. But having to change how everything was set up was time consuming, so it took me most of an hour to complete the required adjustments.[17]

By the end of it, my fingers were sore from unweaving and reweaving the straps, my ears were tired of Romero's endless corrections of the platoon, and my soul hurt because I'd burned an hour of free time that I could have spent doing something other than Dumb Army Shit. In short, I was feeling pretty sorry for myself and having a good old-fashioned pity party.

When my pouches were finally reattached in the Romero-approved manner, I re-covered the Ma Deuce, stuffed my hands in my pockets, and slouched off toward the hooch.

---

17. In a wonderful display of the Army's love of acronyms, the MOLLE (Modular Lightweight Load-carrying Equipment) system uses rows of heavy-duty nylon webbing called PALS (Pouch Attachment Ladder System) to attach pouches like the IFAK (Individual First Aid Kit) to an SM's (Service Member's) IBA (Individual Body Armor).

I climbed into my bunk, pulled out *The Hitchhiker's Guide to the Galaxy*, and prepared to enjoy a bit of escapism to clear my foul mood. I'd only made it a few paragraphs into Arthur's initial meeting with Slartibartfast, though, when the door creaked open and Lussi poked his head in.

"Everybody outside in full kit in five minutes. Romero's taking us to the range to work on reload drills."

Fuck. My. Life.

"Watching dismounts clear that high ground is like watching turtles fuck."

"Slow, ungainly, and dry?"

"Exactly."

There was a short, fifteen minute pause in the conversation as we stared across the valley.

"You probably *like* watching turtles fuck."

"If I saw some right now, I would. At least someone'd be getting some action around here. I'm sure as hell not."

Another ten minutes elapsed.

"You wouldn't be getting any even if we *were* back home."

"Fuck you. You wouldn't either if you weren't married."

Seven more silent minutes.

"Your wife got any sisters?"

"You mean brothers? More your speed, right?"

"Fuck you."

Security missions in Afghanistan were one of the few places on Earth where a conversation could work like this, with short exchanges separated by long empty moments of silence. In any normal environment, something would have happened either to disrupt the flow of the conversation or to force it to progress more quickly, but on security missions two people could carry the thread of a conversation for hours–sometimes even days–with very little information actually being exchanged.

The Charlie Company guys really were taking their time today. There were ten of them, moving up the far side of Lazano's Ridge in a loose wedge. The dismounts were clearing the ridge because we'd seen a group of Afghan men moving across it, so they were expecting enemy contact. This necessitated a movement technique called 'bounding overwatch' in which half of the squad would stop and watch for enemy activity while the other half moved slowly up the steep hill. Every twenty meters or so, the moving element would stop and the other element would pick up and bound past them. Bounding overwatch made sense because it provided better security if the squad was attacked, but it doubled the overall movement time, so it wasn't fun to watch. Even Romero looked bored.

Romero was with the other truck today, parked about twenty-five meters away from ours and set up next to the mortar section we'd brought for fire support. When Stearns left, Lussi moved over to our vehicle, and Romero took over as the *de facto* First Section leader. On patrol, Romero was overflowing with energy. Every time we stopped anywhere, he was out of the truck looking around. He investigated every hill and crevice. He stopped and searched every car and farmer. And he questioned us on everything we did: What are you focused on right now? What would you do if we took contact from that hill? What about that one? How long will it take you to reload that fifty cal? Do you see that guy? He look like he's carrying anything to you? What's the MEDEVAC frequency? What about the fire net? What's the artillery call sign today?

Our answers were rarely good enough, and Romero would respond by telling us what he wanted us to do differently or what we should have been doing instead. He didn't resort to the normal NCO tactics of shouting and smoke sessions. Instead, he just asked questions until you tripped yourself up. Then he told you why what you were doing was wrong. His constant inquiries were exhausting.

As we watched Charlie, though, it felt like even Romero had run out of things to do. He was standing behind the open door of his truck, fidgeting, and looking bored as hell.

"What do you think he's thinking about right now, Lussi?"

"Probably running all of the things I'm doing wrong through his brain, I imagine."

"You've gotten the lion's share of his criticism for sure, man."

"Yep. One of the joys of sergeanting. Responsible for everything the squad does or fails to do and all that."

"Are we that fucked up? I always thought Stearns ran a decent squad."

"Yeah, I thought so, too. Good enough, anyway. Romero disagrees though. Says we were pretty lax. Says you guys know more about video games than combat skills. Wants to fix it."

I thought about the feedback. It rankled.

"He know what he's talking about?"

"Seems like it, yeah. Haven't seen him in action yet, but the dude's smart. Got a Ranger tab."

"Hmm."

I stared at Romero. *Did* he know some shit we didn't know? I couldn't tell. I guess we'd find out the next time something happened.

Which it suddenly did.

The Charlie Company guys were just nearing the top of the ridge when they started taking fire from guys we hadn't seen hiding at the top. From this distance, I couldn't hear any shouting, and the machinegun fire sounded like pops and clicks, but it was obvious the dismounts were in a serious fire fight. The front five guys hit the ground immediately, looking for cover, while the rest of the squad started firing up the hill.

The enemy's position meant we couldn't shoot from where we were–there was too much risk of hitting our guys–so I looked down to Lussi and waited for guidance.

The radio soon provided it.

"Killer One, this is Killer Seven. Reposition your vehicle to get a good line of fire on the enemy position while I work a fire mission, over."

"Killer One. Roger, over."

Lussi turned to Sobon. "Let's take the truck over to the left. Get closer to Romero's position. We should have a good field of fire from there."

Sobon nodded and started to maneuver the truck. As he did, Hogu's MK-19 opened up, lobbing grenades into the dead space behind the ridge and well away from the Charlie guys.

On the radio, I could hear Romero's voice coordinating with the dismounts and calling in an artillery mission.

"Steel Three, this is Killer Seven. Adjust fire, polar, over."

"This is Steel Three. Adjust fire, polar, out."

"Direction, one two zero degrees. Distance, four hundred meters. Over."

"Direction, one two zero degrees. Distance, four hundred meters. Out."

"Dismounts in the open, over."

"Dismounts in the open, out."[18]

---

18. This is the language of a 'polar' fire mission, which is the way you call an artillery strike when you know your location and the distance and direction to the target, but not the target's exact location.

The artillery guys confirmed the mission over the radio as we lurched to a stop a few meters to the right of Romero's truck. I scanned to see where Hogu had been firing, but I couldn't see anyone to kill, so I hollered over to Hogu to see what he could see.

"Where're the bad guys, man?"

"Dunno. I can't see them anymore. They were right on the edge of the ridge, there." He pointed. "But they must have pushed back where we can't see. I gotta cease fire now, though. The Charlie guys are getting too close to where I was shooting."

Indeed, Charlie had made it to the top of the ridge, and now it looked like they were taking more fire from further down the far side of the ridge. There wasn't anything we could do to help them from where we were–the trucks couldn't get down to them without driving into Pakistan.

The radio blared back to life. It was the artillery guys. They repeated the details of the fire mission and announced the incoming rounds.

"Killer Seven, this is Steel Three, shot over."

"Shot out."

We heard the distant blast as the howitzers fired from Sh'kin. This was an adjust fire mission, which meant only one round would be fired so that the guy on the ground–Romero in this case–could confirm that the artillery was on target before calling for more. All of us waited, watching the Charlie Company firefight and anticipating the shell's explosive arrival and its effects on the enemy. The time of flight on a 105mm round is around thirty seconds, so by the time it landed there had been plenty of dramatic tension.

KABOOM

When the shell landed, the explosion took us all off guard for a couple of reasons. First, it had landed much closer than it should have. The target was four hundred yards away. The shell impacted less than a hundred yards from our location–well within the 'danger close' radius of an artillery shell. Second, *the explosion had come from the wrong direction.* It had landed behind us. The enemy was in front of us.

The blast was close enough to rock us all forward, and I watched Romero grab his neck as something hit him. His voice immediately came across the radio.

"*Check Fire! Check Fire!*"

I called down to Lussi. "What the fuck was that?"

"Dunno. Friendly fire, maybe? I think the artillery shot at us instead of the enemy."

The Charlie Company guys had reached the top of the ridge and were firing at someone, but I still had nothing to shoot at. So I looked back over to where Romero was still standing by his truck. His neck was bleeding, and he looked pissed off.

He canceled the artillery mission and instead called the mortar team to start another fire mission. Within a few seconds, he'd gotten the tubes lined up and they had fired a shell over the ridge to land on the enemy's position. As soon as the mortar round landed accurately, he called in a fire for effect mission.

"Cobra One-Five, this is Killer Seven. Confirmed good impact. Fire for effect, six rounds, open sheaf, over."[19]

The mortarmen replied and we soon heard the explosions of the mortar rounds landing on whatever Charlie was shooting at.

Back in our truck I was impressed. "Damn, he got that shit put together fast, even after getting hit in the neck. I don't even know what 'open sheaf' means."

Lussi replied. "I think I remember something about it from EIB. I dunno for sure, though. I told you he knew his business."

"Yeah, I guess you're right."

The mortar fire apparently broke the enemy's will to fight because, within a minute of the fire mission, Charlie had stopped shooting and was moving back toward us. The whole sequence had taken place right along the border, so pursuing the enemy wasn't an option.

"Maybe the Pakis will take care of them," I joked, pointing at the Army checkpoint that had been watching the whole firefight unfold less than three hundred yards from its front gate.

"I think they're more likely to take care of them in the other sense," Lussi replied. "Didn't see those dudes lift a finger the whole time. Probably their cousins or something."

I nodded. It was nothing we hadn't seen a dozen times by this point, and from the same checkpoint no less. I looked around the truck and noticed the crater where Dickey and Collazo's truck had exploded, along with old brass from one of our gunfights a few weeks ago, and the spot where we'd parked the last time we'd come out here to check out

---

19. According to globalsecurity.org, 'open sheaf' is an optional indirect fire command that is normally used to engage targets that are wider than a standard sheaf can cover. With the open sheaf, the distance between impacts of rounds from two or more mortars is half again the distance between the bursts of the rounds in a standard sheaf. In short, it's an obscure bit of infantry nerdery that makes mortars a bit more effective against enemy infantry in the open.

a POO. How many firefights had we had on this stupid ridge anyway? And for what? To secure a random piece of nowhere so rock farmers would quit shooting rockets at us?

Charlie's movement back was slowed because they had to stop and look for a lost piece of equipment. By the time they were loaded up, it was getting late, and I was getting hungry. As we rode back to the base, I thought about Romero's performance.

He'd been as cool as a cucumber out there, and he'd taken charge of the situation and knew exactly what to do. Maybe he *did* know what he was talking about and wasn't just being an arbitrary dick when he made us change how we were doing things.

We pulled into the base without incident, and I jumped off the truck to grab chow before it closed for the night. As I walked off, Romero hollered over at me.

"You're going to clean that gun before you grab chow, right Pace?"

"Um. Roger Sergeant. Of course."

"I figured you were. Figured maybe you just had to grab something before you got started."

I turned around, climbed back into the turret, and pulled out my cleaning kit. My stomach grumbled loudly as I started to wipe the dust out of the Ma Deuce's receiver. One truck over, Hogu was doing the same thing.

"What do you think, Hogu? New platoon sergeant seems pretty squared away, but I guess squared away doesn't come free. No hot chow for our asses tonight."

He nodded. He'd clearly been looking forward to chow, too. "Yeah, that's a fact. Not sure how I feel about it, man. I guess not getting killed is good though. So if he keeps our asses alive, it'll be worth it."

"Yeah, I guess so. I suppose cold goat stew isn't all that bad."

Something crinkled as it hit me in the chest, and I looked down to see a Nutrigrain bar at my feet in the turret. Hogu had thrown it to me.

"Carr gave me a few of these earlier. It'll tide you over."

"Thanks, man"

I unwrapped the bar and wolfed it down before returning to the work at hand. Fifteen minutes later, the guns were clean, and Hogu and I packed up our cleaning kits, grabbed our weapons, and walked over to see what was left for chow.

September 16, 2003
Madison, WI

My dear husband,

I just walked out the door to put my latest letter to you in the blue mailbox at the corner. When I returned, I had four letters from you in my own box! Now you no longer seem far away and unreal. Talking to you last Tuesday was a big help, and your latest letters were so nice – it was almost like hearing you talk. Looking at your pictures on a regular basis helps too.

Have you heard that Collazo was finally diagnosed with a hairline fracture in his spine (along with healing internal bleeding) from the rollover last month? I'm pretty horrified that it took the army three fucking weeks to x-ray him. With another bad jarring, he could have been paralyzed.

My day – especially thanks to your letters – has been pretty good. We had a lab meeting this morning, which was particularly boring. Thankfully, I did manage to find time afterward to work on my current experiment, which has been successful so far.

I've been packing my lunch to save money, but I allow myself to eat out once a week. There are so many places I want to try! Today I just ate my bologna and mozzarella sandwich in the break room for lunch. But for dinner I went to Kabul—an Afghan restaurant which strangely features no goat items on the menu. The food was fabulous. I had lamb and chicken kebabs on saffron rice, unleavened bread (perhaps the same stuff you like?) and a salad with interesting dressing. Perhaps we should go sometime so you can eat some decent Afghan food—this place is owned by a guy with a real Afghan name, so I assume some of it must be authentic.

I had Bacterial Genetics today at 1:00. I emailed the professor ahead of time to let him know my situation (that you pretty much only call on Tuesdays) so he'd understand if I darted out of class to answer a phone call. He was nice about it and then launched into an anti-Bush speech (via email). The important part is that he understands my situation and is fine with me leaving if necessary.

With you gone, I can't focus on anything. The stress is really getting to me — I forget what I'm talking about mid-sentence all the time, and I forget everything—small stuff, major stuff, whatever. I can't concentrate, either. Combine that with a total apathy toward grad school, and it's not pretty. Starting school this year was a huge mistake. I miscalculated how much your being gone would kill my brain and willpower. I think about dropping out every day – not because I don't like it but because I don't care. The only thing I can muster any real feeling about is hoping you're safe and thinking about being together again.

On a lighter note, I'm looking forward to the season premiere of Friends. *It will probably make me sad since you won't be there to watch it with me, but I'll think of you, and we can watch the reruns together.*

Some questions for you now in response to your letters! Do you have any new AT guys besides the platoon sergeant and PL? Did you clean up your toilet accident or leave it for the poor Haji? I shudder to think of cleaning up your crap – particularly if you've had any army food. And I forgot to ask earlier - when your truck flipped, who else was in it? Was Sobon driving? (I'll forgive him, don't worry.)

You're on my mind constantly, sweetheart. I long for the day I can hold you again and know you are safe. Until then, I'll love you hard and write often.

Missing you,
Alycia

I grinned maliciously as my finger hovered over the button that would shower the enemy infantrymen with rockets. Would the attack kill all of them, or just the majority? I couldn't be sure, but, either way, this indirect fire would catastrophically impact their ability to threaten our base again. Fuck them. They deserved death.

What had they been thinking, trying to sneak up on us from across the mountains anyway? Did they think I was a complete idiot?

I pressed the button, and, with a satisfying roar, the indirect batteries fired. Tense seconds passed, and then the rockets impacted the enemy forces, blasting them all to oblivion. I smiled and laughed evilly. All enemy forces: KIA.

Hogu, sitting across the room from me, looked at me warily.

"Why are you laughing so much, Pace. What did you do?"

I handed him the console so he could see for himself.

"Ah, fuck! I didn't think you'd see them coming, man. It took me three turns to get those damned infantrymen over that mountain range. That's bullshit."

Hogu and I were playing *Advance Wars 2* on Hogu's Gameboy Advance. It was a popular game on this deployment, primarily because it was turn-based, which meant it could be stopped when someone had to sprint to the bathroom for a sudden bout of diarrhea or when the base got rocketed. Also, up to four players could play the game at once on a single console. We'd been playing this particular round for a half hour, and the turns were getting pretty intense as the match neared its climax.

I watched Hogu closely to see if I could glean what he was up to. His brows were knitted and he was chewing on his lower lip as he took his turn. His expression told me he was worried about something, so he was probably trying something risky and he was unsure how it would play out.

It's funny, but over the past few months, I'd gotten to know Hogu about as well as anyone in my life. And why not? We spent almost every waking moment of our lives–sleeping, eating, shitting, and patrolling–within fifteen yards of one another. We had no TV. There were no computers. There was no salsa night or Internet lounge. There weren't even any other rooms to hang out in or people to hang out with.[20] There was only our platoon, and because of patrol schedules, there was really just our squad. First Squad.

From a practical point of view, this was a good thing because our platoon's single room was small; whenever the whole platoon was on base the room became crowded and started to stink of farts, body odor, and halitosis. Plus, Second Squad sucked.

This was a truism that everyone in First Squad took for granted. Second Squad's NCOs were shitty dickheads, their soldiers did everything worse than we did, and their shoddiness meant we always had to do extra work to clean up their messes. Second Squad was Them, and We were Us. And, to an infantry private, us versus them is the lens through which all of life got viewed.[21]

The lack of other options and our shared mutual suffering made First Squad tight. This was the kind of tightness that let you notice when one of your buddies was about to throw up, or when he'd just gotten a shitty letter from home and needed one of the

---

20. There was a phone on the base. The phone was a Garmin satellite phone, and each of us was allotted a ten minute phone call once every two weeks. It occasionally worked, although the delay when calling the States was so bad that it rendered anything more complex than "I love you" functionally impossible.

21. It's worth mentioning that Second Squad felt the same way about us. To them, we were a substandard bunch of louts who should have been pulling more guard shifts so they could be properly rested for *real* work. It's also worth noting that when any other platoon–or God forbid another company–was around, all of the distinctions between the squads vanished. It's *also* worth mentioning that twenty years later, nobody cares about any of these rivalries any more.

Slim Jims you kept stashed under your pillow as a pick-me-up, or when he was on the cusp of taking a swing at someone and needed you to help de-escalate the situation.

Or when he was planning some chicanery in *Advance Wars,* I thought, as Hogu handed me back the console. He was grinning like a Cheshire cat. What was he up to?

I glanced down at the screen, and an explosion rocked the room.

Shit. That explosion hadn't come from the Gameboy.

A second rocket impacted almost immediately. Then a third. This time the impacts were close enough to send clouds of dust under our plywood door. I dropped the Gameboy and Hogu and I jumped up and ran to the door to see what was happening. As we pulled it open, we were greeted by the whistle of another rocket and the form of Lussi pushing past us to get into the room.

"That first rocket hit the AHA. There's ammo cooking off everywhere."

As he pushed into the room, I leaned out and looked to my right. Sure enough, the small room that served as both our ammunition holding area and the firebase's operations center had a huge hole in the roof, and smoke was pouring out of it. The headquarters guys, led by Major Willie, were pulling out equipment and trying to douse the flames with blankets, and every few seconds there was the snap-crackle-pop of stored ammunition exploding.[22]

A piece of something whizzed past my head and hit the wall nearby. I closed the door and turned to Hogu and Lussi. Sobon and The Hawk, who'd both been taking naps with headphones on, sat up.

Sobon was cranky as he pulled the watch cap off of his eyes. "Keep it down motherfuckers. I'm trying to sleep."

Lussi replied, "Base is under attack, dickhead."

Sobon looked at Lussi blankly for a second.

"Are we being overrun?"

"No, but some rockets just blew up the CP."

Sobon looked thoughtful. "The satellite phone was probably in there."

Lussi laughed. "Yep, no more calls for us."

The news wasn't sufficiently exciting for Sobon to cut short his nap, so he lay back down and pulled the cap back over his eyes.

---

22. I found out later that the PL was standing at the door to the headquarters when the rocket hit the building. Fortunately, he suffered nothing worse than some hearing damage.

Hogu stood up and stretched. "What should we do now?"

We all shrugged as another huge kaboom shook the ground. What was there *to* do? We were as safe in our hooch as anywhere else, and the guys fighting the operations center fire didn't need us standing around and getting hit by stray bullets.

I looked at the Gameboy. It was still sitting where I'd dropped it. It was still my turn. The cursor was waiting for my input. I sat down and returned to firing rockets at Hogu's other infantrymen.

Romero, kitted out and looking angry, poked his head through the door.

"Get ready, you chuckleheads. We're rolling out on a response mission, and I'm taking First Squad. PL's coming too."

We all responded. "Roger, Sergeant."

Romero left for the ops center in a whirl, the door slamming shut behind him.

Reluctantly, I turned off the Gameboy and put it down. It was a shame to have to abandon a game that was going so well, but we were, after all, at war.

Since Lussi and I were already fully dressed, we threw on our armor and ran out to check on the trucks.

The base looked like an anthill that had been kicked over. Guys were running everywhere in various states of dress. As we ran to the motorpool, we saw a trio of artillerymen in brown t-shirts, shorts, and armor running for the howitzer pit on the other side of the base. They skip-stepped off the path and into a crack between two buildings as another series of rounds cracked past us from the ammo that was still cooking off.

Another shriek came from overhead.

A dozen voices shouted "*INCOMING!*" With nowhere else to go, Lussi and I paused and pressed ourselves against the wall of the building we were nearest to as the rocket exploded violently off to our right.

"Man, Hajj has his shit dialed the fuck in today, doesn't he?"

"Heck yeah, he does."

Lussi and I continued sprinting through the base, dodging past people until we reached First Squad's trucks.

Lussi pointed at the lead vehicle. "Crap, man. Front tire's flat. Something must have hit it."

"Guess we get to do a combat tire change then."

Lussi got behind the wheel and pulled the truck over to the corner of the motorpool where the spare tires were stacked. I grabbed the high-lift jack while he broke the lugs. We'd just gotten the tire off the ground when another rocket screamed in.

More cries of *INCOMING!*

Lussi and I hit the ground and squeezed under the truck as the rockets exploded, causing pieces of something to *tink* off one of the the truck's windows.

We both started laughing at the ridiculousness of it all. We were going to die changing a tire.

"What the fuck?"

"No idea, man. Let's get this done."

He finished spinning off the lugs and slipped the old tire off where it thumped heavily to the ground. I rolled the new tire over, and we both wrestled with it to get it set on the lugs. This took some doing since the tire (with wheel and runflat) was heavy as hell, but after a bit of grunting and profanity, we got it in position, tightened the lugs, and lowered the jack. We packed the jack up and added the flat tire to the stack of other damaged tires in the motorpool.

"We just got our combat tire change badge locked in, Pace."

"I guess you're right. How many people can say that?"

"Not that many, I imagine. You think they're done with the rockets yet?"

"Dunno, man. It's been a few minutes since the last one, though. Maybe."

As we were getting back to the trucks, the rest of the squad arrived with Romero. We huddled around the hood of his truck and listened to the plan for the mission.

Romero started. "Alright, listen up. I'm going to make this quick so we can get out of here before any more rockets hit. There's a suspected POO for this rocket attack, and the commander wants us to get out there and check it out. The Q36 has the attack coming from just past Lozano's ridge, so it'll be a quick drive. Stearns, Lussi, let me know when your PCCs are complete. I want to be ready to move as soon as the PL gets back from his update."

Of course it was Lozano's Ridge, I thought. It's always Lozano's Ridge.

When the brief was over, we all nodded and climbed into our trucks to run through our rollout sequence.

There hadn't been any more rocket impacts during the brief, so I figured they were probably done. How many had that been? Six? Seven? I couldn't quite remember.

Sobon fired up the truck and picked up the hand mike to run through the radio checks. I jumped onto the hood, pulled the dust cover off the Ma Deuce and stashed it under the rear floorboard, then climbed down into the turret and checked my machine gun. The ammo can was still sitting in the MK 93's ammo rack, and I'd set the machine gun's headspace and timing when I'd come out to clean the weapon that morning, so she was in good working order. I racked back the bolt, gave it a quick rubdown with an oily rag, and slammed the bolt back home. She was ready to go. I looked at the other truck and saw Hogu doing the same thing with his MK-19, slathering the insides of it with LSAT, the thick white lubricant that the grenade launcher required to function in this environment.[23]

Everything went much smoother than it had a few months ago when we'd first started patrolling. Back then, last minute sprints to the barracks to retrieve forgotten items was the norm, confusion about what we were doing was common, and the whole process was marked by wasted motion, shouting, and chaos. Now, as I watched Hogu, The Hawk, Sobon, and Lussi get ready to leave the wire, they looked calm and confident. Bored even. What had been an emotionally charged experience not too long ago had turned into a routine, even under fire.

All of the Humvees were already rumbling and lined up when the PL arrived, conferred briefly with Romero, and hopped in his truck. As our four-vehicle convoy rolled through the front gate and past the zig-zag of C-wire immediately outside of it, I opened the Ma Deuce's feed tray cover, fed a belt of ammunition into it, closed the tray, and racked the bolt. I pivoted the turret to the right, hollered "*Test fire*,"and squeezed the butterfly trigger, firing a burst into the hillside near the base to make sure she was working well.

Our convoy wound its way down the hill that Firebase Sh'kin sat on. We passed the two girls' house, waving at them as they leapt up and down begging for candy, and we turned right to head for Lozano's ridge. The drive was short, and in fifteen minutes we were once again passing the crater where Dickey and Collazo's truck had been cut in half.

The road continued past the crater and ran along the top of the ridge, and we lined up and scanned the area. The PL's voice crackled over the radio, "All Killer elements, this is Killer Six. Alright, the POO was right around here. I'm going to take a couple of guys on foot to push up to the edge of the ridge. Gunners, you've got security while we move."

23. LSAT looks exactly like semen. This has led to its use in a wide variety of practical jokes.

"Killer Seven, roger. I'll come with you. One-Alpha, you come too."

From the cab of our truck, I heard Lussi say, "Killer One-Alpha, roger."

Sobon turned to Lussi, "Have fun, man."

"Yep. Another day in the 'Stan."

Lussi got out of the truck and joined the PL and Romero. The edge of the ridge was about a hundred meters east of us. From there, they'd be able to see all the way to the border.

I talked down through the turret hole to Sobon, "Think we'll find anything?"

"Do we ever find anything?"

"Nope."

"I mean, do we imagine that Hajj is just going to light up the base, sit there and eat the counter-battery fire, then wait around for us to show up and kill him? Hell, he could be here now, and how would we know it was the right guy? Maybe that guy over there *is* him."

Sobon pointed to the north, where a farmer was adjusting one of the rock piles that served as some sort of demarcation line between his dirt and his neighbor's.

The guy was a couple of hundred meters away. But from here, he looked just like every other old Afghan (or Pakistani, who could tell?) guy I'd ever seen. He was stooped and slow moving. Every action was deliberate. Patient. Why would he rush? What else did he have to do?

I looked around the ridge. The terrain was wide open, and I could see for miles. The nearest visible building was a Pakistani military checkpoint across the valley. It was at least a kilometer away and the valley between us was deep. It would take hours to get there.

"Where did this guy even come from?"

"Who fucking knows? If the rockets were fired from anywhere around here, I imagine he knows who did it though."

"Yeah. I guess so. Shame we can't ask him."

"I don't think he'd tell us even if we could."

"True."

Sobon killed the truck to save gas, and we lapsed into silence as I continued scanning the area. With the truck's engine turned off, everything was completely quiet. Other than the old man and our three dismounts, I couldn't see any signs of life. It was the middle of the day, and the sun was bright but weak.

"You know, this place has an unchanging quality to it that's unlike anywhere else I've ever seen. I mean, that guy, his rock fence thing, and this ridge have probably been exactly like this for a thousand years."

Sobon was quiet for a moment. Then he replied, "Yeah, I guess it probably has. This place is out of sync with the rest of the planet, man. It's like another world."

He was quiet again before continuing, "And the people. They're just as out of sync. Look at that old man. Let's assume he isn't the guy shooting rockets at us. What does he do all day? Why does he give a shit about that rock pile? Have you ever seen anything growing in that field? No. So, why does he care about it? What's he even doing out here? Where does he live?"

Something about this train of thought must have irritated Sobon, because he started to get a bit wound up as he continued. "I mean, why the fuck does that guy even wake up in the morning? He just does the same thing every day, and none of it fucking matters. He walks from wherever he lives to this shitty field, he moves some rocks around, and he walks back fucking home. He's wasting his life on *fucking nothing.* If I lived like that–just doing the same pointless task every day for no reason–I don't think I'd be able to even fucking wake up in the morning. *I'd just fucking kill myself.*"

He took a breath, pulled out a cigarette, lit it, and took a drag. "I think I'm done now."

As Sobon was ranting, I thought about how many other patrols over the last two months we'd been on just like this one. Drive out. Find nothing. Drive back. Drive out. Hit bomb. Drive back. It put his comments into a context I hadn't previously considered, and I thought about it for a minute before replying, "Are we still talking about the old man, or are we talking about someone else?"

"I don't know what I'm fucking talking about. Fuck this place. I gotta get out of here."

We sat in silence for a while, waiting for the dismounted patrol to wrap up. After a half hour or so, Lussi, Romero, and the PL arrived back at the trucks. Lussi walked our way and climbed back in.

"What's up guys? Anything happen while I was gone?"

"Nope. Not a thing. Just the usual profound discussions about life, the universe, and everything. Find the bad guys?"

"Nah. Nothing. Just some rocks with burns on them. Pakis were yelling something at us, but we ignored them."

The squawk box crackled to life with the PL's voice.

"All Killer elements, this is Killer Six. Negative contact with enemy forces at this time. RTB."

"Killer One-Alpha, roger."

"Killer Seven, roger."

We turned the trucks around and reformed the convoy to head home. As Hogu's truck passed ours, he leaned over in the turret and shouted, "When we're back, it's game on, Pace. You're going down, motherfucker."

I nodded and grinned in reply, then pivoted my turret to the rear to secure my sector.

We drove home in silence. One more patrol in the books.

We were driving further north along Route Saturn than I'd ever been, and I wasn't sure exactly where we were headed or why. Captain Lopez, the Charlie Company Commander, was with us, which was unusual. That, combined with the unfamiliar route, foretold something more than just a normal presence patrol.

Captain Lopez was a huge dude, at least 6'4", and he had an easy confidence that made stressful events seem routine.[24] He got along well with our platoon, and we liked him, too. Wherever we were headed today, it was good to have him along for the ride.

As I bounced along in the turret, I had trouble paying attention to my surroundings. My mouth was sticky and my head ached. I thought about reaching for the bottle of Snapple in my pocket, but I couldn't bring myself to do it. Fuck Snapple. I'd had nothing to drink but Snapple for the last two days, and I didn't want any more. It was better to be thirsty.

Except for the supplies we procured locally, everything in Sh'kin was flown in by helicopter, including potable water. The weather between Sh'kin and Kandahar had been bad for a week, so the helicopters hadn't come in on time, and we'd run out of bottled water. Alternative drinking options had been exhausted pretty quickly, so that all that was left on base was Snapple and Diet Coke. Need something to rinse your mouth out after dry-brushing your teeth? Grab a Snapple. Feeling thirsty after

24. For reasons only the Army understands, Captain Lopez was issued the smallest set of body armor I've ever seen. It looked like a child's vest on his enormous frame.

a workout? Diet Coke's got you covered. Want to wash that morning breath out of your mouth? Have a delicious Snapple. After just one day, I was pretty tired of both beverages, and by the third day, I was ready to face dehydration rather than consume any more.[25] And although having nothing but Snapple for a few days might sound pretty trivial, particularly compared to the sorts of deprivation you read about in books about the Bataan Death March, the Holocaust, or the thousands of other resource deprivation situations humans have faced, in its own small way, the Sh'kin Snapple Drought of 2003 sucked plenty.[26]

I shook off my internal bitch-fest and tried to pay better attention to my surroundings. As Route Saturn wound its way north along the border, it got substantially worse. Whereas the mountains on either side of the road to the south were high but gently sloped so that we could still make our way up them, Saturn here was hemmed in by actual cliffs. I craned my neck to stare straight up at the top of one of them. The Ma Deuce couldn't point that high, so if there was an enemy standing on the ledge thirty feet above us, there was nothing I could do but shout a warning.

I tensed, expecting a bomb to be hurled down on us at any moment. Fortunately none came, and in a few minutes we were out of the constrained area and heading toward an open valley. We found a fork in the road, hesitated for a moment while the PL conferred with Captain Lopez on the radio, then headed east.

The LT called up from the cab of the truck, "Hey Pace, we're in Pakistan."

"Huh? Why? I thought we weren't supposed to go to Pakistan."

"We're usually not, but the boss has something planned. You're going to love it."

"I'll bet."

We continued up the road for a few hundred yards, then stopped. The trucks behind us fanned out in a line, with all of us facing east. The LT got out of ours, and he, Captain Lopez, and several other leaders walked to the front of the line and huddled in a clump.

---

25. I never liked Diet Coke, and I still hate Snapple today. It is right up there with Fanta, Army Coffee, and iodine-purified water on the list of beverages the Army has rendered undrinkable to me.

26. Fun fact: When on mission, soldiers sometimes have to urinate into empty water bottles. The resultant bottles, referred to as piss bottles, are generally saved in a corner until they can be disposed of after returning to base. After two days on a Snapple-pure diet, a piss bottle full of urine and a fresh bottle of Snapple looked identical. This resulted in a few awkward and disgusting situations, and led to much speculation about what exactly goes on at the Snapple factory. Which I'm sure is a fine place. Don't sue me, Snapple.

It looked like they were waiting for something. Captain Lopez knelt down and scooped a handful of dirt into a jar.

"Hey Sobon, what the fuck are they doing?"

"No idea, man. I think the Cappy just picked up a jar of dirt though."

"Like in *Gladiator?*"

"Maybe. Who knows? He's an officer. You know officers do weird shit."

"Very true."

I looked to my right, where Hogu was pulling security behind his MK-19. "Hogu! What're we doing, man?"

"No idea, Pace. I just work here."

I nodded. Yeah, me too.

I looked in the other direction, where the other gunners appeared equally confused. Nobody seemed to know what was up except the leadership, and all the leaders were out front in a huddle.

In a few minutes, we saw a dust cloud to the east that indicated something approaching. I wasn't sure what it might be, but I got down behind the gun to be ready to kill whatever it was. Up and down the line, my *compadres* did the same.

Soon, the dust cloud materialized into a convoy of pickup trucks. Each of them was a white Toyota Hilux full of uniformed guys with a stand-up machine gun mounted in the bed. This was clearly the Pakistani Army.[27] They pulled into the valley and arranged themselves in a parallel line about fifty yards in front of us.

I stared down the barrel of the Ma Deuce at the guy across from me. He was doing the same.

"Man, Sobon, fuck these guys. They look like they want to start some shit."

"They look like a bunch of homeless people to me."

He wasn't wrong. They had helmets and uniforms, but they were a ragged bunch.

"Homeless people with machine guns."

"True."

Sobon climbed out of the truck and braced himself against the armored door to look down the sight of his rifle at the Pakistanis.

"I bet we could take them."

---

27. Or so I thought. Looking back on it, it could have been the Army, the police, or some sort of tribal security force. I still have no idea.

"Fuck yeah, we could. We'd smoke all these motherfuckers."

My fingers hovered above the butterfly trigger of the machine gun. What was the guy across from me thinking? Was he getting ready to shoot? I glanced to my right at Hogu. He was clearly wondering the same things I was, and he looked like he was ready to annihilate the Pakistanis with grenade fire.

The passenger door of one of the pickups opened, and a guy stepped out and started walking toward the group of American officers. Several other Pakistanis joined him.

"I guess they're here to talk after all."

"Maybe. Or maybe it's all a setup."

"Maybe so."

We watched as the group between the two lines of trucks started to converse. We couldn't hear what they were saying, but the meeting seemed cordial enough.

"So, what do you think would happen if one of us just started shooting?"

"Well, they'd shoot back, I imagine. We'd win, but I think they'd kill all the leadership before we could beat them."

"That'd be some crazy shit."

"Yeah, we'd lose *another* PL and *another* platoon sergeant."

"And a company commander, too. Surely that'd have to be some kind of record."

We were sort of joking, but only sort of. The guns pointing at us were real, and my thumbs really were hovering only about a millimeter above the trigger on the Ma Deuce. The situation was a little tense, and as we sat there staring across the no-man's land, the tension continued to mount.[28] I glanced up and down the line. The other gunners clearly felt similarly. It was dead quiet and I felt a bead of sweat run down my neck.

All it would take was a single shot, and this whole situation would erupt into an enormous gunfight. Half of me wanted it to happen. All of us did. There was something crazy about having all of these guns pointing at each other that seemed to make a fire fight inevitable. Shooting would feel good. It would scratch that atavistic need to *do* something rather than just wait.

Validation. As it had during the standoff with the car full of Afghans in August, that's the word that echoed in my head.

"Sobon?"

"Yeah?"

28. Properly, it wasn't no-man's land since there was a group of men standing in the center of it.

"I feel like I'm stuck in that final duel scene from *The Good, The Bad, and The Ugly"*

Sobon whistled the movie's classic theme song, but neither of us laughed. The joke failed to release the tension.

The standoff continued, and the air temperature seemed to rise as well while we stared at each other. Since we weren't moving, there was no breeze to cut the heat. The sweat was running freely down my neck and back by this point.

What was happening? What were they doing? Where was this all going?

These thoughts screamed through my head as I stared across at the Pakistanis.

Crack

I heard the sharp sound from somewhere to my left, and my thumbs twitched reflexively. I caught myself. The sound had been nothing important–I glanced over and saw one of the Charlie Company guys had accidentally banged a piece of gear into a truck door. Shit. That was close.

Right afterward, there was movement in the center, and the group of men broke apart and started making their way back to their trucks.

The PL walked up to our truck smiling.

"Hey gents, ready to go? We're all done here."

For some reason, his easy demeanor and casual comment broke the tension, and I suddenly relaxed. I looked around. With the meeting over, everyone else had relaxed, too. It seemed the moment had passed.

"Yeah. Sure, Sir. It's hot as hell. Let's get out of here."

"Too easy, right? All that driving for a ten minute meeting."

Had it only been ten minutes?

"Wait, you *planned* that meeting, Sir? Did you know those dudes were showing up with an armed escort?"

The PL shrugged as he climbed into the truck. Sobon drove the truck in a circle as we resumed our position at the front of the convoy and prepared to head back to Sh'kin.

"I didn't know quite what was going to happen, but the boss said we were coming out here to meet with some Pakistani commander or something."

"Did he mention that the guy was going to bring a bunch of armed friends?"

"Nope. Sure didn't."

We rode in silence for a while. The tension of the standoff had evaporated, and now I was tired and sort of sluggishly relaxed, so passing back between the cliffs didn't phase me.

"Hey, Sir?"

"Yeah, Pace. What's up?"

"So, do you realize how close you were to death today? We were about a cunt hair from lighting up those Pakistanis, and our best guess was that while we'd probably win, you and the head shed would probably all get smoked."

The PL thought for a moment.

"Then I appreciate you not firing. Pace. I'd rather not be dead just now. I take back all the bad shit I said about you during the meeting."

We both laughed.

"I mean, think about what Tom Brokaw would have said on the evening news. American patrol accidentally drives into Pakistan and gets into fire fight with allied army in unknown valley. All officers and senior sergeants killed. Privates rampage through Pakistani countryside demanding beer, chicks, and hot chow from local population. Situation developing."

"Enlistedmen in charge of a combat patrol? Mother of God. Think of the consequences."

"Yeah, shit might have made sense."

The rest of the patrol went smoothly, which was probably a good thing since I was practically dozing in the turret by the time we got back to Sh'kin.

As Sobon and I reset the truck, we talked about the day's events.

"You know, Pace, it's amazing to me how little we knew about what we were doing today, and about how close we'd been to causing a disaster. And it sounds like the commander knew the whole time! What the fuck? Did I miss something during the brief today?"

"I don't remember anyone saying anything about it. It's possible I wasn't paying attention though, and it's sure as hell possible you weren't."

"That's a fair point. Still, that shit almost got serious. We could have all gotten killed."

"True. We didn't though. And did you see how well the PL calmed everything down at the end? I don't know if he did that shit on purpose, or if he was really just feeling that casual about the whole thing, but I gotta remember that trick if I'm ever in charge of anything. It was like magic."

Sobon laughed. "Sure, Pace. When you're in charge of something. That'll be the day."

I laughed, too. "Yeah, even the Army isn't that dumb."

It was a sunny, dry day that was just starting to turn warm, and as I bumped along in the turret, I soaked in both the heat and the view. We'd never driven northwest before, so within a half hour of leaving Sh'kin, we were in terrain I'd never seen, in a surprisingly different area.

Instead of the dusty plain that Route Corvette crossed to reach the border, or the rolling road that led down to Lozano's Ridge, the track north quickly became circuitous and mountainous. We followed it as it spiraled along the sides of increasingly high ridges, down into riverbanks–dry this time of year–and past isolated mud-brick compounds that commanded the terrain around them.

Apparently, AT Platoon and Charlie Company had either done enough that our commanders felt we needed a break, or the other companies were getting jealous and wanted a cut of the Sh'kin action–I was never clear which. Regardless, after a week spent packing our gear and orienting Alpha Company to the area, which included a few more hairy situations involving anti-tank mines and RPGs, we were now driving north to our new home in Orgun.[29]

The distance was only fifty-six kilometers, but four hours into it, we realized our map reconnaissance estimate about how long the trip would take was wildly off.[30] To say the road was unimproved was an understatement. At its best, it was a rocky, pitted dirt road, wide enough for a single vehicle to traverse slowly and carefully. At its worst, when it merged completely into a river bed, it wasn't a road at all, and our trucks had to pick their way among basketball-sized river rocks to avoid blowing tires or cracking suspensions. For most of its length, though, the road was more akin to a double-wide foot trail, worn into the side of a ridgeline by the feet of 50,000 years' worth of pack animal traffic. Driving along these sections could be quite harrowing, especially when

29. Alpha Company didn't have an organic mounted element, so it had to cobble one together from a dismounted platoon and a bunch of extra trucks. The battalion didn't have enough heavy weapons for them, though, so they ended up with a bunch of captured Russian weapons on their vehicles. I was a little jealous of this until I got to play with one of the DSHKA heavy machine guns. Not only was it inferior to the Ma Deuce in both appearance and function, but the gunner has to link together his own ammo by hand. Go America.

30. Google Maps suggests this drive should have taken an hour and twelve minutes. Either we inadvertently took the scenic route, or their estimate is somewhat off.

there was a hundred-foot drop to a valley below. In these situations, one of my jobs as the lead gunner for the convoy was to peer over the side of the truck to ensure that the road was wide enough for us to use. While we didn't lose a vehicle, there were many instances when our passenger side mirrors scraped the mountainside and our driver's side tires were barely on the road at all. It was even worse for the LMTV cargo trucks behind us.[31]

"At least we're driving during the day. I think this would be impossible at night."[32]

The PL shouted up through the turret to me. "What's up, Pace?"

"Nothing, Sir. Just bullshitting to myself. How long do we have left?"

The PL unfolded his map and cross-referenced it against his GPS. "Looks like we're about half-way there, so, maybe another four hours?"

"Give or take a few more flat tires."

"Yeah. And wrong turns. And ambushes."

"Roger, Sir. And those."

I lapsed back into silence as the trip continued. We bumped along for hours and hours, eating whatever MRE bits were in our pockets, and stopping to piss whenever the terrain was such that we had to stop and wait for one of the trucks behind us. By late afternoon, three vehicles had suffered mechanical problems and were being towed, a half-dozen others had popped tires, and all of us were sick to death of the bouncing dusty Tour d' Afghanistan. We'd even run out of decent conversation.

"Alright, I got another game. I'll throw out an actor. You throw out another actor and the movie that links the two together. We go around the truck until someone loses. First name up is John Travolta."

The PL replied first. "Samuel L. Jackson. *Pulp Fiction.*"

Sobon jumped in. "Bruce Willis. *Die Hard 3.*"

Powers, the forward observer who was with us, went next. "Chris Tucker. *The Fifth Element.*"

"Oooh, nice one."

The next guy was a medic whose name I didn't know. "Jackie Chan. *Rush Hour.*"

It was back to me. I wracked my brain thinking of Jackie Chan movies.

---

31. Whose eighteen to twenty-two year old drivers were some of the most unsung heroes of the Afghan war.

32. I was totally wrong about this. As we'd later find out, driving at night in the mountains was only *almost* impossible.

"C'mon, Pace. Hurry up or you're out."

"Give me a second, Sir. The dust is fucking with my brain."

I thought and thought, but nothing came to mind. Finally, my brain leaped to a technicality. "Jeremy Piven. *Rush Hour 2.*"

The truck erupted in protest. "Bullshit!" "We already used *Rush Hour!*"

"Not *Rush Hour 2,* though. That shit is legit."

"Man, Jeremy Piven isn't even in that one."

"He is, too! He was a salesman."

"Man, fuck that. Sir, your call. Is that a legit reference?"

Everyone quieted down while the PL worked through his decision. He was silent as he chewed on his lip, apparently lost in thought. He seemed to really feel the weight of command on his shoulders–his men were awaiting his arbitration. After a long moment, he broke his silence.

"Nope. That's a bullshit reference. You're out, Pace."

"Fuck that! I appeal the decision!"

"Can't. Captain Lopez is in another truck. No appeals authorized, my friend."

"Damnit."

The game continued below me for quite a while, but I stopped paying attention. Honestly, I wasn't really sure Jeremy Piven was in *Rush Hour 2* anyway. I'd only seen the movie once, and it had been a bit of a guess.[33] Still, losing sucked, and now there was nothing to do but scan the high ground for terrorists and reflect on our time at Sh'kin.

It seemed like we'd been there forever, but as I did the math, I realized we'd only been operating out of the firebase for a few months. What a few months though! When we'd arrived, we'd been so incompetent, that we'd gotten lost and had a flat tire within a kilometer of leaving the wire. Since then, we'd been in multiple firefights, dozens of rocket attacks, had recovered vehicles, called in artillery strikes, and–I thought–gotten pretty good at being infantrymen.

That experience had come with a price, though. Since arriving in-country, three guys had gone home injured, we'd lost all but one of the vehicles we'd arrived with, and multiple leaders who weren't meeting the standard had been rotated out of the platoon.

However, the biggest differences I saw between Fort-Drum-Us and Post-Sh'kin-Us were the internal changes in the guys. Some of them had really taken to combat. They

---

33. Incidentally, Jeremy Piven *was* in *Rush Hour 2*. Such is the fickleness of military justice.

enjoyed the reprieve from garrison bullshit like uniform inspections and police calls, and they found the shooting and rocket-dodging exciting. Other guys seemed to hate war. The rocket attacks and ambushes made them jumpy, they missed their families, and they'd gotten nervous and cynical. Much of the platoon was somewhere in between. All of us, though, had developed enough experience that whether we loved combat or hated it, we had gotten pretty good at it. The blocking and tackling of fire and maneuver came easily now, and many of the tasks that had flummoxed us in July, like working in zero ilum or keeping equipment running in the dust, were now routine.

What was next, though? We still had four months left in the deployment. What would we be like by the end?

As the sun was dipping down toward the horizon, our convoy finally rounded our last corner, passed a bored looking Afghan in a guard shack, and got our first glimpse of Firebase Orgun.

Unlike Sh'kin, which consisted of a single, high-walled building that had been occupied and hardened, Orgun was stacked HESCO barriers arrayed in concentric circles around a few low flat buildings and tents. As we wound through the Afghan-manned entryway, we discovered another difference between the two bases: half of Orgun was occupied by the Afghan Army.

Our trucks rolled past the Afghan section of the base as the Afghans were holding a formation of some sort. I couldn't tell what they were doing, but there was a U.S. guy with a beard standing in front of the formation talking to them. These Afghan soldiers looked sharper than those I'd seen at the border checkpoint at Sh'kin. Their weapons looked better maintained, and their uniforms and gear looked more pulled together and less hodge-podge. I stared at them, trying to decipher the situation as we drove past, but I couldn't make heads or tails of who they were or what they were doing.[34]

We left the Afghan section of the base and followed the HESCO wall along the perimeter of Orgun, passing an artillery pit with a howitzer in it, a bunch of green Army-issued tents, and a wide, rocky airfield. Eventually, we reached a Soviet-style concrete building that was surrounded by its own low-wall of HESCOs.

A familiar face from Fort Drum was leaning against the HESCO barrier waiting for

---

34. I didn't know it then, but I was looking at a U.S. Special Forces Detachment training an Afghan Commando Unit. Over the course of the war, the commando units became the most effective units in the Afghan military, and even after the collapse of the Government of Afghanistan in 2021, many commandos remained loyal to their U.S. friends and trainers and vice versa.

us. We parked the trucks, and I immediately jumped down to greet him. "Nevin! What the fuck have you been doing, man?"

"Not much, Pace. The Commander had me working a bunch of shit up in Kandahar, but now that you guys are finally out of Sh'kin, he said I can rejoin the platoon."

"That's awesome, man. It's great to have you back."

"Yeah, for sure. You guys are going to be holed up in this building for a while, so you can drop your stuff here for now. I heard you'll be moving somewhere else on base at some point, but as usual, it's all a cluster fuck and nobody has any straight answers."

"No surprise there! How's the chow?"

"Hot garbage, of course. Did you expect anything different?"

"No. I guess I didn't. The rooms alright?"

"You'll be snug as fuck on top of each other, but to be honest, it's not that bad. I think your new guys took all the good cots, though."

"We've got new guys?"

"Yes indeed, to include a brand spanking new Staff Sergeant."

Shit. That was a wild card. "How's he?"

Nevin laughed. "He's got plenty of energy. You'll love him."

"What's that supposed to mean?"

Nevin just grinned in reply.

"*Nevin!*"

The PL's voice interrupted our catch-up session, and Nevin jogged to see what the lieutenant needed. I rejoined Sobon at the truck to get our gear unloaded so that we could claim decent cots before all the good ones were taken.

"Well, we've got a brand new base to figure out."

"Sure do. I wonder if it'll be as shitty as the last one."

"It can't possibly be. I mean, it's further from the border, so we're bound to get rocketed less, right? And it's not like you'll be here all that long anyway."

"Less than a hundred days, Pace. That's how close I am to getting the fuck out of here."

I nodded, tugging my ruck out of the Humvee's trunk. "A lot can happen in a hundred days."

Sobon stopped what he was doing and stared at me for a second.

"Yeah, you're right. A lot *can* happen in a hundred days. And most of it's bad."

Sobon returned to his work, and we finished unloading in silence. When we were finished, we lugged our gear into the concrete building to meet our new squadmates and to see where we'd be living.

Charlie 1-19 graduation photo. Author is top row, fourth from left. Powers is to his left.

AT Platoon at Camp K-2, October 2001

Brock getting smoked at Camp K2, 2001

Sobon, The Hawk, Hogu, and the author in Sh'kin

The little girls outside Sh'kin

The PL doing land nav

Guard position atop Firebase Sh'kin

The lovely Sh'kin latrine

The Sh'kin CP, post rocket attack

2nd Section at Orgun

Hajmo the dog, on SGT Radon's cot

1st Section high-centered on a ridgeline

SSG Chadwick at an Afghan checkpoint

AT Platoon on the Margah mission

1st Section truck on a long range patrol

Orgun in the snow

AT Platoon at Orgun

The James Street house. Brandy's house on left.

Part 3

Some Other Base in Afghanistan

You can tell an Airborne Ranger by his dick
By his dick!
You can tell an Airborne Ranger by his dick
By his dick!
'Cause it's long, and it's thick, and it's always in a chick, you can tell an Airborne Ranger by his dick!

Chorus

Glorious, victorious, one keg of beer for the four of us.
Singing glory be to God that there are no more of us,
Cause one of us could drink it all alone.
Damn near.
Pass the beer.
To the rear.
Of the squad.
First Squad!

Army running cadence

12

October, 2003: Firebase Orgun

"I'm twenty-six years old, Sergeant. I think I know how to move some boards."

"Well, if you did, Ramsey, I wouldn't be shouting at you about it right now, would I? You move like old people fuck. Hurry the fuck up or I'm going to die of old age before we get this fucking tent floor finished!"

"Roger, Sergeant."

Staff Sergeant Chadwick and Private Ramsey made a funny pair. Chadwick was tall, scarecrow thin, and overflowing with manic energy. He could switch from laughter to fury in a split second, and he could be laughing again a second later. He radiated intensity. Ramsey was the opposite. Short and stocky, Ramsey was slow and deliberate in everything he did. He did good work, but he did it at his own pace, and this drove Chadwick bonkers.

Our other new NCO, Sergeant Radon, was awkwardly avoiding this interaction. He was a socially uncomfortable guy, and his conflict avoidance–unusual for a sergeant–had us all on edge while we tried to figure him out. A quiet, withdrawn sergeant in the infantry was a strange animal, and we figured strange animals were best handled with care lest they turn out to be rabid. Conversely, the other two new privates, Ross and Williams, were loudly snickering at Ramsey's misfortune as they hauled their own pieces of moldy flooring out of the tent. As privates, their instinct was both to avoid attracting

Chadwick's attention and to laugh whenever someone else drew it instead.[1]

We were clearing the flooring out of the tent on the PL's orders. Our section had been ordered to move into the tent, but, upon initial inspection, the place had reeked of mold and the ancient plywood floor had been deemed the culprit, so we were pulling it out to install new boards. Moldy flooring aside, the tent itself was an old, canvas Army-issued GP Medium, sixteen feet wide by thirty-two feet long, and it looked like it had come straight out of a World War II movie. The effect was completed by the wooden poles that held the whole contraption erect and the black metal pot-bellied stove that squatted in the center of the tent to heat it.

We finished pulling out the old flooring, then got to work with lumber, tools, and our section's collective pre-YouTube carpentry skills. The floor that resulted wasn't entirely level, and you risked splinters walking on it barefoot, but it smelled alright, and it kept us from living in the soft silt that comprised the parts of Orgun that weren't covered in gravel.

With the floor complete, we hauled in our cots, squabbled over who got to be in the corner and who had to be near the flap, and started arranging our gear in the quarters that would serve as our home for the rest of the deployment.

Chadwick put the cherry on top of our new-house sundae when he carried in a ten-inch TV with a built-in DVD player that he'd found somewhere. When the PL showed up that evening, he stood in the doorway nodding, with his hands on his hips as he surveyed our work.

"Well done, men. Shit looks first class." He pointed at the back corner where Radon was setting up his gear. "I'm pulling rank on that corner bunk though."

Radon nodded meekly and shuffled his gear to another cot. The rest of us fidgeted uncomfortably and exchanged sidelong glances. The young lieutenant had trumped a sergeant, and the sergeant had totally caved. In the end, we knew the PL would win. Rank had its privileges, after all. Still, in our hearts we knew the sergeant should have either played the situation off or fought it a bit, if only on principle. Radon had done neither, so in the primitive, pack-like world of the infantry platoon, he'd shown his belly a little too easily, and we'd all smelled weakness. Chadwick had noticed it, too, and he crossed his arms in irritation with Radon.

---

1. Sobon, Hogu, and I were specialists– the full-bird privates in the Section–so our instinct was to sneer at the foolishness of everyone involved while also trying to carry as little of the flooring as possible ourselves. There's a reason they call specialist rank 'the sham shield.'

The moment passed, and the PL walked back to claim his cot. He was followed by a ratty-looking puppy which caught Chadwick's eye.

"Where'd you find that shitty-looking dog, Sir?"

"He's not shitty looking, and he's ours now. He's a friendly little bastard, so I'm making him the platoon mascot. His name's Hajmo. You're welcome."

"Hajmo, eh? Well, he looks pretty shitty to me."

Hajmo was a light brown pup with short hair, a curly tail, and a black snout. He looked absolutely nothing like the photos I'd seen of Afghan hounds.[2] He walked into the tent like he owned it, wandered back to the rear corner of the tent, and jumped onto Radon's cot.

Radon immediately cringed. "Ugh, I hate dogs. Get that thing off my cot."

In that moment, whatever baseline respect we had for Radon because of his rank vanished. Not only had he failed to check the PL's authority, but he was apparently incapable of dealing with a small dog on his bunk. We all stared at him in disgust as he fidgeted next to the cot, staring at the small dog that was happily rolling around on his sleeping bag.

The PL walked over to the cot, picked up the dog, and placed him on his own cot. "It's just a puppy, Sergeant. Relax. It's not a big deal."

"That thing's gross. It's probably got fleas or some kind of disease or something."

The PL looked hurt. "Hajmo is a part of the platoon now. If he's got diseases, we'll get them taken care of."

"Yeah, alright Sir."

Radon returned to unpacking his gear, and the rest of us did the same. I ended up getting the cot adjacent to the potbelly stove, with Hogu on one side of me and Sobon across the small center aisle. I laid my sleeping bag out on my cot and stuffed my rucksack underneath it. I dug out my current book, *The Portable Nietzsche,* and dropped it on the cot. I pulled out my small bag of toiletries and placed it next to my flip flops by the ruck. Unpacking complete. One of the perks of deployed life was that packing and unpacking was pretty simple.

"Hey Sobon, we got a while. Want to go check out the computer lab and see if the Internet is up?"

---

2. He did, however, end up having many of the qualities of an Afghan person, in that he was rugged and a little scruffy, but ultimately pleasant looking, and he could eat anything on the planet without suffering any apparent digestive problems.

"Nah, man. I think I'm just going to hang out here."

"And dream about heading home?"

"Yeah, pretty much."

Since arriving at Orgun, Sobon had become even more morose than he'd been at Sh'kin. In 2003, the Army would keep a soldier deployed until he was thirty days from separating, and since Orgun provided no access to the resources that would have allowed Sobon to research or apply for anything, he knew he'd have only a few weeks after getting home to figure out the rest of his life. That knowledge, combined with end-of-the-movie syndrome, had left him depressed and anxious, and, as a consequence, had left me without my best friend.

"Hogu, what about you? You want to get out of this shitty tent and walk around, or do you want to just lay around on your cot like Sobon over there?"

"Yeah, for sure Pace. Let's go."

We pushed through the flap, and out into the afternoon air.

Orgun had a completely different feel from Sh'kin. Whereas Sh'kin's high walls and rocky ground made the place feel small and a little claustrophobic, Orgun's low HESCO walls meant you could see the entirety of the wide valley the base sat in. Additionally, while the base had some gravel, most of the ground was dusty silt, and this dust created a perpetual cloud that left the sun hazy and weak overhead. If Sh'kin was the Alamo, Orgun was the remote outpost where Kevin Costner's character had been assigned in *Dances With Wolves.*

Hogu and I walked around the outside of our tent, and stepped onto the pallet sidewalk that ran through the center of the U.S. side of the base. We passed the Company headquarters where Captain Lopez and his team ran our slice of the war, then cut through Orgun's small gym, which was equipped with a rubber floor and a bunch of old Nautilus-style weight racks. From there, we cut around the chow hall and ducked into the small dirt building that served as the base's computer lab.

Inside, there was a plywood desk with three dividers that formed cubicles, each of which had an ancient phone and a desktop computer. At the moment, all three were occupied, and a line of guys was waiting for them. Hogu turned to one of the guys we knew, a mortarman named Irizary.

"Yo, what's the deal? How long you been waiting?"

Irizary shook his head. "For fucking ever. Guy over there has been on that phone for a half hour, and the Internet is shit today, so the other guys have spent like twenty minutes just trying to open their email."

The guy (whom I didn't recognize) glanced over and gave us a 'fuck off' look. The other two were visibly irritated as they repeatedly clicked the 'reload' button on their computers' browsers to try to make something happen.

I looked at Hogu. "Yeah, I think this is a waste of time today. Want to get a snack or something instead?"

"Nah, I'm going to hang here. I'll eventually get a shot at the phone, and I gotta make a call."

"Alright, I'll catch you later, man."

I walked out and wandered aimlessly into the open area behind the chow hall. Since our platoon was still split, with Second Section and Romero still down at Sh'kin to help with the hand-off to Alpha Company, things had been slow here for the past week. The furious pace of life at Sh'kin had been exhausting sometimes, but I'd never really been bored. For the first day or two at Orgun, our reprieve had been welcome. But now the lack of activity just made the time drag, particularly with Sobon unavailable. I knew once the platoon was back together things would speed up, and I figured that I should appreciate the reset time, but I really didn't. When things were busy, they were too busy and I wanted a break, but then when I was off, I was bored as hell and wanted to get back out of the wire. Boom or bust, it seemed, was just the nature of combat.

"Hey, Pace!"

I looked up and saw Lussi walking toward me from the tent.

"What's up, man?"

"Grab your gear, we gotta roll out."

I hadn't heard any rockets. "What's up? What're we responding to?"

"Nothing. I think we're doing some kind of civil affairs thing. I'm not sure. Just came up. Get to the truck, and the PL will brief us on it. Have you seen Hogu?"

"Yeah, he's in line at the computer hut. I'll grab him. Also, what the fuck is civil affairs?"

"No idea. Just grab your shit and get to the trucks."

I walked back to the computer lab, grabbed Hogu, and headed over to the truck line, where Sobon was already prepping the radios. He had Williams with him and was explaining how to synchronize the time between trucks so frequency hop would work correctly.

I stepped onto the hood and climbed past them into the turret.

"Hey Williams, you taking over as my driver once Sobon punches out?"

"That's the word."

"Great. Sobon only rolled us once. Think you can do better than that?"

Sobon feigned indignance. "Why are you bringing up old shit?"

"People don't forget, my friend!"

"Maybe I should have hit that rock a little harder and finished the job, although then Alycia would have castrated me when I got back to Drum."

Williams listened to us in silence. He'd been on the ground for less than a week and in the Army for less than six months, and it didn't seem like he was sure whether or not we were serious.

"Don't worry, Williams. Driving over here is easy. Just don't hit any land mines, don't drive off any mountains, and get good at navigating shitty roads in the dark without headlights or depth perception."

Williams nodded. "Sounds easy enough."

We finished prepping the trucks and hopped down to gather around the PL for the mission brief.

"Alright guys, today's an easy one. We're rolling into Orgun proper–the town a few klicks outside the gate–and pulling security for a civil affairs team that's meeting with a local leader. Our role is pretty straightforward. Establish the cordon, watch for trouble, and keep the team alive until they're done with the engagement. Anyone have any questions?"

Ross raised his hand. "Sir, what's civil affairs?"

"They're the folks who hand out soccer balls and shit to the Afghans so they like us."

"Why do we care if they like us?"

The PL thought for a second. "Well, every Afghan isn't a terrorist. Most of them are just normal people like you and me. They've got kids and goats and jobs and shit, and they just want to live normal lives like the rest of us. Most of our operations inadvertently piss those people off though. While we're looking for terrorists, we walk through their fields, hit their cars, and sometimes kill their relatives. That makes the normal people hate us, and when they hate us, they turn a blind eye to terrorists burying land mines in the road to blow us up."

He paused for a second and then continued. "The thought is that civil affairs teams can show the population that our presence here actually benefits them. Ideally, this offsets the damage we do and endears us to the population."

Ross looked skeptical but didn't say anything else, so I chimed in.

"You think that's true, Sir?"

"No idea, Pace. Personally, I don't give a flying shit about making Afghans happy, but I do care very much about keeping you guys alive, so if this helps do that, I'm down."

I nodded. That made sense to me, and I appreciated the PL's candor. There were no further questions, so we piled into the trucks, loaded our weapons, and rolled toward the gate. The civil affairs trucks met us at the entrance, and we rolled through the gate and along the road to Orgun.

"This place looks just like Matamoras. You ever been there, Lussi?"

"Nope. You know I'm from Wyoming. We didn't make it that far south."

"Well, if you ever do, you'll see some striking similarities between a Mexican border town and this place."

Our truck was parked outside a long low government building in downtown Orgun. The building was one story high and made out of poured concrete. The paint on it was peeling, a five foot wall surrounded it, and the whole area had the faded appearance of a place that's been covered in dust for so long that it's forgotten how to look clean.[3] Anywhere else, the building would have looked like a dated, Soviet-style building from a movie set in Central Asia in the '60s. Here, though, surrounded by mud brick houses, herds of trash-eating goats, and a street bazaar with a camel standing in the center, the building looked surprisingly modern. The people, too, seemed relatively modern.

"Look, there's a dude in a sweater. I haven't seen a sweater since we left Drum."

"And that guy has sneakers on. And there's a dude in a tracksuit."

"The women are all still ninja'd up though."

"It's still Afghanistan, man. What did you expect?"

"Terrorists? Guys playing that goat ball game from *Rambo 3*? Women in burkas? I didn't expect to see a dude rocking a gray, cable knit sweater, that's for sure."

"That's a fair point."

We'd been sitting outside the building and pulling security for an hour or so while

3. West Texas has lots of places like this, too.

the PL and the civil affairs guys met with whomever they were meeting with. Normally, perimeter guard was boring as hell, but because this was our first time in Orgun, today was actually interesting.

Aside from the relatively modern appearance of everything, Orgun was *busy.* There were people walking past us, groups of little kids getting shooed up and down the street by their mothers, and far more cars than we were used to seeing. Sh'kin, and the area around it, had been dead. This place was most definitely alive.

Alive was interesting to watch, but it was also a bit unnerving. We'd treated every car at Sh'kin like a potential terrorist threat, but cars had been infrequent enough that doing this hadn't been too taxing. Here, though, with cars driving past us every few minutes, we couldn't search every car, or we'd be doing nothing but searching cars. And there were pedestrians everywhere, too. Who'd search all of them? What about the women? We only had enough guys to man the gun and run one hasty vehicle search. Trying to do anything more than that would have been impossible. Also, if the PL's point about not wanting to overly piss off the locals was correct, then we shouldn't try to do more than that.

Neither the knowledge that we should make friends with the Afghans nor the fact that we had limited manpower made the situation any better, though. When we'd first established the perimeter, I'd definitely worried.[4] As the hour wound down without us being attacked, though, this feeling had faded, and I'd actually started to enjoy people watching.

"Look at that dude, Lussi. How old do you think he is? A hundred?"

The guy in question was a stooped old man with a long gray beard and a face that was as wizened and brown as a piece of beef jerky. He was leading a camel along the road past us, a tall shaggy tan animal with a leather saddle, a nose ring, and a disinterested expression.

"I'd guess about fifty. I think the folks around here age pretty badly."

"Yeah, I guess life's hard in the 'Stan. Just like Wyoming, right? That what you're going to look like in thirty years?"

"Absolutely. Thirty years in Wyoming will make you as hard as woodpecker lips. Most folks don't stick around up there, but the ones that do are some rugged bastards."

The old man and his camel shambled past. An old woman, presumably the guy's

---

4. Except about the guy in the sweater. He triggered zero 'might be a terrorist' indicators in my brain.

wife, straggled along behind the camel. If he looked a hundred, she looked twice that, shuffling along, gumming something with her mostly toothless mouth.

"Man, I still think it's fucked up that the women always walk behind the herd animals here. That's some bullshit."

"Yeah, I guess the Taliban wasn't too big on women's lib. Camels are actually worth something. Women? Nope. Hell, if man-dance Thursday tells us anything, the dudes over here don't even consider the women worth sleeping with."

"The Taliban is gone, though, right? All that shit we saw about beating women and stoning people to death looked pretty sensational on the news, but when you see what the people out here are like, it makes you wonder how much of that was the Taliban and how much was just normal Afghan culture."

"I imagine none of that Taliban stuff was really all that remarkable to the people here. Probably the same stuff they'd been doing for a thousand years."

"Maybe, but maybe it's a little trickier than that. Maybe the Taliban represented an extreme version of normal aspects of Afghan culture. Like, Afghanistan turned up to eleven.[5] I mean, let's look at our country. In America, most folks are Christian. Even the people who don't go to church or anything identify as such and live basically Christian lives. Christian values. Christian holidays. All that stuff. But some folks in America take Christianity to eleven. You've got snake handlers, holy rollers, and Branch Davidians. Not too many of them, but they're out there. If one of those groups suddenly seized power, America would suddenly seem much more dogmatic than it does now, even if that fanaticism didn't penetrate too far into most people's lives."

"Yeah, I get where you're going with that. I guess the question then becomes, where does Afghanistan stop and the Taliban start? Where's that line?"

As I sat behind my machine gun in downtown Orgun, watching the old man leading his camel and wife past a guy in a tracksuit selling fake Rolexes, I wondered which of the two represented a 'normal' person in Afghanistan. Maybe neither of them? Maybe both?

"Maybe the Afghans don't all agree on where that line is themselves."

We lapsed into silence. Eventually, the civil affairs team walked out of the building and climbed back into their trucks. The PL walked over and climbed into ours.

"What's up, gents? Enjoying the sights and sounds of downtown Orgun?"

---

5. A *Spinal Tap* reference, for those who are unsure.

"It's riveting, Sir. You just missed a hot debate on whether or not the Taliban represent Afghan cultural norms."

"Oh, and what did you conclude?"

"We didn't come to a consensus, although we do all agree that lieutenants are both dumb and overpaid. And that they have stupid faces."

"Noted. I'll ensure that opinion is considered the next time I'm making the platoon's shit-burning roster."

"I appreciate that, Sir."

The PL turned to Sobon next. "What about you, Sobon? Any insights into Afghan culture?"

Sobon looked at the PL with a blank expression on his face. He'd clearly not been paying attention. "What was that, Sir?"

The PL shook his head and smiled. "Nothing, Sobon. You're about ready to get out of here, aren't you?"

"Yes, Sir, I really am. Ideally alive."

The PL laughed. "We'll do what we can, my friend. Back to base if you please!"

Sobon fired up the truck, and the PL sent instructions over the radio to return to base. We wove through town and arrived back at base without incident. We parked the truck outside our tent, and as we were climbing out, I called down to the PL from the turret.

"Hey, Sir. By the way, what happened in there? How'd the thing with the civil affairs guys go?"

The PL looked up at me. "We ate some footbread and talked to a bunch of old guys with beards. I guess they were the town elders or something."

"What did y'all talk about?"

"The civil affairs guys asked them if they'd seen any Taliban in the area. They hadn't. Then they asked the elders if they needed any help with electrical problems, or the water supply. They didn't. Then we just sat and looked at each other for a while. I think the point of the visit was to offer them something they wanted in exchange for support against the Taliban, but they really didn't seem very interested in any of it. Frankly, the whole thing was a bit awkward. "

I nodded and wondered aloud, "So, do the people like us better now?"

"No idea, Pace. I mean, they probably like most of us a bit better. You, though? Definitely not."

"That's fair."

The PL left to report to the commander at the company headquarters while the rest of us reset the trucks. Once I was done oiling up the gun, I grabbed a sack of laundry from under my cot and took it to the laundry room, where a smiling Afghan took the bag in exchange for a receipt with something unintelligible scribbled on it.

"Two day, two day," he said.

I nodded and gave him a thumbs up. "Thanks, man."

He nodded enthusiastically. "Good, mistah. Good."

As I walked back to the tent, I thought about the laundry guy. Which type of Afghan was he? Was he the sort of guy who'd knock all his wife's teeth out and lock her in a room before he went off with a group of friends to have sex with a prepubescent boy? Or was he the type of guy who'd put on a sweater and sell watches at a bazaar? Or was he both? Or was he some other type entirely? I had no idea.

Maybe it didn't matter all that much, anyway. It's not like he was coming over for Thanksgiving dinner. He was just some guy who did my laundry and, ideally, didn't pass coordinates to the local Taliban leadership so they could shoot rockets at my tent. When this war was over, he could go back to doing things his way, and I'd go back to doing them mine, and we could happily ignore one another for the rest of our lives.

Was that true, though? Maybe it *did* matter which sort of Afghan the laundry guy was. Did I have some sort of obligation to try to change those parts of Afghan culture that I found so horrible? Could I do anything about them anyway?

I felt like I was onto something, but my brain was tired and unwilling to do any more deep thinking, so I let the thought trail off and walked over to the phone bank to try and call Alycia instead.

I walked into the room and was pleasantly surprised to see it empty. I thought I was in luck until I picked up one of the black receivers and heard no dial tone. It was dead, of course.

"Fuck my life," I muttered, before walking back to the tent and flopping down on my cot to wait and see what would happen next.

October 10, 2003
Madison, WI

Dearest,

It has been thirteen weeks since I saw you last – we're really halfway through now!

It's another crystal-clear, cold night. Camille and I just braved it to take out the trash. It's supposed to get down to 25 degrees. What a switch from just a couple of weeks ago! I'm snuggled up in a sweater right now with a cup of hot chocolate.

Yesterday in the lab, the fire alarm started going off, and, as usual, everyone was inclined to ignore it. Then I saw the hallway fill with smoke and a woman started screaming at everyone to get out of the lab, saying, "It's gonna blow!" I grabbed my phone (not my wallet – see, my first thought was making sure you could still reach me) and ran downstairs with everyone else. Apparently, it was just an autoclave somebody didn't close properly. Autoclaves are used for sterilizing things, and they're basically huge steam-pressure containers. They do explode occasionally, so the woman's screams weren't entirely uncalled for.

You know, every single time I cook (like I did today), I realize that I enjoy making meals so much more than school and science. I loved cooking for you, and I can't wait to do it again. It's so sad that if I'm in grad school or a scientist, I just won't have the kind of time to allow me to make the meals I enjoy.

Additionally, while I certainly like the people here, I find them a tad unidimensional and presumptuous, although I never would have thought so a few years ago. You'll be happy to hear that I too now disdain what you call coffee shop folk – and academia is full of these. I miss the refreshing honesty and vulgarity of army people – people like Sobon, Tyrone, you, everybody who's not a real shitbag (and I'm well-aware that there are more than a few of the latter!). I've changed more in the past two years than I had realized. I don't relate quite as well to these folks anymore... and I don't really want to. It's like they've never been exposed to real life – even the professors. I guess that's where the "ivory tower" thing comes from. At one time, I was built for it, but that is no longer where I want to be. I don't hate or disdain it, mind you. It's just no longer me – although I would have argued this bitterly only months ago.

For many of them, it's like life on the outside doesn't really exist. There's a truly kind, intelligent post-doc in my lab who was shocked that my husband was in Afghanistan because HE DIDN'T KNOW WE STILL HAD SOLDIERS THERE. There's just a gulf between me and the others here that even good intentions and goodwill on all sides can't properly bridge.

I miss you awfully, but I suppose I just have to wait until January for that to get better. I love you more than ever and I can't wait to see you again.

Stay safe,
Your Loving Wife

Rumors of its arrival had been floating around the base for days. Nobody was sure how exactly it would be configured or what wonders it might contain, though. Would it come in a tent on one of the ring-route birds, or would it be driven up in some sort of armored trailer? Some people's visions were downright ridiculous–one guy from another platoon had openly speculated that it would be staffed by Redskins cheerleaders and offer free logs of Copenhagen and Burger King Whoppers to the first ten guys through the door. Others were super-cynical. "It won't have shit by the time it gets here. Everything worth a damn is going to get cherry-picked by the douchebags back at Kandahar." Whatever the truth might be, the news that the Post Exchange, or PX, was coming to Orgun was the most hotly anticipated event of the deployment.

Like everything in the Army, the actual arrival ended up being anticlimactic. The PX ended up coming by helicopter, but the day it was supposed to arrive a dust storm hit, so the aircraft were grounded in Kandahar and didn't make it. The next two days the same thing happened, and the day after that the landing zone got rocketed so the mission was deemed too high-risk and canceled. By the time the PX *did* arrive, half of the guys–myself included–no longer believed it was real at all and instead dismissed its existence as one more wartime rumor.

This skepticism persisted until Lussi walked into the tent one day carrying a gray plastic shopping bag.

"PX is here, bitches. Picked up some beef jerky, a Coke, and a new Discman. Better get over there before everything's gone, though. There's already a line about a mile long."

Like prairie dogs who'd heard something interesting, guys' heads started poking up across the tent. Headphones were pulled off people's ears and watchcaps from their eyes, and wallets were fished out of pockets so their contents could be assessed. Mine was empty.

"How'd you pay, Lussi? I don't have any cash on me."

"There's a finance guy at the entrance. You sign a statement, and he'll give you up to a hundred bucks cash, right on the spot."

"They got anything good?"

"DVDs, CDs, pogey bait, headphones, and other stuff like that."

That was a satisfying enough answer for most of the platoon to put on boots, grab rifles, and walk out to get in line. I thought for a moment. Did I need anything? Honestly, no, and if I bought anything, I'd have to carry it around. Still, shopping *was* something new to do, and it beat the hell out of lying on my cot, so I got up and joined the herd.

Outside, it seemed that Lussi hadn't been exaggerating all that much–the line wasn't a mile long, but it definitely wasn't short. Hogu, the new guys, and I walked over to take our places at the end of a line of about 40 guys. The line started at the headquarters building, strung out toward the motorpool, and wrapped around the outer HESCO wall that separated the base from the airfield. Hogu looked at the congestion.

"Man, I hope we don't get rocketed right now. Haj could kill like half the dudes on the base if they landed one next to us."

He was right. Our capitalist desires had created a pretty juicy target for any enterprising rocketeers.

"Small price to pay for something new to eat, man. Those scrambled eggs this morning were worse than potentially dying by rocket."

"No kidding, right? When I looked in the pan and saw those little islands of scrambled eggs floating in that shitty yellow water, I almost puked. I mean, how do they manage to make 'em so bad, anyway? Seems like you could accidentally make better eggs than that."

I nodded in agreement. "Dunno man. And what about those hash browns? I swear they tasted like Styrofoam. My mouth couldn't even generate enough spit to swallow them. I had to swish them around with a mouthful of Gatorade to get them down."

"I put like five packets of catsup on mine and swirl that shit around into a kind of paste. Only way to make 'em edible, you know?"

The line didn't move quickly, and it took over an hour before we made it to the building which had two doors. The line fed into one, the entrance, while people trickled out the other with shopping bags in hand. A sergeant stood at the entry door regulating traffic. He was an older guy with a regulation mustache in clean DCUs–clearly a visitor from Kandahar brought in to keep the savages in check.[6]

"Only five in the building at a time. One man out, one man in."

In another minute or so, two guys left through the exit, and it was our turn to enter. I walked into the room, half expecting Willy Wonka's Chocolate Factory and half expecting the Department of Motor Vehicles.

The reality, of course, ended up being closer to the latter. The room was low and lined with makeshift shelves stacked with pure Americana, including Slim Jims, copies of *Maxim, Popular Mechanics,* and *Soldier of Fortune*, and DVD classics like *Kangaroo Jack*. The shelves looked like they'd been full once upon a time, but by now, they were pretty picked over. What remained would have been a pretty meager selection under any normal circumstance. However, in my consumerism-deprived state, I was stoked.

Before I could start hemorrhaging cheddar, though, I needed to get some from the finance representative. He was an anxious looking middle-aged civilian guy.

"Good afternoon, Specialist. Would you like to draw some cash today?"

"Yes, Sir. How much can I get?"

"Up to two hundred dollars."

"I'll take a hundred, please."

The guy nodded and directed me to fill out a form with some basic information. I scribbled down my name, rank, social security number, and wrote '$100.00' in the 'amount of withdrawal' space. Easy enough. When I was finished, the guy gave me the yellow carbon copy, opened a small metal cash box, and counted me out five $20 bills.

"Thanks, Sir."

"Of course. Stay safe out there, Specialist."

I nodded, took my fistful of dollars, and started up the aisles. There were only two, and they were each about fifteen feet long, so perusing them didn't take a whole lot of time.

The first section was all consumables. Fortunately, I wasn't a tobacco user because all

6. It's a well-known fact that there's no such thing as a cool-looking moustache that adheres to Army regulations. In fact, the reg seems to have been written specifically to prevent it, since it only allows Hitler-esque toothbrushes or a lip-liner-like skinny-stache. According to the text in Army Regulation 670-1, "Mustaches will not cover any portion of the upper lip line, extend sideways beyond a vertical line drawn upward from corners of mouth, or extend above a parallel line at the lowest portion of the nose." Print a picture of yourself and try to draw a moustache on it that meets those criteria and doesn't look stupid. I dare you. Even if you manage, the first sergeant major you encounter will make you shave it off anyway.

of that product was already gone. There was still a good bit of food, though, so I made a quick survey. Beef Jerky? No. Plenty of that available at the chow hall. Protein powder, Andro, or Hydroxycut? Nope. Not a supplement guy. Twinkies? Hell yeah! Since my diet was painfully lacking in yellow dye number whatever, I grabbed a box and moved on to the magazine section. It, too, was pretty picked over. *Maxim* was sold out, as was *Hot Rodding, Rolling Stone,* and *Men's Health.* Plenty of issues of *Cosmopolitan* and *Glamour* were left though.[7] None of those piqued my interest, and I knew I'd be able to find an abandoned issue of one of the worthwhile magazines in a shitter by the end of the week, regardless, so I wasn't too worried about being short on periodicals.[8]

The digital media shelves came next, but after one look at the prices on the DVDs and CDs, the whole section became flyover territory. The licensed legal movies sold by the PX for twenty bucks apiece couldn't compete with the pirated stuff you could get at the haji-mart for two.[9]

The final section was devoted to electronics. There were a few TVs, a Playstation 2, a dozen Sony Discmans, and a stack of headphones. None of it caught my eye until I saw a set of small battery-powered speakers.

"Hey, Hogu. You think I could mount those in the truck?"

"What? And like, listen to music on mission? That'd be dope–if anyone would let you do that shit."

"Yeah, exactly."

Would anyone go for it? The sergeant major would shit himself if he saw us playing music, but ten hour patrols were boring as hell sometimes, and music would definitely help.

"They're battery-powered, too. You think you can get enough double A's to run them?"

"Yeah, worst case I can steal the batteries out of my NVGs and scam some replacements out of supply. Should be no issue. I imagine they'll last for a while."

"You gotta try that shit, then."

I grabbed the speakers and walked up to the register to pay. I handed the clerk the

---

7. Our base had a grand total of one female on it, an MP, and she was only there for a few months.

8. Any magazine that contained photos of women that one came across in a shitter was best handled with caution.

9. More on this later.

money, and he handed me back a few bills and a stack of cardboard chits. I looked down at the chits. Each chit was the size of a bottle cap and had a denomination printed on it. One said "five cent gift certificate". Another said "twenty-five cent gift certificate". They all had the AAFES logo on them.[10]

"What're these?"

"They're called pogs. Sorry, Sir. We don't use metal coins overseas, so you get cardboard gift certificates instead. Each one has the value of the equivalent coin."

"It's like I work on a banana plantation in South America."

"What?"

"Nothing. Forget I said anything. Thanks, man! Have a great day."

I walked out of the store and watched as another guy got to enter in my stead. Hogu walked out a minute or so later.

"What did you end up with, Hogu?"

"50 Cent album, issue of *The Source,* and some fat pills. And the guy gave me these shitty cardboard quarters. What the fuck is that about?"

"We get paid in store credit. That's how you know we have a great job."

We walked around the line of people waiting to get into the PX–it was still long enough to reach the airfield HESCO wall–and headed back to the tent. I dropped off my bag, grabbed a Twinkie out of it, and walked to the phone room to see if the surge of people at the PX had freed up a phone terminal. I was in luck. There was a cubicle open, so I sat down on the metal folding chair, propped my rifle in the corner, and eagerly picked up the phone receiver. Someone I didn't know was sitting in the cubicle next to me. I tried to block out his mumbled conversation as I started the complicated process of making a call home.

The phone was a military-issued phone–closer to a radio hand mike than a normal telephone. The receiver was made of soft black plastic. A thick, coiled cable attached the receiver to a cradle that was bolted to the plywood wall of the cubicle. Next to the receiver a piece of paper was taped to the wall with a half-dozen numbers and names of places.

---

10. AAFES, or Army and Air Force Exchange Service, is the government organization that runs the PX. Wherever soldiers are in the world, AAFES is there to provide them with the essentials of American life, to include DVDs, sneakers, hair products, and a variety of 'Exchange Select' brand liquors, all of which are bottom-of-the-line rotgut concoctions whose most redeeming feature is that they are sold in a conveniently plastic bottle that won't break when you're too drunk to avoid dropping it.

"Qatar, Kansas, or New Jersey. Which one is my winner today?"

I picked up the receiver, waited for a dial tone, dialed the number for Qatar, and waited to see what would happen. After a few seconds, I got a busy signal. No good. I hung up to reset the connection and pressed the number for Kansas. This time I got a ring, and after a few seconds, someone picked up.

"KSHHHKKKHHK..ow..KSKKSHHSHou?"

The static was deafening. I held the handset away from my ear and talked into the other end of the receiver. "Hi, I'd like to make a morale call, please."

"KSSHHHHSHHHSHHHHKK?"

"Can you hear me?"

"KSHHHKKSHKKSH."

"Hello?"

The line went dead. Whoever had been on the other end had hung up.

I guess they couldn't hear me either.

I dialed the third number, New Jersey, and again heard the phone ring. Again, after a few seconds, someone answered.

"MWR line, how can I help you?"

I waited a few seconds to clear the delay. MWR phones were slow, and words took several seconds to travel to the other end of the line, so if you didn't pause for a while after the other person spoke, you'd end up talking over them.

"Hi, I'd like to make an MWR call please."

"Name?"

"Daniel Pace."

"What state?"

"Wisconsin."

"What's the number?"

I rattled off the number.

"Hold, please."

I waited while she tried the connection.

After about thirty seconds, I heard a whispering hiss, then the operator's voice again.

"Your call is connected."

I couldn't hear anything from Alycia, just the same whispering hiss. The operator came on again.

"You have ten minutes. Thank you."

The operator hung up, and I heard the phone hiss at me again.

"Alycia? Is that you? Can you hear me?"

"Hissssiisssiississ"

"All I hear is hissing."

"Hississsisiss."

"Can you hear what I'm saying at least?"

"Hissisissssis."

"I'll assume you can. I can't hear what you're saying, but I love you, and I'll tell you about how things have been going over here."

I spent five minutes talking into the receiver about everything that had happened over the past few weeks. I described the long dusty drive from Sh'kin and Sobon's problems. I talked about the food, the shower situation, and the arrival of the PX. I told Alycia how my tent was set up and about Radon's problems with the dog. I even tried to describe what downtown Orgun was like, and about how weird it had been to see a guy wearing a sweater. I told Alycia how much I loved her, and how much I missed being with her back at our house. I expressed the hope that she was enjoying graduate school, and I described the happiness with which I greeted her letters. At the end of my soliloquy, I paused and asked if she'd heard anything I said.

"Hssisissssiss."

Had she heard anything? Was I even talking to Alycia, or was it someone else entirely? I was pretty sure *somebody* was on the other end, because if the line was dead it would have clicked over to a dial tone by now, but I could have been describing my day to The Hawk's great aunt for all I knew.

"Your turn. How've things been going there?"

As I listened to the phone hiss in reply, a wave of frustration and sadness hit me. Alycia and I had spoken to each another only a handful of times since I'd arrived in Afghanistan in July. The satellite phone in Sh'kin had been rationed to 10 minutes every other week, and it had almost never worked. The phones here weren't much better, and, in some ways, they were even worse than the satellite phone because they provided the illusion that we'd be able to talk more regularly.

Alycia and I had stumbled through several phone calls like this one–confusing starts that grew into long, one-sided sessions spent talking into the void and wondering whether the other person could hear anything being said or not. I knew from Alycia's letters that she could hear me at least some of the time, though. So, despite the frustration

I kept trying.

The hissing went on for a long while, and as I sat listening to it, I imagined that Alycia was telling me about her day, how the cats were doing, and how her classes were going. After a bit, the hissing stopped, and I just sat there, holding the phone quietly and trying to feel some sort of connection to whomever was at the other end of it. The guy in the cubicle next to me was still mumbling into his handset. I wondered if he'd had better luck with his call. Another minute later, an automated voice cut in and said, "One minute remaining."

That came through clearly enough. Stupid Army.

"Hississsisisss"

"Well, I love you, babe, and I hope things are going well. Take care of yourself."

"Hississisisiss"

I held the phone to my ear, listening to the final hisses and half-heartedly hoping the connection would clear up before our time ran out. The receiver clicked a couple of times. Not until I heard the dial tone did I know the call was definitely over.

I stood up, shaking my head in irritation, grabbed my weapon, and walked back to the tent. It was quiet and so dark when I entered that I had to let my eyes adjust before I could get to my cot. When I got there, I saw that my box of Twinkies had been torn open, and the contents pilfered.

"Which of you fat bastards rat-fucked my Twinkies?"

Sobon spoke up and raised his hand. He had white Twinkie cream on the corner of his mouth. "Sorry, man. I thought they were for everyone."

Hogu, Ramsey, and Chadwick all raised their hands as well.

"Ah well. Fuck it. They tasted like shit anyway."

Lussi tossed me a Slim Jim and some gummy worms. "Here's a consolation prize, Pace."

I thanked him and snapped into the meat stick, wondering what Macho Man Randy Savage would think about us eating Slim Jims in a tent in Eastern Afghanistan.

The shot of dopamine from the shopping trip had faded quickly after the failed phone call, and I felt irritated as I lay down on my cot and picked up the empty box of Twinkies. I grabbed a pen from my bag and drew a moustache on the picture of Twinkie the Kid that adorned the front of the box and asked it: "What the fuck am I doing here, buddy? Why am I in this stupid fucking tent in the middle of nowhere when I could be in grad school in Madison with my wife?"

The Kid only smiled in reply. Smug unhelpful bastard. I stuffed the empty Twinkie box under my bed, grabbed the PX bag and cut open the package of speakers with my pocket knife. These, at least, offered a potential bright spot in an otherwise gray day. Now, if I could just figure out how to rig them up in the truck and keep the Discman from skipping too much.

"When we're back from this ten-day mission, you're going to the board, Pace. Grab a 203 and rig your vest up for it. E-brake is going to be on the gun this mission. Train him up on the Ma Deuce, and he'll go back to Second Section when we get back. Oh, and unless you fuck up the board, you'll be an NCO soon. Start thinking and acting like one."

"Roger, Sergeant."

Sergeant Romero turned to walk off, leaving me to ponder his words as I cleaned and oiled the gun.

"Wait. Sergeant Romero?"

He stopped and addressed me over his shoulder. "Yeah?"

"Can I just stay on the gun?"

"Nope. Sure can't. You've got leadership potential. Time to put it to work."

"Leadership potential? Me? You know I'm Pace, right? You sure you're tracking who you're talking to, Sergeant?"

He only laughed in response and walked off without further explanation.[11]

We'd heard about the ten-day mission yesterday, and we'd been preparing for it since. The mission was to link up with the Special Forces team on the base and to conduct a patrol with them and their Afghan hatchetmen to a place called Margah, about sixty klicks to the northwest. We were to search the town for illegal weapons, then sweep

---

11. Romero and I discussed this conversation twenty years later while I was writing this book. He insists my primary opposition to getting promoted was that once I left the gunner job, I'd have to give up the pistol that came with it. Since we had to be armed everywhere we went, giving up the pistol meant I'd have to carry a rifle to the gym and chow, which was substantially more cumbersome. So, essentially, his argument is that I was too lazy to get promoted. I won't entirely disagree with his assessment, although I will say that the next gunner didn't end up with the pistol either– the PL kept it to take to chow and the gym himself.

northeast to push any terrorists we found into a waiting trap. It was an ambitious plan, and we were looking forward to working with the Special Forces guys and to conducting a mission that wasn't either rocket response or civil affairs support. So far, the preparation for the mission had been mostly unremarkable, except when Sergeant Romero had paused the mission brief to ask Sergeant Radon where we were going, Radon had said he didn't know, and Sergeant Chadwick had proceeded to flip the map board over in a blind rage and shout 'Margah' at him repeatedly, until Radon slunk out of the room.[12]

This promotion news screwed up everything. For one thing, I had to rework my kit, which was always a giant pain in the ass. For another, having E-Brake with us meant we'd have a fourth guy in the truck. That meant we had one more ruck to somehow fit into the packing list, one more guy to wedge into an already cramped truck, and one more ass that would fill up the truck with more nauseating smells after we'd all eaten MREs for a week. Finally, the news meant I'd actually need to spend the mission studying.

During a promotion board, the candidate gets asked dozens of questions about all sorts of obscure bits of Army lore. What's the maximum effective range of an M240B machine gun on a bipod? What's the Army regulation that covers uniforms and appearance? How many seconds do you have to don the M9 protective mask after being exposed to a chemical agent?

As someone who'd never had any plans to stay in the Army past my four year enlistment, I'd never bothered to learn any of this shit, and if there's one thing combat had taught me, it's that none of it mattered anyway. But I couldn't walk into the board and say that. I'd look like a dumb ass, *and* Romero would smoke the shit out of me.

And so I spent the rest of the day before we rolled out on what looked to be a pretty fun mission grumbling, reworking my kit, and trying to bum an *Army Study Guide* off someone. My only consolation was that Sobon was even less excited about the mission than I was.

"Last mission, eh? You make it through this one, and you're home free."

"Yeah. If."

"I mean, what could possibly go wrong?"

"You know how much I hate it when you say that."

"You hear we're getting E-Brake in the truck for this mission? I think we'll have the

12. The rest of us had looked at one another side-eyed and tried not to giggle.

PL with us, too."

"Yeah. I've been thinking about what to do with our rucks. They won't all fit inside the cab or in the trunk, and we can't strap them across the back hatch because we'll need extra spare tires for this long of a mission. Even the tailgate is going to have double-stacked fuel and water cans on it. I think the only way we'll be able to carry the rucks is to strap them to the front grill."

I walked over to look at the front grill. It was a bit banged up from clipping rocks and such, and it was dirty as hell.

"What do you think our gear is going to look like after eight hours of bouncing around on that thing?"

Sobon shrugged. "Don't see how we have much choice. I'll make sure your stuff is on the end, though. That way if something comes loose it'll go right under the tire."

I grimaced. "Dick."

I hated putting my gear at risk, but I couldn't think of a better idea, so I grabbed my ruck, unhooked the shoulder straps and started threading them through the bars of the front grill. Then I walked over to find E-Brake to get him to do the same thing. We finished loading the truck by nightfall, caught a few hours sleep, and by oh-dark-thirty we were assembled in a huge circle with all the Charlie Company guys to conduct a final roll-call before leaving the wire.

First Sergeant Earl, the Company's senior NCO, was hollering out names to make sure everyone was ready to go.[13]

"Sergeant Romero!"

"Here, First Sergeant."

"Jones!"

"Here, First Sergeant."

"Ryan"

"Here, First Sergeant."

Over the next fifteen minutes, he ran down the entire roster of soldiers going on the

---

13. First Sergeant Earl was the epitome of a first sergeant. He was mean as hell, he had a good sense of humor, and he took care of the guys. The characteristic I remember most about him, though, was his hatred of mustaches. He loathed them. He stopped me once when I was working on growing a gross stache to ask me if I knew his favorite thing about them. "What's that, First Sergeant?" I asked. He replied, "The way they get tangled up in my ass hairs. Shave that shit off your face before you get to find out what that's like."

mission–eightyish guys total–then dismissed us back to our trucks to start lining up to roll out.

When we got to the truck, Sobon cranked it and started the radio checks while the PL started his final pre-combat checks.

"E-Brake. You got your NVGs?"

"Roger, Sir."

"Show them to me, and turn them on."

E-Brake did. They worked.

"Pace. You got spare batteries for your NVGs?"

"Roger, Sir." I pulled an extra set of AAs from one of my freshly-rearranged pouches and held them up. I didn't comment on the extra sets I'd swiped for my speakers.

"Sobon. How many MREs do we have in the truck?"

"Four cases, Sir. Enough for four days. Should last us until the resupply with no issue."

"How're we sitting on fuel and water?"

"Six cans of diesel. Two cans of water. All full."

"Good."

"What about motivation? How're you sitting on that? Full?"

"About thirty percent, Sir."

"Good. That's plenty. I think we're ready to go."

The PL walked off to check the other vehicles and to confer with Sergeant Romero. While he was gone, I quizzed E-Brake on gun procedures.

"You got the PEQ-2 ready to go?"

"Roger that. Zeroed it earlier."

"What about the cleaning kit? You have it handy in case you have to clear a stuck piece of brass?"

"Yep. Rod's in my pocket, and the rest of the kit is here." He kicked the can at his feet for emphasis.

"You sweep the turret to clear any stuck links out of it yet?"

He paused for a second. "No. I'll work that now." He leaned out across the turret and started checking for any loose .50 cal links that might be stuck between the turret and the roof of the truck. I'd found out the hard way that links could get bound up in there and cause the turret to freeze, particularly if the truck was sitting on an incline.

I heard a few links *tink* as he extracted them from the turret and brushed them onto the ground.

"Alright. Turret's good to go."

"You got the barrel mitt somewhere nearby in case you need to swap barrels?"

"Yup. In the kit."

"Alright, man. I think you're good to go." I slapped him on the leg. "Just remember, you're the guy the bad guys will try to kill first. Oh, and if you die, try to fall out of the turret so I can get up and take over."

"I'll try to remember that."

I climbed into the back seat to get ready for the long drive. The back seat of a Humvee is no place for an adult to be, much less an adult wearing armor and carrying a gun. The seat itself is a tiny square of hard foam covered in green canvas. This square sits directly on top of a metal storage box that's attached to the frame of the truck without any supplemental suspension or support. This is exactly as comfortable to sit on as it sounds, and if the road is at all uneven, your ass and lower back take a pounding every time the truck hits a rock.

The discomfort doesn't stop there, though. Because of the way Humvees are designed, the back seat also has almost no legroom.[14] This means that, when seated, a passenger's knees are wedged against the metal support behind the front seat, guaranteeing both shin pain and loss of feeling in your feet within a few short hours.

By the time I'd folded myself into the necessary yoga pose to fit in the seat, wedged my rifle and grenade launcher between my legs, and adjusted my assault pack to fit in the crevice between my elbow and E-Brake's legs, I felt like I was in some kind of hostage stress position.

"Hey, Sobon."

"Yeah, what's up?"

"This shit is uncomfortable as hell."

"It gets better. Wait until E-Brake starts blowing ass."

"Man, I miss the gun already. What's that humming noise?"

"Oh, that's the Warlock. The jammer that supposedly keeps IEDs from blowing up. It sits right behind your head in the trunk."

The dull hum was barely audible. It felt more like a vibration in my skull.

---

14. My hypothesis is that, once upon a time, they did. But over the years, as the Army added more features to the trucks, like armor, tracking systems and countermeasures, the only thing the designers felt they could cut was back seat leg room. *Pentagon Wars* substantiates this hypothesis.

"I'm sure that's good for me."

"I mean, it probably doesn't give you any more cancer than all the other shit in here, right?"

"Yeah. I guess we'll probably die of something else before it matters anyway."

"Exactly. Just think, in a month I'll be back home hanging out with your wife and getting ready to start college or something."

I laughed. "Unless you die on this mission, of course."

"Yeah. Unless."

The front door pulled open as the PL arrived back at the truck. He climbed in and told Sobon to roll out.

"We're taking lead on the whole convoy, Sobon. No pressure. Can you see alright?"

"Ilum seems alright tonight, Sir, but we'll see how the roads look as we get farther from the base. We've never gone out this way."

"Very true. What could possibly go wrong?"

"Why the fuck does everyone keep saying that?"

The PL laughed, and we rolled to the front gate. I flipped down my NVGs so I could watch the vehicles start marshalling behind us. The force we were taking on this mission was sliced into two pieces: the Special Forces team with their Afghan soldiers, and our 10th Mountain element. For ease of control, each element planned to move independently during the day and then to link up for clearance operations and nighttime security. The SF guys were moving in a convoy of about fifteen pickup trucks, and they had rolled out an hour before us. Our element was much larger.

One of the biggest challenges in coordinating a mission of this size was managing the logistics. The 10th Mountain was bringing eighty guys for the clearance operation. Cargo Humvees held about six or seven guys each, and gun trucks held four or five. So, just to move the people and provide security we needed almost twenty trucks. Managing a twenty vehicle convoy in Eastern Afghanistan was difficult enough, but because this was a ten-day mission, we also needed to bring supplies like chow, water, ammo, and fuel. That meant we needed to bring LMTVs, the army's medium size cargo trucks, which were much larger than Humvees and could carry substantially more equipment. Our convoy had three of them tonight, each with a water trailer. Also, because the objective of this mission was sixty klicks from Orgun, we'd be operating outside the range of our usual support assets. That meant we needed to bring fire support, communications assets, a medical team, and maintenance guys, which added another seven trucks to the convoy. Throw in a couple of civil affairs vehicles for population engagement and the executive officer's truck because he didn't want to be left behind while the rest of the command team got to do cool shit, and our convoy was over thirty vehicles long. Lining all of those vehicles up under NVGs–we weren't using any white light on the mission–was difficult and took a solid half hour and a great deal of shouting. But all of this turned out to be easy compared to managing the convoy during the movement.

We were the lead vehicle, so much of the difficulty fell on us, and on the PL in particular. The trick to managing a convoy is setting a speed that's manageable by every other truck to avoid the slinky effect that starts to create gaps in the convoy when individual vehicles have to start and stop too often. We didn't have too much trouble doing this initially, as Orgun sat in the middle of a wide valley with relatively straight roads. Once we started winding up and out of the valley on the small roads that zig-zagged into the mountains, though, the troubles began.

"Careful up here, Sobon, the road gets real narrow."

"I can't see over the front corner of the hood, E-Brake. How much room do I have?"

"Wait one. I can't tell. Pace, can you get out and ground guide us?"

"Roger."

I unfolded from the seat and pushed open the door. The gap between our Humvee and the mountainside wasn't more than a foot, so squeezing out took a second. I side-stepped around the truck, walked up to the driver's side, and stared down at the ground through my NVGs. The front wheel of the truck was about a foot and a half from the edge of the road. The drop-off didn't look super steep, but if the wheel went over the edge, the truck would roll a hundred yards to the bottom of the mountain, and since it was a dirt road I wasn't sure how much weight the shoulder could handle.

I held up my hands to wave Sobon slowly forward until he cleared the tight spot. I climbed back into the truck, where the radio was blowing up with leaders from across the convoy asking what was going on.

"Why the fuck are we stopped?"

"What's going on up front?"

"Watch your distance."

"Clear the net. Stop bullshitting on the radio."

The dark and dust meant nobody could see anything more than a vehicle's length in front of their truck, and everyone wanted to know why we were stopped. The PL fielded the increasingly irritated calls as we started rolling again. We hadn't gone twenty yards, though, before E-Brake called down into the cab.

"Slow down, Sobon. We're leaving the convoy behind. We gotta wait for everyone else to make it past that tight spot."

Sobon slowed to an idle, and we inched along the road, slowly making room for the vehicles behind us as each of them stopped to negotiate the obstacle. Being much larger, the LMTVs had a particularly terrible time with it. But they made it past the tight spot without incident, and a half hour later, we were continuing along the trail toward Margah.

The next problem we encountered was a fork in the road. This was before the days of on-screen nav, so to navigate while we drove, the PL had to cross reference the coordinates from our PSN-11, the Army-issued GPS device, with a hand-carried map with the route drawn on it with a marker.[15]

"Which way, Sir?"

"Just a second, I don't see this fork on the map anywhere."

As the PL wrestled with his red-lens flashlight, map, compass, and PSN-ll, the radio blared to life. "Killer Six, this is Cobra Six. Why are we stopped?"

The PL set down his compass so he could free up a hand to click the button on the hand mike. "This is Killer Six. Map check, over."

He hung up the mike, accidentally knocking the compass onto the floor of the truck.

"Goddamnit. Hold up, Sobon. I gotta get out to do this."

The PL opened the door, and I stepped out with him for security. We walked a little way in front of the truck and looked at the two paths.

"What do you think, Pace? Is this the turn we're supposed to take, here?" He pointed at the map. "Or is this just a bullshit trail that's too small to be on the map, and we haven't gotten to the turn yet?"

I looked at the map, then looked up through my NVGs at the terrain.

"To be honest, it's pretty hard to tell, but if we take the turn, we should see it bend to the west in a few hundred yards, and then we'll know we're on the right track. If it curves back east, we're on the wrong track. No idea if that's correct, but that's my read. What's the GPS say?"

---

15. Some of our trucks did have a prototype version of on-screen nav called the FBCB2, (Force XXI Battle Command Brigade and Below Device) but in 2003, they weren't reliable enough to use during mountain navigation. The PL also had a Garmin he'd bought with his own money as a fail-safe, Army technology being what it was.

"Damn thing isn't updating fast enough to tell. The location bounces around enough that we could be at either point."

"Super helpful."

"You're telling me."

The PL continued to look at the map for another minute or so. Behind us, I could hear the radio blaring with requests for information.

"Alright, Pace. I think we take the turn. I'm pretty sure this is it."

"Let's do it, Sir. I mean, what could possibly go wrong?"

We climbed back into the truck, and the PL gave the move order over the net. We turned onto the trail and started down it, with the first half dozen vehicles of the convoy behind us.

The trail curved to the east.

"Fuck. This isn't the right trail. Sobon, can you turn us around?"

"No, Sir. We'll have to back the truck all the way down the road."

E-Brake chimed in. "That's a no-go. We've got trucks behind us. They'll have to back down first to make room."

The PL grabbed the hand mike. "All stations, this is Killer Six. We made a wrong turn, and we need to take the other leg of that fork. I need the convoy to back up to make space, over."

There was an uncomfortably long delay before we heard, "This is Cobra Six. Roger, over."

Nobody said, "What the fuck?" over the radio, but we knew every person in the convoy was thinking it. Disgruntled as everyone may have been, though, the long line of trucks started slowly inching its way backward in the dark, one vehicle at a time, until we were back on the main road and pointed in the right direction. We'd been on the road for hours, and we hadn't gone more than twenty miles or so.

The only good thing about how long it had taken us to get to this point was that morning was coming soon. As the pre-dawn sky started to lighten, our NVGs functioned much better, and driving became a good deal easier until the road merged with a river bed, and we started having problems again.

"Killer Six, this is Cobra One-Six, over."

"This is Killer Six, over."

"Flat tire back here. Wait one while we get it changed."

"Roger."

Sobon stopped the truck and pushed the lever to open the door. Nothing happened. It was stuck. The hinges and latches on the armored Humvees were notoriously finicky because the weight of the doors and the constant banging the trucks suffered wore them out. I smashed my shoulder into the door until it popped open, stepped out, and scanned the terrain around us.

Driving in the riverbanks–or wadis as the locals called them–was always painful. The large round river rocks wreaked havoc on tires and vehicle undercarriages, particularly on armored vehicles. Most of the trucks hadn't been designed for the armor they were now outfitted with, and, as with the doors, the suspensions had trouble with the extra weight.

This was why we'd brought all the extra tires, though, so unless something really serious broke, and we had to sling load the vehicle out with a helicopter, the mechanical problems didn't cost us anything but time.

"And I've got nothing but time."

The PL looked back at me, "What's that, Pace?"

"Nothing, Sir. Just thinking about what I'm going to do when I get home."

"Should be done around Christmas, they say. What's next for you?"

"My enlistment is up in January of '05, so by the time we're back from this trip, I'll be about a year out. With thirty days of leave in the bank, I figure that means I'll be starting to clear by next fall and functionally useless for anything Army related by the summer. My wife's already in Madison, Wisconsin, working on her PhD, so once I'm out, I'll move down with her and start grad school with the GI Bill."

"What're you going to study?"

"Philosophy, Sir."

"So, you plan on bullshitting professionally, and not just doing it as a hobby like you do now?"

"Exactly, Sir. Drink beer. Maybe smoke a little dope. Grow a shitty beard. Try to get fat. You know, live the life of a normal person."

We looked around the surrounding valley. The mountains rose sharply on either side of the riverbed we were driving in, forming long ridgelines that framed the valley, and their rocky sides were sparsely populated by scrubby trees. The tops of the ridges were sharp and grey, and contrasted well with the colors of the dawn sky. It was actually pretty beautiful.

"I can't believe you want to leave all of this behind."

"It has its moments, for sure, but Army life mostly seems pretty shitty. I don't think it's where I'm supposed to be, Sir."

The PL shrugged. "I dunno. I've been talking to Sergeant Romero about it. I think you might be wrong about that."

I shook my head. "Nah, I don't think so, Sir."

The PL nodded and didn't say anything else. We continued scanning the area around us for terrorists, but nothing happened until the squawk box announced the tire was fixed, and we were cleared to continue the mission. We piled back into the truck.

"Hey, Sir."

"What's up, Pace?"

"Can I try out the new speakers?"

The PL nodded. "Let's do it."

I'd rigged up the speakers the day before by zip-tying them to the dashboard of the Humvee. They were attached to my old Discman which I'd wedged into the truck's radio mount as best as I could.[16] The system wasn't exactly blasting bass, and we only had one mix-disc of mp3s, but it was better than nothing.

I reached up to the front and clicked 'play' as the truck started rolling.

"Oh, I, I just died in your arms tonight.

It must have been something you said.

I just died in your arms, tonight."

"Cutting Crew. Nice!"

With helmeted heads bobbing like the protagonists in *Night at the Roxberry,* we continued our combat patrol. The road continued along the riverbed for several miles before splitting off again and turning toward the west. It continued along one side of a broad valley, winding past sprawling walled compounds on the valley floor and the occasional tower-topped escarpment. We'd just passed a particularly large one of these, when E-Brake heard something ahead of us.

"Kill the music. I hear gunshots to our twelve o'clock."

Sobon hit pause on the Discman while the PL responded. "How far? We taking fire?"

"Negative. We're not receiving any fire. I hear a lot of shooting up there though. Machine guns and rifles for sure."

---

16. The Discman was advertised as having 'anti-skip protection,' but the feature was no match for the combined shittiness of our Humvee's suspension and Afghanistan's roads.

"Got it. Sobon, keep pushing. We might not be as far behind the SF guys as I thought we were."

The PL radioed what we knew to the commander while Sobon continued along the road.

"Speed up, Sobon. The commander wants us to push ahead of the main body with the rest of the lead security element to see what's up."

"Roger."

We sped up, bouncing along the road toward a bend where the path wrapped around a steep hill. As we approached the turn, we could hear the fire fight getting louder. There were explosions mixed in with small arms fire now.

"Sounds like grenades. Or maybe RPGs."

"Shit could get hot once we round this bend. Everyone ready?"

E-Brake called down, "Yeah, good to go, Sir."

"Good back here, Sir."

Sobon said nothing. I looked over toward him and saw that he was gritting his teeth. Fuck. I forgot how wrapped around the axle he was right now. I reached across the truck and smacked him on the shoulder.

"It's all good, man!"

He nodded, saying nothing and keeping his eyes locked on the road ahead of us as we rounded the final bend in the road to see a long wide valley. On the south side of the road, the valley floor stayed flat for a quarter mile before sloping up toward a high ridgeline. On the north side of the road, though, the gravelly mountainside butted right up against the road. The slope wasn't too steep as far as mountains in Afghanistan went–probably twenty degrees or so–but the loose gravel meant it would be very difficult to climb on foot and downright hazardous for the truck. Presumably the people who'd initiated the ambush from the top of the ridgeline knew all of this, which was why they had picked this spot.

The target of the ambush was the SF convoy that'd rolled out ahead of us this morning. Their trucks were currently strung out in a long line along the road, and their Afghan and American occupants were exchanging fire with whomever was at the top of the ridge. My initial impression was that the ambush hadn't gone very well for the enemy. None of the SF pickups were on fire, and most of the bullets seemed to be heading toward the attackers, not coming from them.

The PL was enthusiastic about joining the fight. "Sobon! Pull up behind that pickup.

E-Brake, you see any confirmed targets?"

"No, Sir. I can see where the friendly rounds are impacting though."

"Light 'em up."

Sobon pulled our Humvee up and off the road a little, angling it so we didn't block the road for the trucks behind us. Once we were stopped, E-Brake started firing at the ridgeline, hammering it with .50 cal rounds. The other security trucks bounded past us to integrate with the Afghan pickups; their heavy weapons, MK-19s, and 240B machine guns, immediately joined the fight.

I kicked at the door to loosen the stuck handle, then popped it open and climbed out. I positioned myself behind the armored door and scanned the mountainside through my rifle's optic for something to shoot.

Suddenly, the world exploded around me.

BOOM BOOM BOOM BOOM

I'd inadvertently positioned myself right under the barrel of the Ma Deuce, just like the PL had a month ago, and the concussive noise from its barrel was hammering my brain into jelly. Clasping my hands over my ears, I staggered back into the truck to get my bearings, while the .50 continued firing.

The PL was half out of the truck and coordinating with the commander on the radio. When he saw me stagger backward, he looked at me and laughed. His mouth said, "Doesn't feel too good, does it?" But my ears only heard "mumph muffle riiiiiiiiiiiing mumph." I grinned and shook my head. I had a lot to learn about being on the ground during a fire fight.

Our truck didn't seem to be taking any fire directly, so I climbed back out and ran around to the driver's side to lean across the hood. From here, the Ma Deuce was loud, but manageable.

I scanned the mountainside again, but aside from the plumes of smoke rising from the spots where our rounds were hitting, I couldn't see any movement.

"*E-Brake!*"

No response.

I reached in and smacked his leg.

"*What?!*"

"Do you see any enemy?

Pause. "*No.*"

"Then stop firing! Save the fucking ammo!"

"Roger."

Across the line, similar commands were being issued. The security element that had just pulled up had eagerly joined the fight, but it quickly became apparent that the bad guys recognized the futility of their position and had broken contact. As enjoyable as shooting was, any further firing was just a waste of bullets.

"Hey, E-Brake, reload the gun and sweep the turret again, man."

"Mumff mumff riiiiiiiiiiiiiing." I couldn't hear his reply, but it seemed like he was moving in the right direction, so I walked back around the truck where the PL was still on the radio.

"Sir, it seems like they broke contact."

His voice broke through the muffled ringing. "You can stop shouting, Pace."

I nodded. Shit. Voice modulation was tough right now. I looked around at the rest of the convoy. The Afghans were loading back into their pickups, and it looked like they'd be rolling out soon. SF moved fast–our main body hadn't made it around the bend in the road yet.

"We going to clear the high ground, Sir?"

"Negative. It'd take too long. We're continuing mission before they start launching rockets at us."

"Roger that."

I stood outside the truck pulling security while we waited for the slower trucks to catch up. When they made it around the bend, we piled back into our truck and continued our long slow movement.

The rest of the day played out pretty much the exact same way. We'd take an occasional pot-shot or two, return fire, devastate a piece of the countryside with machine gun fire, and then continue mission. Nobody was hurt, and we didn't hit any mines or eat any RPG rounds. So, as with everything in the Army, the mission quickly became first anticlimactic and then boring. Hour after hour we drove, pausing to top off the trucks with fuel from the cans we carried, or to take a piss, and occasionally we took a longer halt when something mechanical went wrong with one of the trucks. It was almost dark when we finally caught back up with the SF guys and got ready to halt for the day.

At this point, the road led through the middle of a five-hundred-meter-wide bowl surrounded by high mountains. The SF guys were huddled around a large fire in the center of the bowl with their trucks arranged around them. Their Afghan partners were positioned on the rim of the bowl in groups of three or four. Each little group had a fire of its own, and each fire had a metal kettle wedged down into the fire itself. They were making their evening tea.

Our convoy pulled up to the SF guys and halted. I climbed out of the truck to stretch my legs and watch as the PL and Captain Lopez got out to talk to them. Romero joined them from somewhere in the back of the convoy.

The SF guys were fascinating. For starters, they all had beards. Lumberjack beards, Viking beards, stringy can't-grow-a-beard beards, spinach-chin goatees–all styles were present. Sitting here in 2025, when every other hipster and redneck has a huge beard, it sounds silly to think that I found the beards so remarkable. But, in 2003, *nobody* had a beard, especially not in the Army.[17]

The SF guys' uniforms were also remarkable. We infantrymen were all wearing standard issue desert camouflage uniforms, and our equipment was all regular Army green stuff. Every one of us looked the same. Not so with the SF guys. Some had desert uniforms. Some had green uniforms. Some had–and this was unbelievable–*civilian jackets on over their uniforms.*[18]

The weapons they carried were equally varied. They carried pistols, assault rifles, sniper rifles, grenade launchers, and rocket launchers. Every guy was an individual, and their combined differences advertised that they were a rugged bunch of bad asses–the sorts of dudes who got to ignore all the dumb shit about the Army and just get the job done.

"Hey, E-Brake."

"What's up, man?"

"That looks like a much better way to do the Army thing."

"I don't disagree. Seems like they get to do all the commonsense stuff and blow off the rest."

---

17. I hypothesize that beards became fashionable again thanks to the GWOT. Images of the 2001 horse soldiers penetrated the brains of larval hipsters everywhere, and, a decade later their desire to appear manly prompted the return of the beard to mainstream society.

18. This may not seem remarkable to a normal person, but to a guy who had just spent the last two years getting dressed down every time a loose thread poked out of a button on his uniform, SF's level of non-standardization was mind blowing.

I nodded. It really did seem that way.[19]

The PL and Captain Lopez finished their conversation with the SF guys and walked back to the trucks. The PL waved the platoon over, and he and Romero briefed us on the plan for the night.

"Alright, we're holding up here for the evening. We're going to set up thirty-three percent security all night, then roll out in the morning. The pre-dawn drive showed us there's not much value in trying to roll out before we can see, so we'll leave at first light. We should make it to Margah for the clearance op tomorrow afternoon."

He turned and gestured toward one side of the bowl, spreading his hands to encompass a large swath of the mountainside.

"First Section, you've got from there, to there. Shit's steep, but you should be able to get the vehicles up there."

He turned and gestured toward another swath of land.

"Second Section, you've got that sector. Everyone, get into position. Once you're set, call it up on the radio. Then, leaders huddle on me for a sync. Any questions?"

We had none. I looked up at the mountainside. It was steep as hell, but the dirt and rock didn't look too loose. Still, thinking about the rollover from last month made my butthole pucker a little.

"Hey Sobon, I think E-brake and I will walk up and ground guide you into position."

Sobon smiled. "What, you don't trust me?"

"Not saying I don't trust your driving skills, but I think I'd feel more comfortable on foot. Also, this way, when you roll down the mountain, someone will be able to pull your ass out."

"When?"

I shrugged. "We can go with 'if,' if you prefer."

He nodded. "Yeah, that works. See if you and E-Brake can find a route that's not too steep."

E-Brake and I grabbed our weapons and started up the mountain. The grade wasn't too bad at first, but after a hundred yards we found that it was much steeper than it had looked. We spread out to pick our way up and find a route the Humvee could negotiate. We paused about half-way up to assess and catch our breath.

---

19. Incidentally, E-Brake, Barrett and I eventually became Green Berets. I fully credit the SF guys we worked with on that 2003 rotation to Afghanistan with our respective decisions to try out. Watching them work was better than any recruiting commercial.

"Hey, E-Brake. My path was terrible. Rocks are too slippery for the truck. How's yours look?"

"Not that good, but I think it'll work."

"Alright. Let's guide Sobon up that way. Then we can hold him there and move further up."

E-Brake nodded and waved his arms to get Sobon's attention. Sobon put the truck in four wheel drive and started crawling his way up the slope toward where E-Brake was standing. The Humvee had no problems with the first part of the slope, but once it hit the steeper grade, it started to struggle. Sobon knew what he was doing, though, and in a few minutes the truck was more or less safely parked on the steep slope just below E-Brake. The two of us then continued our pathfinding trek to the top of the ridge.

As I neared the top, the grade grew even steeper, and I had to pick my way up the last leg on all fours. When I finally reached the summit, I was sweating buckets and breathing heavily. I took a quick look down the other side of the ridge to make sure there weren't Taliban fighters waiting to ambush us, then I turned to look back down into the bowl.

I was higher than I'd thought. The SF guys and their fire looked tiny, and even our Humvee was a long way below us. I could see the entire rim of the bowl around me, and around that mountains stretched as far as I could see, right to the skyline in the far distance. In all that space, I couldn't see a single building. I felt entirely alone–a stranger in a strange land. Where the hell were we? What were we doing out here?

The wind snapped me back into the moment as a strong gust almost blew me off my feet. Apparently the valley we'd been in had been serving as a windbreak because I hadn't noticed any wind at all until I was above the ridgeline. Now, though, the wind was cold, strong, and impossible to ignore. I braced up against a rock and looked over to E-Brake. He'd reached the top, too, and was looking back at the path he'd taken to get there. The wind made conversation at a distance impossible, so I walked over to confer with him.

"What do you think?"

"I think there's a strong possibility the truck will roll down this mountain and end up a crumpled heap at the bottom."

I nodded, thinking.

"Yeah. Possibly. But if we don't get the truck up here, guess what's going to happen?"

"What's that?"

"They're not going to leave this side of the patrol base unsecured, so we're going to have to walk our asses up here every shift and pull security. We'll be in the prone the whole time, freezing our dicks off in the wind. Also, since we won't be able to wake the next guy up without leaving our position for a half hour to walk back down the mountain, we'll probably have to switch from thirty-three percent to sixty-six so we can have two guys in each position to maintain security during our handovers."

As if to accentuate my point, the wind picked up again, sending a knife of cold air down my neck. The sun was going down, and it was going to be a frigid night. I grabbed the watch cap out of my pocket, took off my helmet, and stuffed the cap onto my head before wedging the helmet back on over it.

E-Brake processed my prediction. "That sounds awful. We should definitely avoid that."

"Exactly. So let's see if we can figure out how to get the Humvee up here."

We walked back down the ridge, looking for a navigable way up.

We settled on a zig-zagging route that kept the truck moving directly up the hill but avoided the worst patches of gravel and the larger boulders. I left E-Brake at the top and made my way back down to Sobon to brief him on the plan.

"That's your plan? We're going to get the truck up *that,* just so we don't have to pull guard on foot?"

"Look at how much I'm sweating. You want to make that walk in the dark a couple of times tonight? And did I mention the wind? It's cold as fuck."

Sobon looked back up the hill. I could almost see prudence and laziness battling for domination in his brain. In the end, as is so often the case, laziness won.

"Alright, let's do it. Walk in front of the truck, though. Not behind it. If I roll down the mountain, at least I'll die alone."

"That's the spirit. After you're gone, I'll make sure and mail your porn collection to your folks."

I braced my lungs and legs for the walk back up the mountain and stepped off. Sobon followed in the Humvee, slowly crawling up the slope and keeping the truck pointing toward the crest so that when it lost traction and slipped backward it wasn't at risk of rolling over. There were a few precarious moments when it looked like we were in trouble, but the truck eventually made it to the top of the ridge. However, when it got there, we realized we had another problem to solve.

"Hey, Pace, where do you guys want me to park the truck?"

"What do you mean?"

"I can't leave it on the slope like this. The brakes won't hold all night; it'll eventually slip. I need a flat spot."

I thought for a second and looked around. There wasn't anywhere flat on the whole ridge.

"Can you just pull all the way up to the top and high center it on the point of the ridge?"

Sobon leaned out of the truck to see where I was pointing.

"I dunno if it'll work. Can you dig it out a bit so there's a flattish spot?"

"Sure."

E-Brake and I grabbed shovels out of the trunk, keeping a wary eye on the truck lest it slip and crush us. Once we had the tools, we walked to the crest and started rearranging rocks. In ten minutes or so, we'd moved enough of the big ones that I thought the truck could probably fit well enough.

"That good?"

"Maybe? I'll try."

Sobon revved the engine and let off the brake, causing the truck to lurch up the last few feet to the crest. I stood near the door and guided him forward.

"Slow. Slow. A little more. Another few feet."

Crunch. The front tires slid over the crest of the hill and lurched onto the other side. As they did, the frame of the vehicle smashed down onto the area we'd cleared out.

"Brakes! Brakes!"

Sobon hit the breaks and the truck lurched to a stop. He got out and we surveyed our effort. The Humvee was literally balanced on top of the mountain.

"Awesome. Now we don't have to freeze our dicks off pulling guard tonight."

"She going to hold?"

"Are you kidding? The truck is bottomed out. It's not going anywhere."

"Great. Let's get set up to pull security."

We took turns pulling our rucks off the front grill and getting out enough cold weather gear to stay comfortable for the night. The trucks Chadwick and Lussi were in had apparently been inspired by Sobon's success, because while we worked, they moved up next to us so they could cover their sector from their trucks as well.

When everyone was in position, Chadwick and Lussi worked out the guard roster for the night. E-Brake had the first shift, so I fired back up the tunes and settled into the back seat to try to catch some rack before a long night of guard duty.

October 20, 2003
Madison, Wisconsin

My wonderful husband,

After much thought, it feels important for me to move back to New York after the semester is over, at least for a couple of months. I can't stand the thought of you coming home in January to an empty house. It's pretty good timing for me because my rotations through my three possible labs will be over, and I won't have started in my permanent lab yet. And, of course, it's between semesters, so I will have completed my current classes.

I sent the CMB coordinator an email about taking a leave of absence for your homecoming. Then she had to tell the professor in charge of CMB, and I had to ask my favorite lab P.I. if he would be willing to take me when I come back. He wasn't pleased, but it sounds like he will accept me anyway. However, the CMB program told me that chances are good that if I leave, I won't be able to return.

I've spent a lot of time chewing on this, and I've decided that even if they don't let me return in the spring, I simply must be there when you come home. Worst case, if they throw me out, at least I gave grad school a try. If I hadn't, I would have felt forever dissatisfied. At least now I know what I might be giving up, and I think I can accept that possibility.

Keep me updated on your thoughts about what to do after your ETS. Obviously, my goals and our situation may have changed, so I'd like to know what you think about it. I certainly wish we could just talk about it, but I guess that's impossible. It's been over two weeks since we talked, so hopefully you'll get a chance to call me soon. It feels like ages! I guess we have to continue making major life decisions while communicating only via snail mail!

Speaking of calls, I've put a heart on the calendar for every day you've phoned – including the time from Germany. You've called twelve times total. That doesn't sound like a lot, and when you consider how short the calls were and the fact that sometimes we could only hear static on the other end of the line, it's even worse than it sounds. At least we write each other nearly every day. It's certainly better than nothing.

I called Melissa Butera [the wife of a fellow 1-87 soldier] *tonight and talked to her for about an hour. It was good to talk to somebody who knows you, Ft. Drum, and the other soldiers we associate with. Believe it or not, I miss all that.*

Thanks for mailing me that disposable camera! I got the film developed, and judging from the pictures, you seem to have lost weight. It worries me a bit, but I suppose it's inevitable. I'm looking forward to feeding you back up!

Please let me know if you need me to send you anything – I don't know what you need, and I don't want to burden you with stuff you'll have to carry. Are you doing okay on cameras/batteries?

On another topic, I'm a bit worried about the ten-day mission you mentioned. I don't believe you've done anything like that yet, so I'm quite curious about what it means. I hope all goes well and that you'll be okay. I'll be sending happy thoughts for sure.

I miss you very, very much. I wish I could count down the days until you return, but I'm not exactly sure when that will be.

Until then,

Love,
Alycia

SHHEEEEEOOOOO

KABOOM!

I snapped awake in an instant.

"Whatthefuck?" was the best my brain could produce on such short notice. I rubbed my eyes and tried again.

"Was that a fucking rocket?"

SHHEEEEEOOOOO

KABOOM!

I looked up and saw that Sobon was in the turret.

"Sobon. Are they shooting rockets at us?"

He looked down at me with an irritable expression. "Yeah. And there's not a fucking thing we can do about it. We're stuck up here."

I scanned the area for smoke to see where the rockets were landing. They'd both hit somewhere west of us, well outside the bowl our convoy was set up in.

"They're not even close. E-Brake. You up?"

There was no response, so I walked around to the other side of the truck where E-Brake was racked out in the driver's seat.

"Hey, man. Wake up. Rockets."

His head popped up and he pulled his watch cap off his eyes. "Yeah. Roger. I'm awake. We moving?"

I looked back down the slope.

"No. No point. The odds of one of those rockets hitting us up on this ridge are just about zero. No reason to rush back down and huddle up with the rest of the convoy. We'd just make an even better target for whoever is shooting at us."

Sobon called down. "Hey, E-Brake, once you're up, come relieve me so I can get the truck ready to move."

"Roger."

The radio was alive with reporting by this point. I didn't hear anything remarkable. Mostly, people were just verifying what we already knew–someone was shooting rockets at us. I wasn't sure where the PL was, but I heard Romero come up on the net and tell us to break down our positions and get back on the road to roll out.

SHHEEEEEOOOOO

KABOOM!

"Man. That one was only a klick off, or so. The bad guys are getting better."

"It's probably just some asshole farmer who's pissed at us for being on his land."

"Some asshole farmer with a stash of 107mm rockets?"

"Yeah. Everybody here has a stash of those."

"Maybe so."

Our give-a-shit was running low as we stuffed our gear back into our rucks. Even Sobon–who'd been pretty high strung lately–couldn't get worked up about the explosions. The lack of decent sleep, combined with the morning cold and our inability to do much about the attack, meant the rockets were more irritating than stimulating. We were particularly irate because we would probably have been able to get another hour of sleep if the attack hadn't occurred.

SHHEEEEEOOOOO

KABOOM!

"Stupid Taliban. Had to fuck with us before breakfast."

"Yep. Inconsiderate bastards."

Continuing to ignore the rockets, we finished packing up and prepared to move. It was only then that we realized a flaw in our having driven the truck up to the ridge.

"So, how are we going to turn the truck around so we can get it back down?"

Sobon considered this for a moment as the rockets continued to sail over our heads and explode into the surrounding mountains.

"I can't turn around up here. We'll have to back the truck all the way down."

"Super. That sounds fun."

Sobon's response was to climb into the truck, crank it up, and wait for me to start guiding him backwards. I walked over to his door, and he dropped the window.

"Alright man, you gotta pull off of this ridge to give me the space to walk in front of you."

He nodded, put the truck into gear, and hit the gas. The tires caught, but the truck didn't initially move as the frame was still sitting on the ground. He applied a bit more throttle, and the motor finally triumphed over friction, causing the Humvee to lurch backward. The front end jumped up sickeningly high as the truck started back down the slope, and for a second I was worried the brakes would fail and the truck would careen to the bottom.

The brakes didn't fail, though, and in a few moments I was standing in front of the Humvee, eyes locked onto Sobon's, and using my hands to guide him backward down the mountain. His expression was funny, and I can still remember it today. As the rockets continued to sporadically whistle overhead, and as the truck slipped and slid its way slowly and unnervingly down the hill, he stared at me with a look that was the perfect blend of both despair and resignation. The look said, "I'm doing this thing. I hate every second of it, and I don't want to be doing it. But I'm doing it anyway because, until I get the fuck out of here, I don't have any choice in the matter, so I may as well handle the situation gracefully."

Sobon maintained that expression all the way to the bottom of the hill. When we finally got there, he nodded, opened his door, and lit a cigarette.

"Well *that* was a bit stressful."

"The rockets, or driving backward down a mountain?"

He puffed. "I'm not sure. Probably all of it. Or maybe it was just that I've needed to take a dump for the last four hours but have no idea how to make that happen just now. What are we doing today, by the way?"

Just then, the PL walked up. He'd stayed at the bottom of the hill overnight to discuss the plan for the day's clearance operation with the SF guys and the captain.

"Morning, gents. Nice to see that none of you got blown up this morning. Everybody get his favorite MRE for breakfast?"

I looked down at mine. Beef enchilada. Not too bad. The cold weather meant that eating the jalapeno cheese spread would be like chewing on a congealed chunk of bacon grease unless I heated it up, but at least the meal came with cookies. Sobon had been unlucky enough to get the vegetarian omelet, which was the whammy in the game of *Press Your Luck* we played at every meal. Not only did the omelet itself suck, but the meal came with nothing else beyond a sack of hash inedible browns and a knock-off Pop-Tart.

"All good here, Sir. Where are we headed today?"

"Margah. We start the big clearance op this afternoon. The SF guys are going to reposition on the other side of town and set up ambush positions along all the roads. The hope is that enemy messaging will get confused, and as the bad guys clear town to avoid us, they'll walk into the SF guys' positions."

"Think it'll work?"

"Maybe so. Who can tell? It's better than driving around all day, though."

I nodded. It was indeed. "When do we leave?"

"Soon. Sergeant Chadwick is syncing with Sergeant Romero now. When they're done, Chadwick should get you guys into position. Once the rest of the convoy is in position, we'll get moving."

"Roger, Sir. We'll stand by."

And so we stood by. We were great at standing by. If there's one task I can say with 100% confidence that we did well, it was standing by. The trouble was, we never knew how long we'd *be* standing by. A minute? An hour? A day? Did Sobon have time to take a shit? Did I have time to eat a meal? Who could say?

Still, fortune favors the bold. So while we waited I cut open my MRE, popped the beef enchilada and cheese spread into the heater, poured a bit of water into it to activate the heating element, then propped it on the hood of the truck to wait for it to cook. For those who have trouble picturing what this setup looks like, here is the instructional illustration provided in each MRE. In my case, the 'rock or something' was the hood of our Humvee.[20]

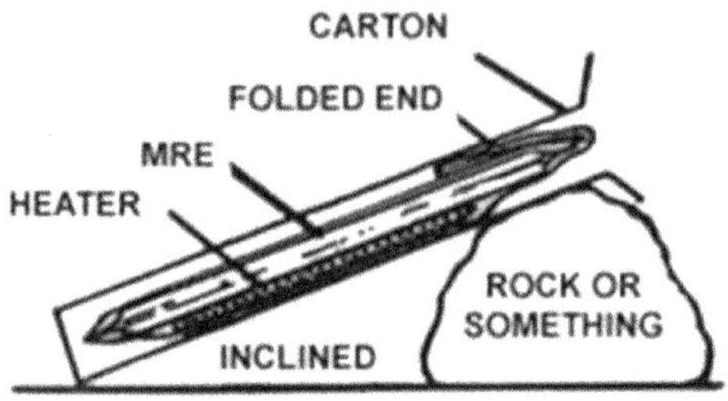

"Hey Pace, you know the second that's done cooking, we're going to have to roll out, right? No way in hell you're going to have time to eat that."

"Maybe, Sobon. But maybe I will have time."

"Nope. Not a chance. That's why I'm not taking a dump right now. The second I got the first kid dropped off at the pool, we'd get the order to move."

"You're too cynical, man. We'll have plenty of time. Besides, you can't let that poison build up in you. That dump is 100% MRE. Think of the mess of chemicals piling up in your colon right now. You gotta purge it, man."

Sobon shook his head and puffed on his cigarette. "Nope. No way."

I poked at the cooking enchiladas. The package seemed hot enough, so I picked up the cardboard package and dumped the water out. I had just pulled the meal out of the heater and torn open the package when Chadwick walked up.

"Just finished with Romero. Why the fuck are you eating, Pace? It's time to move out. And what the fuck are you smiling about?"

"Nothing, Sergeant. Just laughing at an inside joke between me and Sobon."

"Well you two ass hats can laugh at shit on your own time. Let's go."

"Roger that."

Chadwick stomped off to the other trucks to issue similar guidance. Smiling ear to ear, Sobon ground out his cigarette on the hood of the truck.

"You going to say 'I told you so' motherfucker?"

---

20. The actual graphic does, in fact, say 'rock or something.' I've always suspected this was the result of an inside joke by the Oompah Loompahs who work at the MRE factory that somehow succeeded in making it through all of the government's fun filters. On a tangent, I still sometimes eat MREs, and if I ever find a golden ticket in one of my meals, I'm skipping the tour.

"Do I have to?"

Sobon climbed into the truck and cranked it back up as I looked at the beef enchilada. The package was already open, so trying to save it for later would be messy. But the meal was also scalding hot, so shoveling it down would hurt. And I was hungry. Who knew when we'd actually stop somewhere long enough to eat today?

"Ah well, fuck it."

I braced myself for the pain and crammed the first third of the enchilada into my mouth.

"Fuck. Fuck. Fuck." I cursed and swore with my mouth full of blazing hot chow. I gulped down the chunk, then shoved in another. This one went down better since, presumably, all the nerve endings in my mouth were already melted. By the time Chadwick returned with a why-the-hell-are-you-eating look on his face, I was stuffing the MRE trash into a pile on the floor of the truck and folding myself into the backseat. I gave him a thumbsup and an enchilada-covered grin through the back window of the Humvee to indicate I was good to go. Scowling, he nodded in reply and walked back to his truck.

"So, how was breakfast?"

"Delicious, Sobon. It was incredible. You really missed out."

"I'll bet. You sure it was that good? You're talking a little funny. Mouth hurting?"

"Nope," I lied. "Not a fucking bit."

Sobon laughed. "Whatever you say, man."

Our conversation was interrupted as the PL's door suddenly opened and he climbed inside.

"You guys ready to roll?"

"Yessir. Good to go."

"Alright, let's move out."

Sobon put the truck in gear and we rolled through the bottom of the bowl and back onto the road. Behind us, the rest of the convoy started lining up, and in fifteen minutes we were back on the road to Margah. I flipped the music back on, and the speakers came to life with the sounds of NOFX's "Please Play This Song on the Radio."

"I hate these guys, Pace. Change it."

"Can't, Sir. It'd be bad luck. We gotta listen to the disc on shuffle in the order God intended."

The PL nodded and considered my response. "Fair enough. By the way, did you guys get breakfast before we left?"

Sobon's sudden bark of laughter startled the PL.

"What? What did I miss?"

I replied. "Nothing, Sir. Just an inside joke. Beans and bullets. Enlisted guy shit. Nothing you gotta worry about."

"Carry on then, men."

We drove all morning, winding through the hills and valleys of Paktika province without incident. Eventually, around midday, as we climbed out of a river valley, I saw a broad mountain-rimmed plateau open up in front of us with a large village spread out along it.

The PL checked his map, then said, "This is it. Margah. Sobon, keep going. We'll continue along the road until we hit our designated blocking position."

The PL grabbed the hand mike for the company net and radioed his report up to the commander. He then grabbed the other hand mike for the platoon net and radioed instructions to the rest of AT Platoon. Our section fanned out to the right side of the road and took up positions where we could look down on the village. Second Section did the same thing from the other side of the road. Simultaneously, the Charlie Company guys dismounted from the cargo trucks, broke down into squads, and started clearing their assigned streets from one side of the village to the other.

The clearance went pretty smoothly. The dismounts went through the whole village, house by house without encountering any resistance. Normally we would have set up an element on the far side of the village to prevent anyone from leaving. But, since the point of this operation was to push the bad guys out of town so that the SF element could catch them, we deliberately left one side open. This meant that instead of starting a running gunfight in defense of their hometown, any Afghan who was worried about being detained could just leave before our guys reached his house.

Our platoon's role in the operation was pretty dull. Primarily, we were to serve as mobile heavy weapon platforms to support the dismounts if they got into a fire fight. Our secondary job was to provide an evacuation platform if anyone got hurt, and our tertiary role was to serve as a collection point for detainees and any enemy materiel the commander wanted to take back to base. The first two jobs didn't lead to any excitement, but the third generated a few moments, like halfway through the mission when the dismounts dropped off a particularly feisty detainee.

From our vantage point, we watched two Charlie Company guys drag this one particular Afghan out of his house and escort him down the street to our truck. The guy had his hands flex-cuffed behind him, but every few feet he tried to break free and run away. A slew of women poured out of the house behind them as he was dragged away, and they were now following the group down the street, shouting and weeping. The whole procession took almost fifteen minutes to get to our position, and the Charlie guys looked irritated. The PL, Chadwick and I met them for the prisoner handover, and The Hawk and Sobon stepped past them to make sure the group of wailing women kept well away from the trucks.

The detainee was wearing Army goggles that had been spray painted black and pulled down over his eyes so he couldn't see, and his hands were bound behind his back with plastic zip ties. He was dressed in the standard knee-length Afghan tunic and baggy pants combo, and he was barefoot.

The Charlie Company sergeant who was holding him by his upper arm looked sweaty and pissed off. "Watch this shithead. He's fucking trouble."

"What'd he do?" Chadwick asked.

"When I came in the house, the motherfucker rushed me. Dude was screaming and waving his arms and shit. I smacked him in the mouth, but he still didn't calm down. So I had to get a couple of dudes to hold him down so we could wrap him up and pull him out of there."

The sergeant wiped his forearm across his brow before continuing. "Then, to make matters worse, when we bagged him up, these crazy bitches started shrieking at us and shit. Like, what am I supposed to say? Your boy is a crazy son of a bitch. If you didn't want us to arrest his ass, you should have kept him in check. Probably a Taliban psycho or something."

"You guys got a terp?"

"No. He's with the other squad. Commander wants you to hold this dude here until he can get the terp up to question him."

"What are we supposed to do with the women?"

The sergeant shrugged. "Fuck if I know. Take 'em out to dinner. Do whatever you fucking well want. I gotta get back down to my squad."

Chadwick nodded. "Yeah. Gotcha. We got it from here."

The sergeant nodded, and he and the other dismount started back toward the village to continue the clearance operation.

Chadwick pondered the situation. "Alright. Hawk, Sobon–hold what you got and keep the women over there and away from the trucks. Lussi, Pace–let's take this shitbag over to the other side of the truck and sit him down somewhere he won't cause any trouble."

We all replied with "Roger."

Chadwick, who was still holding the detainee by his upper arm, tugged him to follow us. "Let's go, guy."

The Afghan reacted poorly. He started shrieking incoherently and thrashing around. Chadwick kept his grip on the guy's arm and wrapped another arm around his neck. "Calm down, motherfucker!" Lussi and I stood nearby, uncertain about what we should do. Do we get ready to shoot this guy? Do we hit him? What was the right answer here?

As though the women could read our minds, their wailing escalated. Sobon called back over to us. "Uh, Sergeant Chadwick. They seem pretty excited. What do we do with this?"

Chadwick looked irritated and flustered. "Fuck if I know. Just keep them over there and don't shoot anyone unless you have to."

Sobon nodded, and The Hawk started to try to placate the women. "Now, if you'll just calm down, everything will be okay." His words didn't impact the group of distraught women a bit from what I could tell. But why would they? The women didn't speak English. Who knows what they thought we were going to do to this guy?[21]

And what *were* we going to do to him, anyway? The detainee was still squirming ineffectively in Chadwick's grasp, and Lussi had moved to assist, grabbing the Afghan's other arm which was still loose. The two wrestled the prisoner to the other side of the truck, and now they were trying to figure out where to put him.

"*What the fuck? This dickhead is biting me.*" Somehow the prisoner had managed to get his mouth onto Chadwick's gloved hand and was chewing on it.

"Err, do we hit him now?"

"Uhm. No. Not yet. This dude doesn't have many teeth, so he's sort of just gumming me. Wait. Is he crying now?"

The Afghan, his mouth full of black glove, and both of his hands restrained, had started weeping and wailing loudly. This seemed to thoroughly confuse Chadwick.

"What the fuck is wrong with this guy? I think he's fucked up in the head or

21. There was an urban legend that ran around the unit which claimed that many Afghan villagers thought we were Russians and that we were still here from the last war. I myself never heard an Afghan say it, although some of the villages were remote and isolated enough that it sounded plausible to me.

something. Just help me get him on the ground."

Chadwick pulled his hand out of the guy's mouth, and he and Lussi pushed the Afghan down into a squatting position.

"Stay, motherfucker. Just stay there." Chadwick wiped the slobber from his glove onto the side of the truck, then turned to Lussi. "Lussi. Keep a hand on this dude's shoulder and don't let him move. Pace, stand in front of him and pull guard."

"Roger, Sergeant."

Squatting seemed to calm the guy down, and his wailing subsided into subdued sniffling. This, in turn, seemed to calm the women down, which seemed to help Chadwick relax. Nothing changed for a few minutes, so the lot of us settled into a sort of uncertain detente while we waited for something to happen.

Something did happen about fifteen minutes later when E-Brake called down from the turret that a cargo Humvee was headed our way from the village. When it reached our position, an Afghan wearing a ratty U.S. military uniform climbed out of the passenger seat. The PL quickly moved over to greet him.

"An interpreter. Finally. Come help us out, bro."

The interpreter nodded and followed the PL over to the detainee.

"Ask him his name."

The interpreter translated the PL's question, which provoked a weird, high pitched mish-mash of something that might have been words in response.

The interpreter seemed confused and tried again with a different Pashtun phrase.

More weird noises. I didn't speak ten words of Pashtun, but even I could tell the detainee wasn't answering in any way that made sense. The interpreter tried a few more times before turning to the PL.

"Sir, this man is, how do you say, troubled in the head."

"You mean he's mentally retarded."

"Yes, that. You see, he has the mind of a child."

The PL thought for a minute. "Okay. That sort of makes sense. Here, come talk to these women."

The interpreter followed the PL over to the group of women and had a quiet conversation that was out of my earshot. When the PL walked back over to us and told us to cut off the guy's flex cuffs, it was obvious what conclusion had been reached.

Chadwick grabbed the guy's arm and lifted him to his feet, and Lussi used his pocket knife to pop the flex cuffs off the guy's wrists. Chadwick removed his goggles, and gestured to go join the women.

"Za. Go. Go!"

When we saw the guy's eyes rolling about in slack confusion, it was obvious the interpreter was correct, and for a second he just looked around uncomprehendingly. After a few moments, though, he realized that the group of women was calling to him, and he looked toward them, smiled hugely, and shuffled off to join them. The women surrounded him, patting him with their hands, talking to him soothingly in Pashtun, and pulling him away from our trucks and back toward the village.

We watched them go, relieved that the situation was over.

"Man, I'm glad we didn't smoke that guy."

"No kidding. I would have felt terrible killing a retard."

"Right? But how were the Charlie Company guys supposed to know he was fucked up? To them, he was just some guy who came at them when they entered the house. He's lucky he's alive."

I nodded, and we all walked back to our positions to continue pulling security for the dismounts clearing the village.

That *had* been a close one. It had been sort of like the time back at Sh'kin when we'd almost shot the dudes driving toward us at the night checkpoint. Nobody had been doing anything really wrong, but the situation itself created the sort of uncertainty that could easily lead to bad outcomes for everyone involved. It was a weird aspect of war that had never previously occurred to me: there were the good guys (us) and the bad guys (the Taliban) and we were fighting each other because the Taliban had helped people blow up the World Trade Center. But out here in the middle of nowhere, there was this whole other group of people who didn't really seem to be on either side. They were just a bunch of people trying to live their lives. Did they help the Taliban sometimes? Probably so, and that pissed me off. But I suspected the Taliban were probably just as pissed off because the people weren't helping them as much as the Taliban thought they should have been. What would I do if I were one of the people who lived here?

I wasn't sure, but as I watched the women walk back down to the village, I *was* sure I'd be pretty pissed if I came home to find out that while I was away a bunch of foreigners had broken into my house, scared my wife, and arrested one of my mentally handicapped relatives.

But then I realized something else. "Hey Lussi. There were a bunch of women there, but no dudes except the retarded guy. Where were the guys in that family?"

"They probably fled town before we got here because they're Taliban, I imagine."

I nodded. Man, nothing about this was simple.

The rest of the clearance operation was boring.

In the end, we detained one old guy who seemed to have ties to the Taliban, blew up a big pile of illegal RPG rounds and old rusty machine guns, and confiscated a bunch of Taliban flags. The dismounts piled back into the cargo trucks, and we all left the village, traveling in the same direction we'd hoped the fleeing Taliban would take. The SF guys and their Afghan partners went the other way to continue whatever cool shit they had planned.

There wasn't much daylight left, so a few hours later we stopped for the night, set the trucks in a perimeter, ate some chow, and pulled another long night of guard.

"I have a picture
Pinned to my wall
An image of you and of me and we're laugh-
ing,
we're loving it all

But look at our life now
All tattered and torn
We fuss and we fight –"

The metallic plink-plink-plink of machine gun rounds hitting the improvised armor of our Humvee door interrupted the sound of Tom Bailey's rendition of "Hold Me Now" that was blaring from the Discman's speakers.

The radio blared to life with reports from the other trucks in the convoy.

"Killer Six, this is Killer Seven, contact, 3 o'clock"

The PL had the bubble guts today, so he was simultaneously vomiting out of his cracked door, keying the radio hand-mike that was attached to his helmet with tape and a plastic spoon to respond to Romero, and fumbling for the Discman to turn it off.

"Hold me now, whoa
Warm my heart
Stay with me
Let loving start"

"Turn that shit off!" he shouted, wiping his mouth with his sleeve.

From the back seat, I leaned around E-Brake's legs to grab at the Discman. My M203 vest, with its load of 40mm grenades, hindered my efforts, though, and I couldn't reach it.

"Sobon! Grab that shit!"

"So perhaps I should leave here
Yeah, yeah, and go far away
But you know that there's nowhere that I'd
rather be than with you here today"

Mirroring the truck in front of us, Sobon pulled ours off to the side of the road so we could engage the Afghans firing at us from the cliff overlooking the convoy. As he mashed the brake pedal, the truck lurched to a stop, and the Discman dislodged from underneath the radio mount where Sobon grabbed it and hit the pause button.

I slammed my shoulder into the perpetually stuck rear door of the truck. As it sprang open, I stumbled out, hanging onto the door to use it for cover from the continuing fire. While the PL shouted coordinating instructions to the other units to flank the enemy element, I loaded a grenade into my M203 and fired at the cliff.

The grenade exploded violently against the cliff face, dislodging a rock and startling a huge black vulture that took to the sky with an irritated shake of its feathers. I reloaded the M203 and fired again, this time successfully hitting the top of the cliff.

The enemy fire stopped, and I scanned the cliffs for further activity. Our PL was out of the truck by now and starting to heave bile into the dirt. "Man," I thought. "The Hajji crud got him good today."

"E-Brake, you see anything?"

"Nope. No movement."

"You fresh on ammo?"

"I'm good so far, but can you hand me another can in case I need it?"

"Yeah, no problem."

I leaned over to unstrap the cans of ammo we had stacked at E-Brake's feet. I passed it up to him, heard E-Brake slam the new can onto the roof of the truck, and felt a .50 cal link hit me on the shoulder as he swept the turret clear of the expended brass. I hadn't had to tell him to do that this time. He was learning well.

"All good."

The lieutenant had recovered by now, and he looked back at me, smiling under his mirror shades.

"That was some good shit, Pace. We killed the fuck out of those douchebags." He looked down at the puddle of vomit at his feet. "I think I left a piece of my soul in the dirt, though. The tea and footbread that guy gave me yesterday are wearing my ass out."

The radio speaker confirmed that the Charlie Company guys were clearing their way across the top of the cliff now. It would take them a while to finish the job, so we had some time to kill.

"Yeah, it looks that way, Sir. I never thought I'd have a fire fight with the Thompson Twins playing in the background. That was crazy."

The PL and I laughed for a while at the weirdness of the situation. Of all the potential songs I could imagine for the soundtrack of a search and destroy mission in eastern Afghanistan, "Hold Me Now" wasn't on the list. Maybe a little Drowning Pool, or if you were in the One-Hundred-and-Worst Airborne maybe even a little Phil Collins, but definitely not the Thompson Twins.

As I continued to scan the cliffs around us from left to right and top to bottom, two old Afghan men walked by our truck, their sharp eyes peering out from under their weather-beaten brows to meet ours as they passed. When our fire fight had started, they'd both squatted down alongside the road to wait, and now that the firing had stopped they'd gotten up to keep moving. We stared at each other, no signs of comprehension passing in either direction, as they passed the trucks, continuing on their slow walk from wherever to nowhere, shoulders stooped, and hands folded behind their backs as always.

Today was the last day of our ten day mission, and it'd be good to get back to Orgun. The clearance operation had been a big success. Over twenty bad guys had been killed or captured, and the SF guys had recovered a bunch of weapons and bomb-making material. I wondered whether the retarded guy's dad or brothers had been among the enemy fighters.

The PL's feverish voice interrupted my train of thought. He sounded terrible. "Hey, Sobon."

"What's up, Sir?"

"I'm tracking you're supposed to be on the next rotator flight out of Orgun. Looks like this'll be your last mission."

"That's great news, Sir. When do I leave?"

"The day after tomorrow, I think. Weather dependent."

Sobon nodded. It looked like he was finally going to get out of here.

It took another hour for the Charlie Company guys to finish their sweep of the cliff top. Eventually, Captain Lopez's voice came across the radio, directing the convoy to continue its movement. Our truck's tires crunched across the gravel as it rolled back onto the road to fall in behind the truck in front of us.

"Mind if we fire back up the tunes, Sir?"

The PL opened the door to vomit again and leaned back into the rolling truck. He closed his door, wiped his mouth again and took a swig of water.

"Absolutely, Pace."

I reached forward to set the Discman back onto the radio mount and hit play. As the weak sunlight shone through the turret onto my gloved hands, Tom Bailey's voice resumed singing the '80s lyrics I was so familiar with. It *would* be good to be back in Orgun to sleep on a cot and eat something besides MREs.

In the Army, good-byes are generally dumb. Everyone thinks they're bidding someone farewell. They hold a ceremony and make a big deal about it, only to find out twenty minutes after they thought the person was gone that he's still there because of some random circumstance nobody could have foreseen. This can happen a dozen or more times, and each occasion ends the same way–with the guy and all of his friends standing around awkwardly. It's a lot like two people who, having said good-bye, then find themselves still walking together in the same direction for another five minutes. Then, when the individual actually does depart, no one bothers saying goodbye because everyone is sure he'll just come back in an hour anyway. But then he doesn't. That's what it was like to say goodbye to Sobon.

We'd put together a small ceremony and presented him with his going away award. There'd been hearty hugs, damp eyes, and a few half-assed speeches. Then, we'd helped him carry his gear to the truck that was going to take him out to the airfield, and we'd all stood around waving as he left–the first of our group of prisoners who'd been paroled. Then, an hour later, he was back. The birds hadn't been able to fly in because of bad weather.

This same thing happened a half dozen times over the course of the next day. Sobon packed his gear, we helped him carry it to the truck, and he came back an hour later to drop it all in a heap and flop down dejectedly on his cot.

By the time Sobon actually left, nobody imagined he was really gone. It was late morning, and the few people still left in the tent were too engrossed in their own diversions to bother helping him yet again with his gear. He thus walked out of the tent alone, muttering something about what a pain in the ass it was to carry everything he owned. I remember him stumbling over someone's boots and the spring-loaded door slamming into him on his way out, knocking him sideways and provoking a burst of profanity. When the door closed with a bang, I was sure I'd see him again in an hour. But it's been twenty-two years, and I haven't seen him since.

13

December, 2003: Firebase Orgun

I was standing outside the tent at parade rest, waiting with two other guys for my run through the gauntlet. In theory, I'd been preparing for the promotion board for almost two months. In reality, studying had mostly fallen behind other more important priorities, like conducting combat operations, lifting at the gym, or watching Season One of *Sex in the City*, a DVD we'd accidentally procured from the Haji-mart while trying to buy a copy of *The Seven Voyages of Sinbad*.[1] As I ran through the list of potential questions the board could ask me–none of which I knew the answers to–it became painfully obvious that my chickens were about to come home to roost.

"*Next!*"

The guy in front of me took a deep breath, straightened his uniform, and pushed through the tent flap. I stepped forward and took his place, straining to see if I could hear what the board was asking him. Despite my best efforts, I could hear nothing through the walls of the canvas tent.

What would they ask? Questions about the maximum effective range of weapons systems? If so, which ones? For the most part, I was alright at those, but what if they wanted to know the names and numbers of the various Army manuals? If that was the

1. That's right, an infantry platoon in Afghanistan watched every single episode of what was most certainly the chickiest chick flick of its day. We watched it in a tent, squatting unshowered on cots that were oriented in a ring around a ten-inch TV that was propped up on an MRE box next to a pot-bellied stove from the sixties. Every single dude made at least five comments per episode about how gay the situation was, but every single dude also sat through each episode from start to finish.

case, I was screwed because I didn't know any of them. And who was in there, anyway? I had a vague sense that senior NCOs were the people who sat on promotion boards, but which ones? Did I know them? The only person that I knew was definitely *not* on the board was the one I'd have most liked to see on it–Sergeant Romero. I'd seen him walking around the base just a few minutes ago, unfortunately.

What was I even doing here? Why was I bothering to try to get promoted? I was getting out of the Army to go to grad school in a year, right?

The problem was, I wasn't as sure about that plan as I had been a few months ago. Despite all of the silliness and discomfort that came with Army life, I was actually starting to enjoy combat. It was confusing, and it often seemed as though the specifics of what we were doing didn't make a lot of sense. But it felt *real* in a way nothing else ever had. I'd dodged rockets and shot machine guns at people. I'd seen buddies get wounded and guys die. I'd frozen my ass off, eaten slop, and shit my brains out in a third-world nowhere. How could I go from that to sitting in some classroom somewhere, or smoking pot and talking about esoteric nonsense? After combat, everything else seemed pretty trivial.

Alycia seemed to feel similarly. Our limited communications revealed that she was starting to feel that life outside the Army felt like a veneer-version of real life–seemingly critical problems trussed up to seem important but with none of the substance of the kinds of concerns that deployment left us grappling with.

What would a career in the Army look like, though? Would we stay at Drum or move somewhere else?

"*Next!*"

The call from inside the tent snapped me out of my head and back into the here and now.

"Well, here we go."

I straightened my uniform and pushed my way through the tent flap. It was dark inside the tent and smelled like mold. My eyes struggled to adjust to the dim light provided by a single overhead lamp. To my left, I could make out a row of folding tables. Behind these tables sat three uniformed figures. I couldn't make out faces in the gloom, but three individuals were clearly staring at me and waiting for me to do something.

I'd been instructed to report after entering, so I walked across the dirt floor, stopped in front of the tables, came to the position of attention, and awkwardly saluted.

"Specialist Pace reporting as ordered."

I stood in silence as the middle figure shuffled a few papers laid out in front of him.

"Relax there, Specialist Pac-tee."

I recognized the voice's thick Hispanic accent. It was the Alpha Company First Sergeant. I'd been on a couple of missions with him, and he'd never once gotten my name right. Since Pace is a short name, I figured this had to be on purpose, but I'd never seen any indication that he disliked me any more than first sergeants seemed to dislike everyone.

"We're going to ask you a series of questions today. Answer to the best of your ability. You understand, Specialist Pac-tee?"

"Roger, First Sergeant."

"Sergeant Ganoway, you have the first question."

Ganoway was the Platoon Sergeant who'd been with us during one of our bigger firefights at Sh'kin. I knew his name, but I'd never interacted with him personally.

"Alright there, Specialist Pace. I'd like you to tell me the maximum effective range of an M203 grenade launcher."

I stood blinking for a second. I was carrying an M203 and had been doing so for more than a month. I hadn't done much preparation for the board, but even I knew the range of my own weapon. I could see Ganoway smiling. He clearly knew this as well. Had he given me a softball on purpose?

"350 meters, Sergeant."

Ganoway nodded.

"Correct. What's the kill radius of a 40mm grenade?"

"A 40mm grenade has a kill radius of five meters and an injury radius of fifteen meters, Sergeant."

Ganoway nodded again. "I have no further questions, First Sergeant."

"Sergeant Philpot, you have the next question."

Philpot. That's who the other guy was. I worked with him a bunch of times at Sh'kin.

"First Sergeant, I saw everything I needed to see about Specialist Pace's potential as an NCO when we were in a firefight together. I have no questions for him."

Wait. Did I just hear that right? Philpot has no questions for me? What the hell was going on here?

The First Sergeant nodded and made a few marks on his paper before looking back up at me.

"Specialist Pac-tee, please exit the tent. We will inform your chain of command of the results of the promotion board."

"Roger, First Sergeant."

I started to turn to leave, remembered I needed to salute, stopped myself abruptly, rendered an awkward salute, and turned on my heel to leave the tent. I pushed through the canvas flap, squinting as I stepped back out into the bright sunlight. Blinking, I walked past the next guy in line and toward the chow hall to grab a Gatorade.

Was I done? Was that it? I'd only gotten two questions. Was that normal? Had the board gone well or poorly?

I got my answer later that day from Chadwick. I was lying on my cot reading when he stomped into the tent in a huff and walked up to my cot. He had his rifle slung over his shoulder as usual, but he was carrying a second M4 in his hand.

"Hey, Pace. You passed the board. Congratu-fucking-lations."

"Thanks, Sergeant."

"As a future NCO, I've got something for you to address. Ross forgot his fucking weapon in the shitter. Here it is."

Chadwick dropped the extra rifle on my cot.

"Make sure he never fucking forgets it again. I'll leave the specifics of how to accomplish that task to you. You're a smart guy. I'm sure you'll figure something out. Meanwhile, I'm going to go pull Romero's dick out of my ass. He's the one who found the weapon, and he's been fucking me up all morning."

"Roger, Sergeant."

As Chadwick left to find Romero, I peered across the tent to where Ross was sitting on his cot watching me nervously. He knew the pain was coming; he just wasn't sure what form it would take.

I looked around me. Why was the tent so quiet all of the sudden? Shit. Everyone had heard that conversation, and they were waiting to see what would happen. My role in the Army's disciplinary process had suddenly and dramatically changed. Instead of being the victim or an amused bystander, I'd become the hand of justice. I thought about Radon and how quickly his authority had evaporated when he'd shown weakness. Then I thought about all of the successful NCOs I'd known and how they'd always seemed so angry in these situations.

I wasn't angry at all, though. On the contrary, I was mostly amused. Had they been amused, too? Was the anger something you just had to fake to provoke the necessary response?

I tried to make myself mad, but it wasn't a role I was familiar with. I struggled for a second, trying to tap into some sort of inner pool of rage, but I couldn't do it. I couldn't fake the anger. I'd have to try something different instead.

Rather than leaping to my feet and shouting, I stood up and smiled.

"C'mon, Ross. Time to earn back your weapon."

"Roger."

He got up warily. Despite the lack of anger, he smelled a trap. The tent watched and waited to see what I was going to do.

"May as well take your top off and leave it here. Don't want you to soak it through with sweat. Buying this thing back is going to be expensive."

His face relaxed a bit as he unbuttoned his shirt and stripped down to his T-shirt. He knew what was coming. So did the rest of the tent. The expectation in the room evaporated. There were snickers at Ross's fate as the platoon returned to its normal off-time activities. Ross and I walked outside, and I started running him through an Army smoke session.

"Front."

He flopped onto his belly.

"Back."

He flipped onto his back.

"Go."

He jumped up and started running in place.

I had Ross do front-back-go's for a long while, then ran him through every other Army exercise I could think of. It was cold outside, but Ross was quickly soaked in sweat. As I stood with my arms crossed, holding his weapon and watching him work, a variety of people walked by on their way to wherever they were going. Nobody stopped to watch us, but everyone noticed. The privates grinned openly, but kept a wide berth to avoid being inadvertently sucked in. The NCOs gave the situation the side-eye and nodded approvingly. The officers walked by without even a glance: a private was getting smoked by a new NCO for losing his weapon. The Army standard was being upheld. The system was functioning as it should.

When we were finished, Ross was dripping sweat and exhausted. I clapped him on the back and handed him back his rifle.

"To make sure you don't lose that shit again, I want you to dummy-cord your rifle to yourself. Get some five-fifty cord, tie one end to your rifle, and secure the other end to your wrist."

"Roger."

We walked back into the tent and went to Ross's cot to tie off his weapon. I walked back to my cot to sit down and pick up my book. Across the tent, I could hear Ross's friends have a mumbled conversation about what had happened.

I laughed to myself. Maybe being a sergeant wasn't going to be that bad.

The day the command announced that the deployment was being extended, First Section was on laundry detail. The Afghan workers were on strike again, so between missions we all had extra work to do. Around the base, privates were burning shit, picking up trash, shuffling construction material around in the motorpool, and, in our case, doing everyone's laundry.

The firebase's laundromat was a 20 foot by 20 foot, single-room mud hut with a single door. Normally, all we had to do to get our laundry cleaned was to drop off an Army-issued sack full of dirty clothes with the smiling Afghan at the door, pick up a receipt, and come back to recover said laundry in a few days.[2] What happened between those two points in time was a little mysterious. For one thing, nobody had ever heard the sound of a working washer or dryer. For another, based on what we *could* hear, there were at least ten guys working in the building, and they seemed to be having a perpetual boombox fueled dance party. What exactly *was* going on in there, and how the hell were they cleaning the clothes? We were curious, but not all *that* curious, so nobody ever investigated, and the laundromat remained a black box.

Since picking up the laundry mission, however, the inner workings of the laundromat had become clear. The reason we'd never heard the sound of running washing machines was because none of the washing machines functioned. The Afghans had solved this problem by taking the lids off of the top-loading machines, manually filling them with water, pouring in powdered soap, and stirring with broom handles until the laundry was clean. They'd then hung the laundry on lines strung across the room to dry. To make it warm enough for this to work, they cranked up a cherry-red stove in the corner to eleven. All of this was (of course) labor intensive, so what we'd thought might have

2. With at least one sock missing, guaranteed.

been a dance party was actually just a steamy room full of sweaty guys listening to music while they stirred tubs full of our wet, dirty clothes. Which was now our job.

All day, First Section filled tubs, stirred clothes, and hung them out to dry. As the future sergeant on site, my role was to keep everything moving. Our actual sergeant on site, Sergeant Radon, should have been the one to do this, but Sergeant Romero had deemed him useless and put me in charge instead.[3]

"Keep firing, assholes!"

"How many assholes we got on this ship?"[4]

"Yo!"

"Yo!"

We didn't have a boombox–the Afghans had taken it with them–but we did have an endless string of movie references to run through and enough bitching to fill the whole day. Surprisingly, morale was good. Relatively speaking, laundry detail wasn't that bad. It was thirty degrees outside, and we were acutely aware that some of the other guys were out in the cold burning half-frozen buckets of human shit, so stirring laundry in a steamy room seemed like a pretty good gig.

"How long you got left on that load, Bobby?"[5]

"Five minutes, maybe?"

"That shit dry yet, Ross?"

"Nope. Needs more time."

"We gotta fold this shit, Pace?"

"Fuck no. Just stuff it in the bag. They can fold their own shit."

By the end of the day, we were all sweaty and tired, and after spending hours in a

---

3. This was a super-awkward moment. The final decree had come at the end of an operations order brief during which Radon had miserably failed to demonstrate that he knew his part. Romero had pulled us both aside and said, "Pace! You're in charge of this useless motherfucker. He does whatever you fucking say. Keep him out of my fucking sight." I'd been pretty astonished by this flagrant disregard for military rank structure, but, on the whole, Radon seemed fine with the change. It's worth noting that Romero remembers this somewhat differently. He swears that he was friendly throughout the conversation. I ascribe our differing perspectives to what we mean by the word 'friendly.' In Romero's world, calling someone a useless motherfucker *is* friendly.

4. A reference, of course, to Mel Brooks' classic movie, *Spaceballs*

5. Williams was nicknamed Bobby because he bore an uncanny resemblance to that character in *King of the Hill.*

ninety degree steam room, the walk back to the tent was unusually cold. When we got there, everyone dispersed to their cots to dry out. The rumor was that negotiations with the Afghan workers had succeeded, so we didn't think we'd be on laundry duty again the next day.

"That means we're probably rolling out early, so have your shit ready before you go fuck off somewhere."

"Roger."

There was a shrill squeal from the far corner of the tent. Radon had found Hajmo curled up on his cot again. The dog had grown into a cheerful Afghan mutt that sort of looked like Benji, but with an upwardly curled tail. Despite–or maybe because of–his generally agreeable nature, he seemed to enjoy sleeping on the cot of the one guy in the tent who hated him.

"Somebody get this dirty mutt off my cot."

I looked across the tent toward Radon. "Just push him off, Sergeant. It's fine."

"I'm not touching him. He's probably got some kind of disease."

"He's just a dog, and I think he likes you, man."

"Well, I don't like him."

Hogu chimed in, "Maybe he thinks your cot smells like a bitch."

Radon stared back at Hogu impotently as laughter rolled around the tent. Getting *de facto* demoted hadn't done Radon's popularity any favors.

The moment was interrupted by Chadwick and Lussi walking in. They'd been at a meeting with Romero and the PL.

"Alright shitheads, listen up. You know how you all thought you'd be home by Christmas? Well, you won't. We're staying another three months. Or maybe more; If Sergeant Romero gets his way we'll be here for twelve. So write your families or your boyfriends or whatever you gotta do and get ready for some more fucking combat."

The tent's only response was to stare blankly back at Chadwick. After ten seconds or so, he said, "Well, carry on," and left the tent. The silence continued.

We'd heard rumors about the extension for weeks, but now it was official; the six month tour would be at least nine.

After a bit, Hogu broke the silence. "Man, fuck. I was hoping to be back in The City by New Year's."

"Yeah, I was looking forward to getting home too, man."

Everyone else felt similarly, and there wasn't much more to say about it, really. I picked up a pad of paper and started scratching out a letter to Alycia. She'd probably hear about it from the FRG before my letter arrived, but I figured it made sense to tell her myself as well. The letter didn't take long to write, and once it was done, I changed into shorts and zipped myself into my sleeping bag. The last thing I did before going to bed was to turn the pot-bellied stove up a few clicks. It felt like it was going to be cold as hell outside tonight, and I didn't want to wake up in the middle of the night freezing my ass off.

November 23, 2003
Madison, WI

Dear Daniel,

I just got through with a wine and SpaghettiOs dinner in front of Sex and the City. *I didn't even think about what a ludicrous meal this was until I was putting away the leftovers!*

The rumors you heard just got confirmed - we finally got official notice that you guys won't be back until May, and I saw that the 25th Infantry Division won't be out there to replace you until April. Why does the army start out by telling everybody six months and then lengthen it? Wouldn't it be better to tell people it will be twelve months and then surprise them with six or nine? It sucks that, just as I was starting to feel like we were getting close to the end, it was all yanked out from under me. I imagine you feel pretty crappy, too – worse, probably.

I made my plans to go home for a bit after this semester based entirely on the fact that you were supposed to come back in January! Now that I've gotten the ball rolling, given everybody notice, and have my heart set on going home, I can't bring myself to reverse course. It's a good thing that you're leaning toward staying in the army, because this means that if I go home in December, it's going to be for good rather than for a few weeks. I'm slowly coming to terms with the fact that the deployment extension has pushed me to make a more permanent break with this place than I originally intended. It's probably for the best.

In response to the question in your latest letter, I do believe that I would have been perfectly happy in grad school if we had not gotten married and I had come here in 2001 as originally planned. But my experiences in the past 2 ½ years have changed me - and my interests, expectations, priorities, and goals as well. I have absolutely no regrets; in fact, I'm thrilled things have gone the way they have.

I was upset to see that five guys died in a helicopter crash today, but I kept calm because I doubted you were doing any air assault stuff near Bagram. However, at about 7:00 p.m., somebody walked up our stairs and knocked on our door. This was completely unexpected, and I found myself petrified, shaking, and nauseous. The moments it took for Camille to walk to the door and open it felt like an eternity. I started shaking even harder with relief when it turned out just to be the maintenance guy. Why he came on a Sunday evening to look at our roof, I have no idea. I hope you never have to feel that way about me–thinking somebody's coming to tell you I'm dead. I assume you never will. I can't believe how quickly and violently I responded–it's amazing how well I keep my fears bottled up most of the time. I'm always taken by surprise when they bubble over.

I've been worrying lately about how the whole Afghanistan situation will be resolved. I have a bad feeling that it won't be, and all of this pain, stress, and separation will have been in vain. I seriously hope this won't be the case.

As usual, I'm longing to talk to you in person or, at the very least, on a real phone! There's just so much going on right now! Talking to you is always wonderful no matter what, but it sure is hard to carry on a normal conversation because of the weird time lag and static. We're constantly interrupting each other. But at least we can tell each other, "I love you," and communicate anything important. I need to hear your voice tonight.

Until we can be together, and I can tell you in person (or show you), know that I love you with every particle of my being. You are my world, and I will be here waiting for you no matter how long it takes.

Missing you,
Alycia

I was groggy, and my head was throbbing as I slowly gained consciousness. I opened my eyes. They were watering profusely and my throat hurt. The tent was dark, and the only light I could see was the orange glow of the pot-bellied stove. Something was wrong. I unzipped my sleeping bag and sat up. Damn, it was hot in here. I was soaked with sweat and I sat blinking for several seconds. Why couldn't I think? What was going on?

I looked around the tent. It was hazy. Far hazier than it should have been, even in the near darkness. Nobody else was awake. I picked up the plastic water bottle by my cot and fumbled with the lid. I couldn't get it open. What was wrong with me? My brain could make no sense of anything. My bladder throbbed. That, at least, penetrated the haze.

"I gotta piss."

I stood up and almost fell over. The tent swam as I struggled to regain my balance. Was I sick or something? I didn't exactly feel feverish, but I couldn't think or see clearly, and my balance was terrible. I stumbled toward the exit, stopping for a second to stare at the pot bellied stove. It always glowed orange when it was hot, but it was really cooking tonight. Why, though?

I started coughing. Something was making my throat and lungs hurt. Once I caught my breath, I continued my walk across the tent, tripping twice on the scattered debris on the floor. I finally made it to the door, pushed it open, and stepped outside.

I was immediately met with a blast of frozen air. I looked around. It was way below freezing tonight, and snow had fallen. I took a deep breath and stood blinking for a second. The fresh air cleared my mind, and I immediately felt better. Something was wrong in the tent.

Ignoring my bladder, I pulled back open the door and peered inside. The air was thick with haze; that *hadn't* been my imagination. Why had it made me feel so bad, though? And why was it so hazy in the tent? The stove! That had to be it.

I let the tent door slam shut and walked around outside so I could see the spot where the stove's smoke-stack poked through the tent's roof. The piece of black pipe was there, as always, but there was no smoke coming out.

"Shit. It's venting straight into the tent."

I ran back inside and turned on the light.

"Everybody up. Now!"

I propped the door open and pushed across the tent to turn off the stove. The guys started to slowly come around. They were clearly having the same problems I'd had. Chadwick got to his feet and started rubbing his eyes.

"What the fuck are you doing, Pace. Why's it so fucking hot in here?"

"The stove's clogged, Sergeant. It's dumping carbon monoxide into the tent. We gotta fix it, or it's going to kill everyone."

Chadwick shook his head and blinked.

"Step outside for a second, you'll see what I mean."

He slipped on his boots and walked outside. I started shaking the other guys to get them up. About half of them were out cold, but Hogu and Lussi woke up enough to understand what I was saying. Chadwick pushed back into the tent.

"Yeah, I get you now. We gotta take this thing apart and fix it."

I nodded. "It's cold as hell, but if we don't, we can't run it tonight, and we'll freeze our asses off."

We grabbed three sets of heavy gloves and started taking the exhaust pipe off the stove. It was hot, and we burned ourselves a few times, but we ended up getting all the pieces outside the tent and dumped them in the snow. I picked one up and looked through it, but I couldn't see the other end.

"This thing is clogged with soot."

I held the length of pipe in one hand and started kicking it. Big piles of soot fell out. The other guys did the same with the other pieces. In a few minutes, we had the pipes cleaned out well enough to function again, and we took them back into the tent. By then, the fresh air coming in through the door had somewhat cleared the place out, and everyone was awake. The PL walked up in his shorts and watchcap.

"Why the fuck is it so cold in here? And why are the lights on?"

We explained the situation as we put the stove back together.

"Will that fix it?"

"I think so, Sir."

Once the job was complete, we turned the stove back on. It seemed to be running fine, so I walked outside to verify smoke was coming out of the stack again. It was, so I walked back in. "I think it's running fine, now. We can rack back out."

"Keep the door cracked to let the rest of the bad air out."

"Roger, Sir."

I started back toward my cot, but my bladder reminded me that I still had to pee. So I turned around and walked across the base to the piss tubes to take a leak. The urine steamed vigorously as it hit the freezing air. It was way too cold to be standing out here in my shorts. I quickly finished so I could get back to sleep. By the time I was back in the tent, the lights were off, and it had quieted down. I picked my way across the floor and zipped back into my bag. The stove was getting hot again, but it seemed to be working correctly now.

As I dozed back off, all I could think about was what a stupid way that would have been to die.

The road outside the gate had disappeared under the thick blanket of snow that had fallen overnight. Lussi and I stood staring at it as we prepared for our upcoming mission. Chadwick would be back from the final pre-mission update session with the PL soon, and, once he arrived, it would be up to us to figure out how to make the mission happen.

The mission itself was simple enough; we were going to conduct a counter-rocket patrol. This involved driving the roads for several klicks around the base, then dismounting and hiking to the top of a few of the more prominent ridgelines that overlooked Orgun and were commonly used as launch sites for rocket attacks. The logic behind these patrols was that our presence would deter terrorists from using these sites in the future, in much the same way that moving your trash cans into a better-lit area deters raccoons from rooting through them.[6]

We'd conducted dozens of these patrols, and normally they were routine–if dull–operations. Today, though, the snow presented a challenge we hadn't encountered before.

"Where's the road?"

"No idea, man. The snow erased it."

Neither of us was new to snow. Fort Drum got an obscene amount of it every winter.[7] Drum had plows and salt to keep the roads open, though, and even before the plows came through, tall stakes driven into the shoulders of the roads let you estimate where the pavement ended. Afghanistan had none of that.

"Yeah. That's not gonna work. I can't tell where the road ends at all."

"Do you still have that old Russian map with the minefields marked on it? Any chance we'll hit a mine if we go off-road?"

---

6. The one is just about as effective as the other.

7. Syracuse, just to the south, frequently tops the 'snowiest cities in America' lists. http://niche.com reports its average snowfall at 123.8 inches per year.

"Map says this section is clear, for what that's worth. What I'm worried about is the rocks."

Lussi was right. While Humvees were built for off-road work, the terrain around Orgun was covered with large, sharp rocks. Normally, we could see them well enough to navigate around them, but the snow made that impossible.

"Yeah, that's a good point. We'll have a flat tire before we make it a hundred yards. Think we can just eyeball where the road is, though? We know it runs just to the right of that guard shack. Can we estimate well enough to make mission?"

Lussi thought for a minute.

"Maybe. We can try. Don't think it's going to work, though."

"Yeah, I don't think so either."

We heard the sound of crunching gravel and turned around. Chadwick was here. "You two ready to roll?"

"Roger, Sergeant. Not sure how it'll go, though. We can't see the road."

Chadwick looked out across the field of snow, thought for a moment, and scowled. "Can't cancel a combat operation for weather."

"Yeah, we know. Don't think we're going to make it very far before we pop a tire, though."

"We got spares, right?"

"Yeah, only one per truck though."

"Well then, we're good to go. We're the 10th Mountain Division for God's sake, I think we can handle some fucking snow."

We nodded and climbed into the trucks. Lussi was on point with The Hawk. I was in the second truck with Ross.

"You good, Ross?"

"Roger, Specialist. Hooah."

I grinned and stared at him. 'Hooah' sounded different now that I was in charge. How many times had I said that word to a sergeant and made it sound just like 'fuck you?' Was that what Ross was doing now? Who could tell.

"Hey man, I'm not sure how this is going to go. The road's invisible. Just do the best you can to stay in the front truck's tracks."

"Roger that."

I heard Chadwick's voice on the radio calling our departure time to the TOC, and we rolled out through the gate. The Afghan guard manning it gave us such an expressionless stare as we passed that it could have been anything from a casual greeting to a look of incredulity that we'd be dumb enough to try driving anywhere right now. Even after all of these months, I still couldn't read them.

I watched Lussi's truck through the windshield. It was moving slowly and uncertainly, clearly trying to sniff out where the road was. The Hawk was an experienced driver, but he was clearly unsure about where to go. Suddenly, the front right side of the truck pitched up sharply before lurching back down. They'd clearly hit a big rock and now they were high-centered on it. The Humvee's back wheels spun helplessly for a few seconds before The Hawk put the truck into four wheel drive and dragged the truck's frame forward and off the rock with an audible squeal.

"Man, I'm sure that was good for the undercarriage."

The lead truck's efforts had cleared enough of the snow that I could now see what they'd hit–a 12 inch boulder. They'd drifted off the road and were now in the adjacent field. The Hawk seemed to notice and steered back the other direction to get back on the road, but almost immediately lurched to another stop. After a few seconds, Lussi's door popped open, and he climbed out to see what was wrong. I opened my own door and stepped out as well.

"What's up, man?" I hollered.

"Blew a tire."

"Fuck. I'll come help."

I radioed our predicament to Chadwick and walked over to help Lussi and The Hawk swap the tire. Our patrol hadn't made it a hundred yards out of the gate, and we'd already killed a truck. This didn't bode well for the mission.

As we wrestled with the tire in the snow, Chadwick walked up.

"What the fuck happened?"

"Can't see the road, Sergeant. Hit a sharp rock and blew a tire."

Chadwick looked around, irritated. The field of snow was flat and featureless. We could see nothing but a sheet of white from where we were to the Afghan security checkpoint building five hundred meters ahead of us. The road curved around from here to there, but we had no way of determining exactly where it was.

I brushed the snow off my pants and stood up. "Blowing a tire is about the best thing we're going to get out of this patrol, Sergeant. If we hit one of the small ridges or ditches wrong, we'll flip a truck. This is dumb. There's no way we're going to get anything meaningful done today."

His face screwed up angrily as he prepared to lay into me, but then seemed to reconsider. "Yeah. That makes sense. We gotta cancel this shit. Driving around in this snow when we can't see anything *is* dumb."

Chadwick walked back to his truck while we finished swapping the tire. I could hear from the tone of the commander's voice on the radio that he wasn't thrilled with our conclusion, but in the end he had come to terms with it just as Chadwick had. Accepting that we couldn't overcome something as simple as the weather was a tough pill to swallow. But American technology was frequently defeated by the conditions we encountered in Afghanistan. Dust destroyed our electronics, the cold sapped our batteries, the darkness foiled our night vision, and the winding mountain roads ate our vehicles.[8] The American military had spent decades procuring equipment and training to operate in the worst conditions imaginable, but the unique blend of challenges that Afghanistan presented invalidated much of that preparation.

Once we were finished changing the tire, we turned the convoy around and started the crawl back to the gate. As if to further demonstrate Mother Nature's triumph over American technology, Chadwick's truck blew a tire on the way in. After much swearing and two hours of cold, wet work, our platoon ended where we'd started, pulling back through the gate to Orgun. As we drove past the Afghan gate guard, he gave us the same inscrutable stare he'd given us on the way out.

This time, though, I was pretty sure I knew what he was thinking: "What a bunch of dumb asses."

It snowed for almost a week, canceling all of our missions and leaving us with entirely too much time on our hands.[9] Guys responded to the boredom in a variety of ways.

---

8. Even the baseline altitude of the country was a problem. Afghanistan's air is so thin that many of our helicopters couldn't operate in parts of the country, and those that could did so with reduced range and lift capacity.

9. The enemy's operations were apparently also impaired, because we didn't suffer a single rocket attack the entire time. I attribute this to Afghans' choice of footwear. I never saw an Afghan wearing anything but flip flops or sneakers, neither of which is much fun in the snow.

Some, like Sergeant Skerret, one of Second Section's NCOs we'd picked up back in Sh'kin, took the opportunity to double up on gym time. Skerret was already a huge dude, whose enormous legs exploded out of the Ranger Panties he rocked around the FOB, creating awkward moments for everyone.[10] Throughout the snow break he could be found either inhaling copious quantities of the chow hall's terrible scrambled eggs double stacked with androstenedione, or shouting loudly at himself in the mirror as he crushed reps at the squat rack. Skerret was usually a pretty easy-going guy, but woe to the private who ran afoul of him after he'd crushed his preworkout.

Other guys got a bit wilder as all of the adrenaline that was normally dissipated during combat operations sought a new outlet. A whole group of people from across the company somehow acquired Airsoft guns from the Hajimart and started having nightly wars. Squads of soldiers of all ranks staged raids on one another's tents, ambushed each other outside the chow hall, or had full-on *Gangs of New York*-style battles in the snowy open areas of the firebase. These were tacitly supported by the command, at least for a while–it seemed like as healthy a pastime for a bunch of infantrymen as any other–until someone took a pellet in the eye and had to be MEDEVAC'd. That ended the fun for everyone.[11]

Yet other dudes, feeling a burst of holiday spirit, got whimsical. Williams and a couple of other guys spent a whole morning building a snowman that they dressed up in captured enemy equipment, complete with a bayonet and old fifty-cal brass for his face. The Snow Haji, as we dubbed him, guarded the entrance to our tent for days. Like any other broke-ass private, he even (illegally) smoked a few cigarettes on his shift–at least when someone was willing to let him bum one.[12] Before the Snow Haji melted, Williams got a couple of Afghan guards to pose with him for a photo, and to this day, the resulting picture captured one of the very few instances in all my time in that country when I saw an Afghan smile.

Hogu and I took advantage of the slowdown to work on our chess game. Hogu

10. Wherever Raimond is today, I suspect he still has PTSD from waking up on his cot with Skerret's barely-covered crotch and rhinoceros legs looming over his face. Their cots were adjacent to one another.

11. Although it still continued covertly until the Sergeant Major decreed that all guns must be physically destroyed and said that if he saw another one, the owner would be on extra duty for a week.

12. Smoking on guard is strictly *verboten*. The orange cherry at the end of the cigarette is visible in the dark for miles.

had grown up in Brooklyn and had fond memories of watching old guys play chess at the park. He wanted to learn how to play better, but he'd never had time. I'd played casually growing up and enjoyed it, so I offered to help him improve his game. Throughout the snowy week, we spent hours pushing pawns, rooks, and queens around the black-and-white marble set I'd bought off an Afghan for ten bucks. Neither of us was all that good when we started, and despite the hours of play, I don't think either of us got much better. Still, the games were fun and they helped occupy our brains and pass the time.

Guys played spades, took naps, rewatched DVDs, or, in Lussi's case, continued to rework the lyrics to doo-wop songs from our parents' childhoods For most of us, the slowdown was a much-appreciated break. However, one individual hated the time off from minute one–Sergeant Romero. After just one day of inactivity, Romero started to get irascible. After two, he became a rampaging avatar of pure irritation. And, much to our chagrin, his solution to being bored and irritated was always training, training, and more training.

I remember one of our chess games getting interrupted when Romero pushed through the entrance to our tent, letting in a blast of frigid air and greeting us with a scowl.

"Get off your necks. It's time for a Kim's game."

About half the section was in the tent. We looked up at him blankly until Ross spoke up, "Who's Kim, Sergeant? And is she hot?"

"Kim's not a girl, numbnuts. It's a memory game, Ross. You chuckleheads drive around all day looking for terrorists. Kim's game will make you better at finding them."[13]

"How do we play it, Sergeant?"

"I'll show you. But first, get your asses out of the tent while I set up the game."

Romero wasn't a guy to be slow around, so we all hustled to put our boots and shirts on, grab our rifles, and file outside to wait in the snow for whatever surprise he had in mind.

"What do you think he's got in store for us, Lussi?"

"No idea, man. Probably some weird Marine shit."

Williams chimed in, "I bet he smokes the fuck out of us."

---

13. The Kim's game was named after an identical memory exercise played by the main character in Rudyard Kipling's book, *Kim.*

"Maybe so." It was always a possibility that we'd somehow pissed him off.

We had been outside for about five minutes when Romero popped back out of the tent.

"Alright, here's the game. When I say 'go', you're all going to file into the tent and line up around the empty cot that's just inside the entrance. It's currently covered with a poncho liner. When I pull off the cover, you'll have sixty seconds to study the objects on the cot and memorize them. Then, I'm going to cover the cot back up and ask you questions about the items you saw. Any questions?"

"Can we write anything down, Sergeant?"

"No. Nothing. Just memorize what you see."

We all nodded.

"Go!"

We filed into the tent and lined up around the cot as directed. The cot was covered with random objects. An open MRE with the contents organized on top of the package. A disassembled M9 pistol. A *Maxim* magazine opened to a page with a hot chick in a swimsuit. A boot that was half unlaced. A handful of tent stakes. A rifle. A small pile of 5.56 rounds. My eyes scanned these and the twenty or so other items in front of me. When the sixty seconds was up, Romero threw a blanket back over the cot.

"Alright everyone, time for the next part. Come outside."

We stepped out into the snow and Romero smoked the shit out of us for fifteen minutes. As we pushed and squatted, I muttered to Hogu.

"What bug crawled up his ass?"

"Dunno, man. Maybe it's part of the game."

We finished the smoke session, breathing hard and sweating, and stood in the snow waiting for our next orders.

"Everyone, take out something to write with."

We all fumbled in our pockets for pens and paper. Ramsey didn't have a pen, so he had to buy one from Chadwick with twenty more pushups.

Once we all had our pens and paper ready, Romero continued, "Question one. How many rounds of 5.56 were in the pile?"

Oh shit. I didn't remember that. I was pretty sure there were 5.56 rounds on the cot, but as to how many, I had no idea. The expressions on the faces around me suggested nobody else did either. I wrote down 'six' and waited for the next question.

"Question two. What color was the girl's swimsuit in the Maxim article."

Uh. Red maybe? Blue? I guessed again, scratching my answer down in my notebook.

"Question three. Which item was to the left of the M9?"

I tried to picture the cot layout in my head. How had the items been arranged? I thought I remembered the boot being next to the pistol, so I scribbled that answer down.

The test continued, with ten questions in total. Once we were finished, Romero walked us back into the tent so we could check our answers. I'd gotten two answers right. How the hell had I gleaned so little information from staring at the cot for sixty seconds? I sucked!

Romero agreed. "Ten pushups per wrong answer, chuckleheads."

As we pushed, he explained the importance of the game: "As infantrymen, reconnaissance is one of the most important missions you will receive. Reconnaissance, for all you cherry-ass fucks out there, is the the fine art of looking at things that higher tells you to look at and then communicating what you saw. But how can you do that if you don't remember it? You may not have a radio in your observer position. You may not be able to communicate because you're in a hide site and don't want anyone to hear you. That means you have to be able to remember the things you saw *accurately*.

"That means remembering how many people you saw. What equipment they were carrying. What their uniforms looked like. Whether or not they looked disciplined. It means being able to count the number, make, and model of every car that passes through a given intersection in a given time period. It means being able to notice the number of wheels on a piece of enemy artillery and whether or not the tracks are suspended.

"Being good at reconnaissance means being able to do all of that and more. And based on the number of pushups you're doing, you guys currently couldn't find your asses with both hands in the dark. Kim's game is one of the ways you're going to get better at it, though."

By the time he was done with his monologue, we had all finished our pushups and recovered. Some of the guys had done the full hundred, and nobody had done fewer than fifty. Clearly, we weren't very good at reconnaissance.

"The snow has kept us from conducting combat operations, and that's bullshit. The snow *isn't* keeping you from improving yourselves as soldiers, though. Instead of laying on your necks and playing fucking Gameboy, why don't you try bettering yourselves professionally and learning how to be decent infantrymen. It might keep your asses alive."

His point made, Romero looked around the room, making eye contact with each of us, then walked back out of the tent. We all stood around quietly for a few minutes, letting what he said sink in. Did he mean we needed to start training right now? Was he going to come back into the tent and ambush us if we lay back down on our cots? With Romero, particularly cranky no-combat-today Romero, nobody wanted to risk it.

Raimond spoke up first. "Man, Sergeant Romero didn't get his fix today, so now he's in withdrawal. It's like he's a junkie."

Lussi laughed. "A combat junkie. An *angry* combat junkie."

A lightswitch flipped on in Raimond's brain. "Bro, that's his call sign now, Sergeant Lussi. Angry CJ Seven."

The call sign stuck, and twenty years later, we still call him that.

December 11, 2003
Fort Drum, NY

My dear Daniel,

Sweetie!!! I love our home! I'm so happy to be back that I'm in heaven... it seems like I never left. I am so much happier here, and I feel much more connected to you than I did in Madison. I didn't realize just how unhappy I was there until I unlocked the door here – then the contrast slapped me in the face. Everything here is perfect, and everything reminds me of you. It was so wonderful to find everything exactly as I left it–even your water glass is still on the table! This place is really, truly, completely home to me now. I even feel that way about Watertown–I still hate it, but I've grown to really love hating it. I can't stop smiling.

I carted Sobon around for a few errands this morning. Clearing looks like a huge pain in the ass for him since he doesn't have a car. Sobon and I reminisce about you almost incessantly... he really cares about you, babe, and he's worried about you guys, too.

It felt incredibly good to be back on post – I had forgotten how much smilier/friendlier everyone is here. I was light as air and sweet as sugar to everyone whose path I crossed. I may bitch about the army community, but after this sharp contrast to the civilian world, I will never again take it for granted. I feel much more at home here than in the civilian world... isn't it funny how we and our preferences change over time?

I talked to the mother of the boy who took care of our yard this semester. Apparently, there was so much trash in our yard all the time that her husband confronted Brandy about it. There was even underwear here! He's some kind of area coordinator, so his voice actually carries some authority, thank goodness. Not that things are much improved, mind you–our yard is still full of trash and even one of her window screens.

In other neighborhood news, Misty says she's finally leaving Tyrone for good. He's restricted to the barracks, and she's gleefully doing a number on their house. She drilled a big hole in the wall between the bedrooms so she can run a cable into her room since it doesn't have a cable jack (she wants to be able to watch TV in bed). Worse (and more comically), she has been stacking every one of her trash bags against the big garage door rather than taking them out to the can for a couple of months. There is a wall of them waiting to fall out on whatever unlucky person opens that garage next. I suppose she wouldn't be able to enact this particular scheme if it weren't winter, but it's been so cold that the trash doesn't even stink.

I am the new key caller in the FRG as of today. Nelly was sick of it, so she asked me to step in. I don't really have anyone to call at this point. Dickey told me today that he is getting divorced, The Hawk is rumored to be on his way home, and Chadwick's wife gave us no phone number. That leaves Raimond's wife, and that's it until you guys get new people who are married.

At 1600, I went back to Clark Hall to pick up Sobon and brought him home to lift all the heavy stuff out of my car, which he obligingly did. We then picked up Dickey and ate dinner together at Applebee's. I had a really great time talking about Sh'kin with them. It made me feel much more connected to you. It was also good to hear their versions of your stories–they were pretty identical, but it was nice to get extra details and to be able to ask questions. Dickey seems to be taking everything fairly philosophically, and he's making some pretty smart plans for his near future. I'm impressed with his outlook and attitude. Sobon and Dickey both say they feel very guilty for being here right now, although they are simultaneously grateful for it.

At this point in my letter, I had to take a break because you called, and the phone actually worked! How wonderful that you get a few days of R&R!! It's incredible to be able to talk to you without the omnipresent time lag and static! I hardly know how to react. I'm so used to whittling my conversation down to a few short details–a week's worth of important stuff into a few short minutes. It's very odd to shift gears. I wonder if being with you in person again will feel strange for the same reasons? I wonder also how long it will take for us to feel normal again.

This war has been going on for over two years now, and I don't really see an end in sight. I don't feel like we have clearly defined goals, so I'm not sure how or when we'll decide it's over. I just want for us to do this right and be done with it all! I fear that you will be doing too many more OEF and OIF rotations, and that they will be far too long.

Well, it's bedtime now. I spend time every night and morning here dreaming about you, my love, and tonight will be no different. I love you, and I always will.

Yours always,
Alycia

The pallet had been dropped off in the darkness of Christmas Eve like a gift from Santa Claus. It was an eight-foot-long by seven-foot-wide by four-foot-high package of what appeared to be hot chow. Who'd brought it and why? Nobody knew. But, we all knew (or thought we knew) who'd ordered it: The Special Forces guys. It had to have been them; who else could have finagled such a brick of luxury in this dump?

Our ability to inspect what was on the pallet had been limited because it was taken away by a forklift at first light, but that hadn't stopped rumors about its contents from spreading across the base like we were kids speculating about presents on Christmas morning.

"I heard it's free Benchmades for everyone and Christmas dinner. I tell you, man, I smelled something good in there."[14]

"I bet it's stacks of porn and dip."

"Fuck it, I'm going for gold. I bet it's booze. Two beers and a shot of Jack for everyone."

The speculation continued all morning. I don't think anyone actually expected to guess correctly, but the act of guessing did provide a bit of diversion on what was otherwise a pretty depressing Christmas day since, so far, Christmas on Firebase Orgun felt exactly like every other day we'd spent in Afghanistan. There'd been guard duty to pull, bad eggs for breakfast, and a counter-rocket patrol to conduct.[15] The only thing that

---

14. The Benchmade was shorthand for a popular switchblade knife that had somehow made it into the Army's supply system.

15. Shouting 'Merry Christmas' to the people we drove past hadn't elicited much response.

made the day feel any different was the knowledge that six thousand miles away–literally on the other side of the planet–our families were celebrating the most magical time of the year without us.

By lunch, with the day's patrol behind us, even speculation about the pallet had grown old, and our platoon spent Christmas afternoon trudging around the base listlessly and half-heartedly rattling off seasonal movie quotes like, "Merry Christmas, shitter's full," to one another as we went about our normal deployed lives.[16]

I'd been hoarding a sack of Christmas cookies from Alycia, but as popular as they were, even week-old cookies couldn't capture much of the Christmas spirit. On the contrary, they only highlighted just how little of it there was to be had. Nor did the day change until almost dinnertime, when the PL poked his head in the tent and said, "Our chow block is at 1800. That's in fifteen minutes. No PTs. Come in uniform."

I picked my head up. *That* was something different. Chow was never served in block times.

"What's up, Sir?"

"The battalion commander has something planned for Christmas dinner."

The lower-enlisted-guy radar in my head started buzzing; that sounded dangerously rank-heavy.

My eyes narrowed, "What's that mean, Sir?"

From somewhere across the tent, someone shouted, "It's a trap!"[17]

The PL rubbed his forehead. "It's not a trap." Then, after thinking for a second, "Alright, it *might* be a trap. But," he added with a smile, " it's a trap with dinner."

We were skeptical. "Yeah, I'll bet. Same dinner as always, I imagine. Maybe they'll break out some extra catsup packets if we're lucky."

"I heard there was ham and turkey."

Ham and turkey. I hadn't seen ham or turkey since last Thanksgiving. The optimistic part of my brain thought ham and turkey would be awesome. The skeptical part of my brain thought otherwise.

"*Army* ham and turkey? What the fuck does that look like? I bet someone shot a buzzard and roasted that shit."

"At least the buzzard would be fresh, though. And I heard they flew in apple pie from

---

16. A quote made famous by Randy Quaid's character in *National Lampoon's Christmas Vacation.*

17. Another movie quote, of course. This one was from *Return of the Jedi.*

Bagram."

My mind leapt back to the pallet. Maybe it wasn't just the SF guys getting hooked up. Could it be that the Army had actually done something pleasant?

The PL left the tent, and we all started to get suited up in our uniforms for chow. Like everyone else in the tent, I remained skeptical, but the skepticism was tinged with a little bit of hope. Besides, we had to eat chow anyway.

Ten minutes later, we were lined up outside the door to the chow hall, waiting our turn to enter.

"You ready to be disappointed, Hogu?"

"Always, Pace. Always. The Army could fuck up a wet dream. Don't think there's much hope for Christmas dinner."

I nodded. Nope, there really wasn't. We waited for another few minutes. Then, when it was our turn, we pushed through the door and walked in.

My initial impression was that everything was different. The chow hall had been transformed into Christmas Town. There was tinsel hanging from the ceiling. Instead of the usual sad line of green plastic boxes full of Army food pressed against one wall, there was a buffet line built out of folding tables with a cloth spread over them. Behind the line stood the battalion commander, battalion sergeant major, and two of the first sergeants. One of them had a Christmas hat on, and all of them were holding serving utensils.

The sergeant major greeted us. He was standing behind a big foil-wrapped ham, and he had a fork and a carving knife in his hands. "Merry Christmas, men. Grab a plate and get some chow."

I looked around the room. Upon closer examination, it wasn't as different as I'd initially thought. The tables and chairs were the same as they'd always been, the tinsel was sparse and didn't cover the cracked dirt ceiling, and the bottles of Texas Pete were still crusty and mostly empty. Still, the Army had clearly tried to make the setting festive and, weirdly, it worked.

Smiling, I grabbed a cardboard cafeteria plate and lined up.

The sergeant major looked downright cheerful as he asked, "Ham or turkey there, Specialist Pace? You're pinning sergeant in a week or so, right?"

"Turkey please, Sergeant Major. And roger, I do pin soon."

"Congratulations, son."

"Thanks, Sergeant Major."

I moved down the line and was greeted by the battalion commander.

"Hey soldier, you want sweet potatoes and green beans?"

"Roger, Sir. Both."

He scooped both onto my plate. "Your platoon is doing good work out there. Keep it up."

"Thanks, Sir."

I continued to shuffle down the line, receiving a roll, gravy, and a slice of apple pie. Each item was accompanied by a positive comment from someone high ranking enough that I rarely saw them unless I was in trouble for something. By the time I made it to my table and sat down with Hogu, I felt genuinely festive. Most of the other guys seemed to feel similarly.

We ate our meal, talking through our respective Christmas traditions and wondering what the dudes we'd sent back to The States–Gilbert, Sobon, Dickey, and Collazo–were up to today. The crowded chow hall echoed with laughter as guys speculated about what–or who–their wives and girlfriends were doing back home, and everyone contrasted this Christmas dinner with previous ones.

Dinner was usually a pretty mechanical process, but somehow, the command had transformed this one into an enjoyable event. For a half hour or so, all the realities of our wartime lives were forgotten. All of the dust and diarrhea faded into the background, and the Army's rank structure and its associated awkwardness were ignored. For a short while, we got to feel like normal humans again.

Unfortunately, the feeling began to fade as I threw my tray away and pushed out of the door to walk back to our tent, and by the time I'd stopped by the piss tubes and made it back to my cot the feeling was gone.[18] I was back in Afghanistan, and we had another mission scheduled for the next morning before the sun came up. Still, things weren't as grim as they'd been before dinner. The dinner *had* made a difference. The *effort* had made a difference. It showed that the command recognized the shittiness of the situation and had an appreciation for what we were giving up to do what we were doing over here in Orgun.

Although the magic itself hadn't lasted long, the positive effects of that dinner reverberated for days afterward, and took the edge off a cold, dark, dusty period in the

18. There's nothing quite like the smell of pickled piss wafting out of a piece of PVC pipe to suck the magic out of the air. It's like Febreeze for Christmas spirit.

deployment. In hindsight, our firebase Christmas wasn't as remarkable as the World War One celebrations I've read about, where the combatants climbed out of the trenches and paused the fighting to celebrate with one another–there was no way in hell that was going to happen with the Afghans. Still, our Christmas dinner, small as it may have been, penetrated our thick candy shell of cynicism and turned out to be pretty special–special enough that I still remember it today, twenty-one years later.

It was a crisp clear January day, one of the first of the new year. The winter sun was prominent overhead, and its weak rays were just warm enough to make the heaps of snow on the ground start to think about melting. AT Platoon was formed up in ranks in the open space between the motor pool and our tents. I was standing in front of the formation at parade rest, and Sergeant Romero was standing next to me. Romero started things off.

"Platoon, *attention*!"

We all snapped to attention.

"Men, we are assembled here today to recognize one of our own who's been selected for promotion to sergeant. Pace has been with the platoon for more than two years, and you all know him well. As of today, though, he'll no longer be one of the boys–he'll be a sergeant. Joining the ranks of the NCO corps is a big deal and an honor, and it comes with a great deal of responsibility.

"One of those responsibilities is to lead. Sergeants lead from the front. But they don't tell people what to do, they show them how to be through their own example. Leadership isn't about barking orders. Be the first to do a thing, and the men will follow.

"Another of those responsibilities is to uphold the standard. Never let the standard fall. If you see a thing that's fucked up, make the on-the-spot correction right then to fix it. The men will rise to your expectations if you maintain a high standard, but they'll fall to the lowest level you'll let them get away with if you don't.

"Finally, remember that being a sergeant means things are no longer about you. They're about your men. Everything you do should be with them in mind. They eat first, they sleep first, and they get guidance, even when you don't have any yourself. They're always watching. Don't be the guy who takes his cut and leaves his men to fend for themselves. You'll lose their trust and your authority immediately."

Romero paused for a few seconds to let his words sink in.

"Now, Sir, can you come up here to administer the oath and issue the promotion."

The PL had been standing behind the formation. He walked to the front and stopped, facing me.

"Raise your right hand."

I did.

"Now, repeat after me: I, Sergeant Pace, do solemnly swear..."

"I, Sergeant Pace, do solemnly swear..."

"...that I will support and defend the Constitution of the United States against all enemies, foreign and domestic..."

"...that I will support and defend the Constitution of the United States against all enemies, foreign and domestic..."

"...that I will bear true faith and allegiance to the same; and that I will obey the orders of the President of the United States and the orders of the officers appointed over me..."

"...that I will bear true faith and allegiance to the same; and that I will obey the orders of the President of the United States and the orders of the officers appointed over me..."

"...according to regulations and the Uniform Code of Military Justice."

"...according to regulations and the Uniform Code of Military Justice."

"So help me God."

"So help me God."

With the oath of enlistment complete, we lowered our hands, and the PL gestured for Sergeant Romero to join him.

"Now, publish the orders!"

From behind the formation, I heard Chadwick's voice shout the promotion orders:

"The Secretary of the Army has reposed special trust and confidence in the patriotism, valor, fidelity and professional excellence of Specialist Daniel Pace. In view of these qualities and his demonstrated leadership abilities and dedicated service to the United States Army, he is therefore promoted from Specialist to Sergeant effective immediately with a date of rank 1 January, 2004. By Authority of the Department of the Army."

Sergeant Romero spoke again. "Now, for the fun part. Sir, if you'll help me pin this young buck sergeant."

"Of course."

Each of them took a set of metal sergeant's insignia out of their pockets and removed the metal caps from the sharp pins that fixed the rank to one's uniform. With Romero on my left and the PL on my right, they first tore the old, sewn-on specialist rank off my collar, then pushed the pins of the new rank into place. As was tradition, they didn't replace the metal caps.

Romero stepped back and announced to the platoon. "Congratulate Sergeant Pace!"

The platoon laughed and whooped while the PL shook my hand. Everyone lined up behind the PL as he spoke quietly to me, "Congrats, man. You earned this." As a formality, he thumped the newly-pinned rank into my collar bone.

Sergeant Romero was next in line. "Good job chucklehead. Now it's time to get to work." He punched one of the two pins into my chest hard enough that I felt it lodge into my skin. I grinned, "Thanks, Sergeant."

Chadwick was even more enthusiastic. He placed his hands on my shoulders and used his thumbs to grind the two emblems into my collarbones. I kept a smile on my face, gritted my teeth, and tried not to wince as he offered a bit of traditional wisdom, "Being an NCO isn't comfortable, man. *Remember that.* And congratulations."

The rest of the platoon congratulated me in a similar fashion. Tradition dictated that if you outranked a guy, you got to 'pin' him by driving his new rank into his chest. Some people took this more or less seriously, but by the end of the line, the two emblems were lodged deeply into the meat below my collar bones.

After filing through the line, everyone peeled off to mill around in a loose pack. The tension built as we all knew what was coming next. Once there was no one else in line, the tension broke as Sergeant Lussi shouted, *"Get him!"*

I was dogpiled by laughing people leaping in from all directions. I made a token effort to resist, but it was twenty to one, and everyone knew I'd lose. This was the last time all of my old private buddies would get to kick my ass without consequence, and they were going to make the most of it. They ground me into the snow, heaping themselves on top of one another until I was buried under half the platoon. The struggle continued for a minute or two. Then guys started picking themselves off the pile until I was the only one left. Lussi grabbed me by my hand, helped me up, and handed me my headgear, which had been lost in the scuffle.

"Congrats, man. Might want to clean those up."

I looked down to where he was pointing. Spots of blood had leaked through my uniform and stained the area around the new rank on my collar.

"Yeah, I guess I need to. Good times."

"Yup. I still have the marks from when I got pinned. Good times, for sure."

I brushed off the snow and walked back to the tent. As I pushed through the door, Hogu jokingly called, "*At ease!*"

I smiled and shook my head. "Fuck off, man."

"Roger, *Sergeant*. Fucking off."

I laughed. This NCO thing was going to take some getting used to.

January 16, 2004
Fort Drum, NY

Dearest,

I've just had a good, hard cry because I miss you so terribly tonight. Sobon gave me some pictures he took of you, which I combined with the 3 Doors Down song "Here Without You." This combo proved to be very combustible, so here I sit, mascara running down my face and my heart aching in my chest. I know we're six months closer to seeing each other than we were, but we still have too much longer to go.

There's just something about being past the six-month mark of this deployment that makes me feel like shit. I suppose a big part of the reason is that we thought you'd be coming home right about now. It seems that every time we're approaching the finish line, the army is going to say, "Just kidding!" and move it back a little more. When will it end? This would be so much easier if I had a reliable mental countdown going, but even if they gave me a date, not one fiber of my being would trust it. As much as they pretend to care about morale, you'd think they would factor this in–this kind of uncertainty is hard on everyone.

Sobon, Dickey, and I went out to dinner to celebrate Sobon's ARCOM with a V. Both of them were depressed–Dickey because he's getting divorced and Sobon because he's Sobon–so I mostly just listened to their troubles. I like doing this for my friends, so I didn't mind at all. They do make me feel guilty for being so happy with you, though. How did we get so damn lucky?

Unfortunately, I wasn't so very lucky tonight–somehow, LTC Howard's wife was seated at the table beside us, and there is no doubt she recognized me from the FRG.

There I was with not one guy, but two! (Maybe one would have been worse?) She gave me quite the eyebrows-raised look and whispered to her friends, who turned to look at me. It made me feel all dirty and guilty and like I would have to sneak around to see our friends to avoid this sort of thing. It sucks to be in a place where my every move is suspect. I never really thought about it until now, but I guess I should be more careful about the appearance of infidelity even if you yourself trust everyone involved and aren't worried about it. The gossip machine here is unreal (probably because it is typically fed plenty of accurately salacious fodder).

Congratulations on finally getting pinned! It's so weird you're really an NCO now. I can't wait to see you in your new stripes!

Your wife,
Alycia

14

February, 2004: Firebase Orgun

The battle between Radon and Hajmo over Radon's cot had ended in a draw sometime after the start of the New Year. Radon had been fired, loaded into a helicopter, and pushed into whatever oubliette the Army used for storing disgraced sergeants, and Hajmo had been shot when the theater commander decided pets were unhealthy and ordered them all to be exterminated.[1] We'd been on mission at the time, and the PL hadn't taken the news well. Still, Hajmo was dead, and there wasn't anything to do about it but continue to fight the war.

January had been a boring slog. We'd battled the snow, ice, and mud more than the enemy, and I'd worked to settle into my new role as a sergeant. Being an NCO was considerably different than being a private had been, and the guidance Sergeant Romero had provided in his speech about accountability and leadership turned out to be spot-on. All of a sudden I was no longer supposed to shut up and wait to be told what to do. Instead, I was expected to see what needed to be done and do it. Also, I wasn't supposed to do it alone; I was expected to put other people to work *with* me, which meant I had to start ordering around my old peers.

This was easy with the soldiers who joined us late in the deployment–Gardea and

1. The only survivor of this purge had been Dennis, the Charlie Company XO's cat, who'd escaped notice until the situation had blown over.

Jimenez; they'd never really known me as anything but a sergeant.[2] Even the guys who'd been with us for a few months, like Williams or Ross, didn't have much trouble adjusting. We were buddies, but I'd always been pretty senior. However, telling the guys I'd been friends with for years what to do, was weird.

"Hey, Hogu. C'mon man, we gotta go give the trucks a final check."

Hogu was lying on his cot, immersed in a game of *Advance Wars.* "Yeah, just a second, Pace. I'm going to finish this up."

Hogu's response gave me pause. It was two hours before we were scheduled to roll out on the latest big mish, and I wanted to make sure everything was good to go before Chadwick and Romero did their walk-throughs. Should I let Hogu finish his game? Or should I tell him to get off his ass and come with me now? What about the fact that he hadn't called me 'sergeant'? Was that a casual slip, or did it represent something I needed to deal with? Hogu had been my friend for a long time, at least in Army terms, and I didn't want to dig into his ass. I just wanted him to do what I said so that we could finish what needed to get done.

I stared at Hogu for a moment thinking about how to handle the situation. He must have noticed something was amiss because he looked up, had a moment of realization, and said, "Shit, Sergeant Pace, I forgot you got promoted. My bad. Yeah, man, I'll get ready to go."

"Thanks, man. Tracking it's weird. Just gotta get this done, ya know?"

"Yeah, I get it. Takes some getting used to, right?"

"For sure."

I breathed an internal sigh of relief. The moment had passed without incident, and I walked across the tent to round up the rest of the section to get ready to roll out to some nowhere-ville north of us called Sharana.

A few minutes later, we were all outside working on the trucks. It was a chilly but otherwise beautiful day, without a cloud in the sky. The snow was gone, and during the OPORD the intel guy said the temperatures would get up into the fifties, which was a hell of a lot warmer than it had been for weeks. The Section split up and got to work.

"Hey, Sergeant Pace, why are we going to Sharana?"

Williams and I were working on filling one of the radios in the PL's vehicle. As usual,

---

2. Jimenez, or Jimmy as we called him, was a chubby easy-going transfer from Korea. Gardea was a high-speed dude who'd been at the Ranger Battalion until he got booted for getting a DUI on his motorcycle.

the timing was off between the trucks, so the frequency hop feature wouldn't let the two radios talk to each other. I missed Sobon. He'd have had this fixed in no time. The new guys required more hands-on attention.

"We're going to recon the area, Williams."

"But *why* are we going there? Is it an important area? Taliban headquarters? Anything like that?"

I pushed more buttons on the radio to bring it back to the screen that would allow me to reset the timing. Fuck, these things were irritating. I missed the relative simplicity of the Ma Deuce.

"No idea, man. I think we're supposed to recon the route up there so somebody more important than us can do something in Sharana in the future, but if you want some sort of cosmic explanation for the importance of Sharana, I couldn't tell you."

"That's bullshit, Sergeant. Doing things when we don't even know why, I mean."

I looked up from the radio, irritated. "Does it matter?"

"What? Does what matter?"

"Why we're doing it. Does it really matter?"

"Hell yeah, it fucking matters, Sergeant. I'm risking my life out there, and I'd like to think it's for something that matters."

I thought back on the last five months' worth of patrols. The vehicle stops. The random searches. The harassment and interdiction fire. Why had we done any of it? The 'why' section of our OPORDs always mentioned something about 'denying the area to enemy forces,' but had we done that? It seemed like there were still just about as many enemy forces out there as when we'd started. Were we making an impact? Then I thought about all of the rocket attacks on the base. *Those* were certainly making an impact. They'd never killed anyone. But every time a rocket hit, I had to stop whatever I was doing and drive off the base to look for the assholes who'd shot it. How many hot meals and how much sleep had rocket attacks cost us over the last five months?

Hmm, *here* was a 'why' I could get behind: "We're doing this to get rid of the terrorists in Afghanistan, and we're doing that so they stop shooting rockets at us, and we're doing *that* because the rockets fuck up chow."

Williams thought for a moment as we continued fiddling with the radio. "So, I'm risking my life every day so we don't miss the opportunity to grab a cardboard plate and eat some of the worst food I've ever had in my life?"

"Yeah, pretty much. And sleep. We're going on this patrol today so we can eat regularly and sleep through the night without interruption. That's what it boils down to, Bobby. "

Williams grinned and laughed. "Alright, fuck it. I can work with that."

We finally got the clock on the radio set to the correct time and date, and I left Williams to finish synchronizing it with the other trucks. I walked over to where Hogu was lubing up the MK-19.

"Hogu! How's she looking, man? Ready to kill more bad guys?"

"Sure, yeah, let's do it. More bad guys! What the fuck is Sharona or Sharana, or however you say it, even for? Why are we going there?"

"Hot chow, man. We're going for hot chow. The more bad guys we kill, the more often we get to eat hot chow without getting interrupted by rockets."

"With motivational material like that, they should make you a fucking general or something."

"Absolutely. General Pace. I'd probably be like that crazy admiral out of *Hotshots.* Senile and ridiculous. For now, though, I'm barely even Sergeant Pace, and I don't know why we do much of anything, so I gotta make shit up."

Hogu laughed and kept wiping down the gun.

I walked past his truck to check in with Sergeant Chadwick, who was using a fine-tip marker to draw phase lines and routes on his new map. Sharana was far enough to the north that we'd needed new maps, and we'd spent the previous evening cutting them out and laminating them to get ready for this mission.

"Hey, Sergeant, PCC's complete. Need help with the map?"

"No. Is yours done?"

"Yeah, I finished mine this morning."

"Let me see that shit."

I handed my map to Sergeant Chadwick, and he compared it to his, making adjustments as necessary.

"So, what's special about Sharana, Sergeant? Why're we going there?"

Arts and crafts made Chadwick irritable. There was something about getting the lamination straight and getting the squares from the maps to line up properly that pissed him off. "Fuck if I know, Pace. Who cares? Does it really matter?"

"No, I guess not. Just curious. I wonder if the PL knows."

"Probably. That sounds like a bullshit officer question to me; I'm sure he knows the answer. All you need to worry about is getting your truck ready and making sure you don't get your ass lost if you end up running point."

"Roger, Sergeant. Makes sense."

"Oh, and CJ's riding with you today. Don't fuck up and make me look bad, man."

Fuck my life. Having Romero in the truck with me meant he'd be up my ass about security for the whole mission, and I'd get bumped to the back seat.

"Roger, Sergeant. I'll go find him and make sure we have his gear loaded up."

Chadwick nodded and focused his attention back on his map.

I walked further down the line of trucks to Second Section's vehicles to find Sergeant Romero in conversation with Nevin.

"I'm telling you, Sergeant. That intel from the OPORD was total bullshit."

"Why do you say that, Nevin?"

"'Cause the motherfuckers have never been right once about anything. Tips. Enemy locations. None of it. I swear to God those tit-heads must be back there calling Madam Cleo at the Psychic Friends Network or something for their information. That's how fucking bad it is."

Romero laughed. "Yeah, maybe so. You think you could do better at it?"

"Hell yes, I could. I can make shit up with the best of them. If I got to sit in that comfortable ass room looking at shit on the Internet all day, I'd spin you a fucking yarn that would blow your mind. It probably wouldn't be any more accurate than the nonsense the current guys spew, but it'd at least be more entertaining."

"Well, maybe we'll give you the chance to do that someday." Romero turned to face me. "What do you need, Sergeant Pace?"

Nevin chimed in, grinning. "Oh, did I forget to call at ease for you there, *Sergeant?* My bad. Need me to do some pushups?"

I laughed. "I'll overlook it this time, motherfucker." And then to Sergeant Romero. "Hey Sergeant, I heard you were riding with me for this mission. You got any shit you need me to load in the truck?"

"Just my ruck. I'll bring it over later. No issue. Your guys ready to go?"

"Roger, Sergeant. Everyone's good to go. Three days of food, water and fuel, as directed."

Romero nodded. "Good. Once I finish with this chucklehead, I'll come by and double check."

"Roger, Sergeant. Oh, and the guys were asking, what's up with Sharana? Why are we going up there?"

"Didn't you listen to the OPORD? What'd the commander put out?"

"He said we were conducting the patrol to deny the area to enemy forces."

"There you go, then. There's your why."

"Can't we do that around here? Why drive two days north? Is Sharana special?"

"I don't know, Pace. I guess we gotta find new bad guys to kill. Maybe the ones here gave up."

I nodded. "Got it, Sergeant. They gave up. Makes sense."

I turned to leave. Nevin called out as I was walking away, "You know the real reason we're going up there, right?"

"No, why's that?"

"The commander got tired of seeing ole Angry CJ Seven here stomping around the base all cranky and shit, so he's sending us north to get rid of his ass for a few days."

Romero laughed. "Angry? Me? What're you talking about? I'm the most cheerful motherfucker in-country."

Now it was our turn to laugh. "Right, Sergeant. You're a regular Mr. Congeniality."

"I think you nailed it, Nevin. That's gotta be the reason."

I walked back to my truck, pulled my gear out of the front seat to clear it for Romero, and started packing it into the small uncomfortable back seat. It was going to be a long couple of days.

The rest of our prep for the mission went smoothly, and by the designated time of 0300 we were lined up and ready to roll out. Our convoy was relatively small for this trip. Just AT Platoon, a few trucks' worth of Charlie Company guys, and a section from the Battalion Mortar Platoon. Since we were rolling out of range of Orgun's artillery detachment, we had to take our own fire support.

I heard the PL call our start time over the company net. "Cobra Base, this is Killer Six. SP time now. Ten vics, forty-two PAX, over."

"Roger, over."

We left the wire and rumbled north through the town of Orgun and up the dry wadi until it linked up with a winding highway that led to Sharana.

"Here be monsters," I murmured to myself.

"What was that, Pace?" Romero had heard me.

"Just bullshitting to myself, Sergeant. They used to write 'Here be monsters' on the parts of old maps nobody had been to. It was a way to fill out the margins with something more interesting than blank space, and I think it captured folks' sense of fear surrounding the unknown. That's us, today. We're rolling into the unknown. There be monsters, you know?"

Romero didn't respond for several minutes. The dark terrain, rendered green by my night vision, rolled by, and I assumed the conversation was over until he suddenly said, "That shit is crazy, right?"

"What do you mean?"

"That there are still places around here we know next to nothing about. When was the last time anyone but the people who live in Sharana went to Sharana? Hell, when was the last time anyone but an Afghan even heard of it?"

"Yeah, it's super crazy. It's hard to imagine there are places as disconnected as this still left in the world. I don't think people back home would believe me if I tried to describe it."

Another hour or so of dull, kidney-rattling driving along the rocky roads of Paktika province passed before the conversation continued.

"You know, I think there's an opportunity here."

"Really? What opportunity is that, Sergeant?"

"So, parts of this country are actually pretty beautiful, right?"

"Yeah, once you get past the dust, the lack of infrastructure, and the religious fanatics."

'Yeah, put all that to one side and consider the mountains. You know what people pay to ski mountains like these in Colorado?"

"Not really, no. A lot, I imagine? Skiing was a pretty bougie pastime for those of us who grew up in Central Texas."

Romero laughed. "Take my word for it, then. They pay a lot. Imagine how much it would cost to run one of those places here? Nothing at all, right? Labor is cheap. Land is functionally free, and Afghanistan has that whole 'untouched wilderness' thing going for it."

"I don't know if I'd call it untouched. I think the Taliban deforested the shit out of most of it. And you'd definitely have to pay an assload of money for security, or all your clients would get executed for conduct unbecoming a radical Muslim."

"Yeah, ignore that. The trees will grow back, and Americans won't know that shit anyway. We would have to do something about security, though."

I thought for a bit. "You know, why overlook that stuff at all? Why not build it into the model? You could set this place up as a combat-action theme park. Sure, folks could ski, but they could also run patrols, defuse land mines, arrest mentally retarded people, shoot terrorists, and suffer from major digestive problems. I'm sure there are bored corporate executives who'd pay a fortune to do all of that."

Romero nodded. "You know, you're right. That's a hell of an idea. We could sell people slots to participate in the Afghanistan Experience. It'd be the whole nine yards, just like you described. Then, we could sell *other* people slots to participate in the rescue operations we'd have to conduct on the first people. We'd make a fortune."

"Sure. There must be literally dozens of wealthy lunatics who'd give us money to risk death for no better reason than that they feel their lives don't have enough excitement. I'm pretty sure we'd spend most of the money we made on lawsuits, though."

"Nah, we'd just need iron-clad legal disclaimers. Maybe we could even work some of them into signing us up as their life insurance beneficiaries."

"Now you're talking!"

Our conversation about the entrepreneurial opportunities that patrolling in Afghanistan represented lasted through the remaining hours of pre-dawn darkness and well into the morning. Romero didn't believe in listening to music in the truck, but this was almost as good for passing the time. It was the best truck conversation I'd had since Sobon left.

The day was turning out to be clear and cold, the sun was high overhead, and we were just discussing the relative merits of whether to arm our clients or let them scrounge for weapons locally, when we saw a convoy of trucks rolling down the road toward us. They were Hilux pickups full of uniformed Afghans, and we quickly realized they were the Special Forces guys from Orgun out on patrol.

"What are they doing here?"

"Dunno, Williams. Pull off to the side and stop. Pace, wait here with the truck. The PL and I will go see what's up."

"Roger."

We pulled over and stepped out of the trucks. The drivers behind us peeled off to either side of the road in a staggered column formation, just as they were supposed to do, and the PL walked up to wait with Romero for the Special Forces guys to arrive. They did so soon, and once their trucks had stopped two bearded men with long wild hair, mirror shades, and mismatched desert uniforms climbed out to talk to Romero and the PL.

"Hey, Williams."

"What's up, Sergeant?"

"The SF dudes look like they have a hell of a lot more fun over here than we do, don't they?"

"That's for fucking sure. No uniform standards. No shaving standards. Just walking around looking bad-ass and doing whatever the fuck they want. Don't look too bad at all."

"No, it sure doesn't. Maybe I need to try that out someday. It definitely looks more fun than being in the infantry."

"Thought you were getting out and going to grad school or something."

"Nah, I've been chewing on it, and I think I'm going to stay in for a bit longer. The PL has been trying to convince me to go to selection to be an officer, and I think I might give it a try in another year or so. Maybe I'll go SF after that."

"Pssh. I get why you'd want to go SF. But, an officer? Why'd you want to do some gay shit like that, Sergeant? Just to make more money?"

"The money'd be nice for sure, but I think it has more to do with what you were asking about earlier."

"You mean why the hell we're here?"

"Yeah, that. I feel like that whole aspect of our deployment is lacking, and I don't get the impression people care all that much. To be honest, to this point, I guess I haven't really cared all that much either because, you know, we're in combat. And that's sort of been an end unto itself. I mean, I get the whole September 11th thing, but I don't really see how what we're doing is connected to that. I thought I got the connection when I first got here, but that was when I thought everyone over here was Taliban. I don't know about you, but I haven't seen anyone that was obviously Taliban the whole deployment, and I'm not sure I ever even understood what the Taliban was in the first place. Don't you get the impression we're just over here sort of fucking with a bunch of random tribesmen who pretty much just want to be left alone?"

"Sergeant, I'm a driver. I have no idea how my actions connect to what we're doing day to day, much less how they support any sort of grand plan to achieve whatever the fuck we're trying to accomplish over here in Afghanistan. As you so accurately put it, I'm on this patrol so people quit fucking with my chow schedule."

I laughed. "I guess I feel like if the Army is going to go through all the trouble of flying a bunch of us over here to kill people, it should all make a little more sense to the people doing it. And I think, maybe as an officer, I can figure out a way to make that happen."

Williams lit a cigarette and puffed on it thoughtfully.[3] "Yeah. To be honest, Sergeant, I think there's a pretty fat fucking chance you're going to achieve that, officer or otherwise. But at least you'll get paid more."

I nodded and shrugged. "At least I'll get paid more."

Sergeant Romero and the PL finished their conversation with the SF guys and started walking back toward our vehicle.

"What's up, Sir n' Sergeant? Any change to what we're doing?"

Romero responded. "Yeah, FRAGO. We're going to help these guys finish the mission they're working on. They got a tip there's a shithead around here, and we're going to help them bag him."

The PL continued. "Shouldn't take more than a few hours. Call Chadwick, Skerrett, and the C Co and mortar leadership up here, and we'll talk through the plan."

When everyone was assembled, the PL laid out his map on the hood of the truck and talked through the new mission.

"Alright, gentlemen. Here's the situation. The SF guys have confirmed there's a bad guy on their target list holed up in a village just east of here. They want to bag the guy, but they don't have enough bodies to isolate the village during the raid, so they need our help. There are only two avenues of approach into the village, so we shouldn't have much trouble locking it down. We'll push east to about a klick outside the village as a convoy. From there, First Section AT platoon will push through the village and block the far side. Second Section will stop short of the village and block the near side. Mortars and C Co, you guys will establish a support by fire position at this point to provide fire support throughout the operation. The SF guys have Cobra support coming in, though, so once they're on station you'll have to hold fire until they leave. Once the SF guys are complete, we'll collapse back to the firing position and continue movement to Sharana. Does anyone have any questions?"

Nobody did. It seemed like the sort of thing we'd done a dozen times. We piled back into the trucks and were ready to roll when the radio crackled to life.

"Killer Seven, this is Killer One Alpha."

"Go ahead, over."

"My truck is dead. Going to need a minute to set up recovery, over."

"Roger, we'll come to your location. Killer One, take point, over."

3. His resemblance to Bobby Hill made smoking look a little incongruous.

"This is Killer One. Roger, over."

Romero turned to Williams. "They're two trucks behind us. Pull us back there so we can get them hooked up for a tow. We're too far out to take the truck home, so we'll just limp it along through the rest of the mission."

"Roger, Sergeant."

Williams backed the truck up to Lussi's stalled vehicle. The Hawk, Lussi's driver for the mission, was already standing in front of the truck and getting it ready to be towed. I climbed out to help rig everything properly.

"All we've got is a sling leg? Who's got the tow bar?"

"Two of the pins are busted in the bar, so we can't use it until the mechanics finish fixing it."

This was unfortunate. The tow bar was the best way to recover a vehicle. Heavy and stable, it made steering the dead-lined truck along the bumpy terrain relatively easy. Sling legs, on the other hand, were primarily designed to hang Humvees from the bottoms of helicopters. They *could* be used for towing, but they weren't ideal because the springiness and length of the leg made it difficult to control the rear vehicle.

We finished hooking up the leg and took a second to assess our work. The sling leg stretched between our truck and Lussi's and was chained to both Humvees' integrated tow pintles. The setup looked solid enough, but driving was going to be challenging for a number of reasons. First, since the pintles were designed to be attached to a tow bar, which was shaped like a triangle, the single sling leg had to stretch between the two trucks at an angle, which caused the front truck to pull on the rear truck unevenly. Second, since the sling leg was springy, the transfer of force between the vehicles was very uneven, causing the rear truck to lag behind while going up hills, then surge ahead and risk a collision while descending. Finally, the whole arrangement had to be conducted without power steering or power brakes in the rear vehicle since neither of those systems worked because the truck was dead.[4]

Driving this way took a great deal of practice for both drivers. The front driver had to manage his speed and turns carefully enough to allow the rear driver to maintain his distance, and the rear driver had to wrestle an unwieldy multi-ton metal hippopotamus up and down the mountains of Afghanistan without getting in a wreck.

---

4. If you recognize a great deal of comedic potential in this setup, you're not alone. The entire dick-dance is regularly recreated for the amusement of crowds in a sport called 'chain racing.' YouTube it. You won't be disappointed.

"Well, that's the best we can do. Hawk, you ready for this? It's going to be a long, shitty drive."

"Roger, Sergeant. I got it."

The Hawk *looked* ready for it, which was a little unusual. He was one of those guys who didn't look anything like a recruiting poster. He was short and stout, and his combat equipment always looked like it was the wrong size. It hung off of him in an ungainly way no matter how often he tried to straighten it out. His facial expressions, too, weren't exactly John Wayne movie material. He was usually smirking, and often actually giggling, as he negotiated whatever Afghanistan threw at him. But, despite his appearance, he was a solid driver who'd been negotiating windy mountain roads for months. In The Hawk's mind, today was just going to be another day at the office.

I shook my head thinking about the upcoming drive, clapped The Hawk on the shoulder, wished him luck, and climbed back into my truck.

"Hey, Williams."

"Roger, Sergeant?"

"You ready for this tow job, man. You done a few of these before?"

"Only with the tow bar, Sergeant."

I sighed. "Okay. Well, take it slow. Driving with the sling leg takes some getting used to."

"Roger, Sergeant."

The PL gave the signal to start movement, and Chadwick's truck pulled onto the road. Williams revved the engine, and we pulled up behind him. After about ten feet, though, the sling leg reached its full length, and our truck suddenly lurched to a stop as the weight of the vehicle behind us killed our momentum. In response to the sudden change, Williams instinctively hit the brakes, and I immediately felt the heavy thump of the Humvee behind us smashing into our rear bumper. I gritted my teeth. It was going to be a very long day.

"Don't hit the brakes, man. You gotta just get used to the accordion effect."

"Roger, Sergeant."

We started rolling again, more slowly this time, and I could feel the change in our truck's momentum as the weight of the Humvee behind us evened out and we started rolling smoothly. In a minute or two, I relaxed and looked down at my map to see how far away our updated target was. After a few minutes of beer math, I concluded it would take us about an hour to get there.

"Hey Sergeant, when do the Cobras come on station?"

"They should roll up with the SF guys, and they're scheduled to do the hit a few minutes after we're in position. I need to figure out how–"

WHAM

Sergeant Romero was cut off mid-sentence when the truck we were towing crashed heavily into us, causing my head to smash backward into the metal strut behind me. Romero looked pissed.

"Williams! What the fuck, you chucklehead?"

"Sorry, Sergeant. Had to brake hard to roll through that ditch without bottoming us out."

"Brake more smoothly, man. You're driving for two now, and the guy behind you can't see what's coming up."

"Yeah, I got it, Sergeant. My bad."

As it turned out, Williams didn't get it, at least not for several more hours.

By the time we reached the target village, our two trucks had collided at least a dozen times and everyone's nerves were on edge. As we drove through the village–still bumping and jerking along like a couple of clown cars–the Afghans gathered to stare at this American circus. They didn't laugh, though, and there were no smiles on their (as always) expressionless faces. Still, I gritted my teeth as I imagined how ridiculous we looked.

We finally made it to our designated blocking positions, and once the truck stopped Romero and I hopped out to pull security. I looked behind me to where The Hawk was still sitting behind the wheel. Even from here, he looked sweaty and wide-eyed. Clearly, the drive had taken its toll on his nerves, too.

Sergeant Romero must have noticed the same thing because he walked over to The Hawk's door and offered him a few words of encouragement about his performance. Nuking The Hawk in place wouldn't make the task he'd assigned him go any smoother, and we still had a long drive ahead of us after the SF mission was complete. As Sergeant Romero consoled the frazzled Hawk, Lussi and I looked up the road to watch for any cars trying to leave the village. We were above the village, so we could see pretty much everything in the area.

"So, how was it in the rear truck?"

"Sucked ass. We're getting dragged along with zero power in the vehicle, and every time your truck slows down, we rear-end you. Nothing we could do about it, though.

With the power brakes out, this beast doesn't exactly stop on a dime. I must have smacked my arm or knee on the dash about fifty times."

"Yep. It was pretty awesome up front, too. Any idea what's wrong with the truck?"

"Alternator, I think. All the new gear they wired in draws too much power for the old system. Same with the suspension and steering. The truck wasn't built for what we're asking it to do. Think that's why we're getting so many breakdowns."

"Gotcha."

We killed the trucks to save gas, so the next thirty minutes were quiet in the way rural Afghanistan so often was. Nobody in the village tried to leave, and no cars approached from the other direction either. Nothing happened at all until I started to hear a distant humming sound. At first, I thought I was imagining it, but the hum got louder until it became a mechanical buzz that my brain finally recognized. Helicopters.

Romero heard it, too, and picked up the radio.

"Killer Six, this is Seven, over."

"Roger, go ahead, over."

"Audio contact on helicopters, over."

"Roger, stand by."

The radio was quiet for a few seconds until the PL's voice returned.

"Killer Seven, this is Six. Two Cobras en route to our location at this time, over."

"Roger, over."

Romero turned to us. "That means the SF team is on the way."

"We need to do anything?"

"Nope. Hold what you got."

It wasn't long before I saw the helicopters. First, they were two dots that appeared low on the horizon. Then, as they got closer, they came into focus as Huey Super Cobras. They approached from the same side of the village that the SF team would, but being much faster, the gray helicopters flew circles around the mountains that lined the road, looking for terrorists. I'd never seen a Cobra before. The only military helicopters that had ever supported us were Apaches, Blackhawks, and Chinooks. In contrast to these, the Cobras that were now overhead looked ridiculously dated–like childhood toys come to life.

"Holy shit, those look exactly like the G.I. Joe helicopter I had when I was a kid."[5]

---

5. The G.I. Joe Dragonfly was, in fact, based on the AH-1 Cobra attack helicopter.

Romero responded. "Yeah. They're Marine Corps birds. And the Marines only use old shit. Those Cobras have probably been in service since Vietnam."

"Are they worth a shit?"

"Oh hell yeah, they are. In my experience, Cobra pilots are a hell of a lot better to have on station than Apache pilots. Cobra pilots don't give a fuck about risk. They'll get right down in the shit and stir it up. Apaches tend to hang around in a high hover and work from a distance."

As if to emphasize Romero's point, the Cobras swooped low over our position. They couldn't have been twenty feet overhead; it felt like they were close enough that I could grab one of them.

"Yeah, I'll take a Cobra any day."

The helicopters continued flying circles around the village until the SF convoy eventually arrived, driving fast, plowing through the village's narrow roads, and forcing people to hurry out of the way. They exhibited a disdain for local concerns that I envied. In seconds, the trucks had pulled up to a house in the center of the village and disgorged a pile of BDU-wearing Afghans who ran into the house. From my position, I couldn't really see what they were doing, but no shots were fired, and a few seconds later the Afghans swarmed out of the house and climbed back into their pickups.

"Was that it? Did they get the guy?"

"Don't know. Can't tell from here. We'll hear one way or the other soon."

"How long do we maintain our positions?"

"Until the SF team is off station. Once they leave, we'll collapse and head up to Sharana."

"We've still got four or five hours of driving left, yeah?"

"Yeah, that's my guess."

SHHEEEEEOOOOO

"Ah fuck. Rockets."

KABOOM

We couldn't tell where the rocket came from. But, it had landed in a field a bit north of the village–nowhere near any friendly forces.

"Do we move?

SHHEEEEEOOOOO

SHHEEEEEOOOOO

SHHEEEEEOOOOO

Romero thought for a few seconds before replying. "Nah. We'll hold what we've got. We're as good here as anywhere else."

KABOOM. On a ridge a few hundred meters east of us.

KABOOM. In the middle of another field south of the village.

KABOOM. I heard this one, but I couldn't see any smoke. Must have been on the other side of the mountains.

"Who knows what those jackasses are even shooting at."

At this point in the deployment a rocket attack no longer elicited much excitement in any but the newest of new guys. The helicopters, on the other hand, took the rockets very seriously. Like angry hornets, their relatively slow looping pattern quickly gave way to aggressive diving maneuvers.

"Think they can see the bad guys?"

"Dunno. They seem pretty fired up, though."

The rockets stopped, but the diving and whirring around continued until the SF team wound its way back out of the village. Once the convoy made it to the main road, the Cobras abandoned their search for the rocketeers and flew off to clear the route to wherever the trucks were headed, leaving us alone on the ridge once more. In another few minutes, the helicopters were reduced to specks on the horizon, and soon afterward they vanished altogether.

"Well, that's that."

The PL's orders to collapse the position came across the radio, and after another embarrassing clown-procession through the village, complete with no fewer than three two-truck collisions, we were once again assembled and on the road to Sharana.

WHAM! "Sorry, Sergeant."

WHAM! "Sorry, Sergeant."

WHAM! "Sorry, Sergeant."

This sequence repeated itself every few minutes for all four remaining hours of the drive, since our trucks collided pretty much every time we had to take a sharp turn, cross a ditch, or slow down to avoid a large rock in the road. By the time we finally made it to our checkpoint outside Sharana, Romero looked like he was about to reach across the truck and throttle Williams. Everyone in the truck was visibly relieved when the PL finally called across the radio and said that we'd reached our stopping point for the day.

It was almost dark as we pulled off the road, arranged the trucks in a perimeter, and put together the security plan for the night. Once the guard roster was done, the mortar firing position established, and the dismounts had emplaced their observer position, I joined the other sergeants in a huddle in the center of our perimeter to get guidance from Sergeant Romero and the PL.

"Alright gents, after a degree of pain and suffering, we made it to our designated checkpoint for the night. Well done. Tomorrow starts the fun part. A Co will show up before first light and, together, we're going to recon a bunch of spots around the area. A Co is tasked to check out a group of large buildings outside the main village. We're going to recon a big field and a couple of other spots. We'll be done with all of our tasks by tomorrow evening, and once we wrap up, we'll push as far south as we can get by nightfall and finish the drive back to Orgun the next day. Squad leaders, tracking you've already got the plan. No change from what we already discussed. Anyone have any questions or issues?"

I did, but I didn't want to ask my question in front of everyone else, so I waited until the PL had dismissed the group.

"Hey, Sir."

"What's up, Pace?"

"No questions on the plan, but why are we doing this? What's important about Sharana?"

"Shit, did I not put that out? Higher wants us to check out Sharana because they want to build a new FOB there. Sharana is the capital of the district, so I guess they want a base closer to where the Afghan government is making decisions."

"That makes sense. A couple of the guys were wondering why we were up here, and this'll help put their questions to bed."

The PL nodded. "I guess I take it for granted sometimes that the guys know why we're doing what we're doing."

"Sir, to be honest, nine times out of ten nobody gives much of a shit, but every once in a while somebody gets curious, and it's good to have something to tell him."

"Yeah, that makes sense. Unrelated question. How're you liking being an NCO? Romero says you're thinking about staying in the Army."

"I am, Sir. Alycia and I want to do a few more years before we call it quits. Becoming an NCO has taken some getting used to, but it's not bad. At least I'm one rung up from the bottom of the totem pole now, right? Better to be the guy watching people burn shit than the guy burning it."

The PL laughed. "Leadership at its finest. Have you thought about trying out to be an officer? Romero and I think you'd be a good fit, and you've already got your degree, so the process would be pretty straightforward."

"I put in my paperwork a year ago just to see what would happen, and the Army assigned me to the Chemical Corps, so I withdrew my application. I took it as a sign from God that being an officer wasn't for me. Handling nerve gas and inventorying gas masks isn't my bag."

The PL nodded. "The Chemical Corps is probably not the best fit for you, I agree. Want to try again this year? There's a board coming up in March, right after we redeploy."

I thought for a moment. If I was going to stay in the Army anyway, being an officer would probably be a better bet. On any given day, I'd almost certainly have a better idea what was going on, and the pay was definitely better.

"What'll I have to do if I accept?"

"You'll resubmit your application, then you'll have to stand in front of a board and answer a bunch of questions about duty, honor, country, and that sort of thing. If they don't think you're a douchebag, they'll send you to the Officer Candidate School at Fort Benning. I saw those dudes marching all over the place with a drum while I was at Airborne School. You'll also get to wear an ascot, organize your wall locker and desk drawers, and sing songs before bed every night."

"That's quite the sales pitch, Sir."

"It's gay as hell. You also get to spend a lot of time scrubbing out the bathroom. You're going to love it."

"Alright, Sir. You sold me."

"Wonderful. I'll let Captain McCarron know, and we'll get the paperwork together. In the meantime, let's finish up this reconnaissance operation and get back to the land of hot chow."[6]

"Absolutely, Sir. I've been trying to avoid taking a shit until we get back to base. The shitters aren't great, but dropping a deuce out here on a rock in the freezing ass cold is worse, for sure."

"It's the little pleasures that make life worth living."

I walked back to the truck where Gardea was trying to fall asleep on the hood and

6. Captain McCarron was our company commander. Because he ran the whole headquarters company, though, he was stationed back in Kandahar so we only saw him a few times during the deployment.

Williams was standing in the gun pulling guard.

"Hey, Williams, you awake up there?"

"Roger, Sergeant."

"How are you doing man? You stop sweating from that drive yet?"

"Barely, Sergeant. I was white knuckling the shit out of the steering wheel for hours. That shit took a year off my life, and I thought Romero was going to fucking murder me."

I laughed. "I'll bet it did. I think Sergeant Romero finally calmed down a bit, but I'm not gonna lie. Having that truck smash into us every five minutes got pretty fucking old."

"You're telling me. No way to avoid it, though."

"Yeah, maybe not." I turned to walk off and get my gear ready to rack out, but then stopped when I remembered something. "I got you an answer on the 'why' from earlier, by the way."

"No shit? We're not just doing this to keep chow from getting interrupted?"

"Nope. Apparently, they're going to build a big ass base up here because Sharana is the capital of the district."

Williams nodded. "Okay, that's an answer I can live with. That explanation makes this shit make at least some kind of sense. Now I can stand up here all night knowing I'm not just freezing my balls off for no reason. You've officially motivated me, Sergeant. Thank you."

"Carry on, soldier. I'll pretend that wasn't sarcastic. And don't thank me, thank your recruiter."

"Ha! Fuck that guy twice in the ass, Sergeant."

I spent the next half hour walking around the perimeter and checking on the guards to ensure everyone was good to go. It was a cold clear night, and the temperature was dropping rapidly, so by the time I was done trooping the line, I could clearly see my breath. When I got back to the truck, Gardea was snoring on the still-warm truck hood and Williams was immersed in the weird solipsism that is an infantryman's night guard shift.

I dug my poncho liner out of my gear, climbed into my truck seat, and wrapped up in the woobie as best I could for a few hours of shitty, cramped truck sleep. Tomorrow would be long and dull, most likely, but after that, we'd be back at Orgun and, after *that,* some day in the future, I'd be back home with Alycia trying to figure out this officer business.

January 31, 2004
Fort Drum, NY

Sweetheart,

Well, Sobon is out of here for good. He wanted to spend his last night chilling and reminiscing (and fretting/freaking out) at Dickey's house. He drank a lot of beer and, as usual, started getting very sentimental about you and his other army friends. He really seems to be considering returning to the army after college. He wants to be a part of this organization with his two best friends–you and Ed Mikkelsen. You would not believe how much he loves and misses you–it's touching.

It's since warmed up a little, but two days ago, I went on one of the most awful runs of my life. The track at the gym was closed, so I had to run outside. It was bitter cold, snowing, and very windy. My snot froze in my nose, the wind made my eyes water, and then the tears froze. I was in agony, and I could barely see. Is this what PT is like for you poor bastards all winter?

Today I went with Melissa Butera and her mother to the Salmon Run Mall for a yellow ribbon rally. We brought the posters we made yesterday and met up with Mrs. Howard, Mrs. Watson, and some other, lowlier 1-87 wives like us. It was predictably long and boring, with speeches from two generals and local and state politicians, including Hillary Clinton. More than one of them excitedly stated that the 10th Mountain deploys more frequently than any other division and that you are sure to be called on again soon to fight in the war. Are they nuts? This is not what you say to pump up a bunch of wives who haven't seen their husbands in over six months. Clinton even said something like, "When we need troops for a mission, we ask, 'Is it difficult? Is it dangerous? If the answer is yes, we say 'Send in the 10th Mountain!'" Several of us wives kind of looked at each other, and over the applause yelled, "Send someone else!" I hate politicians. Every last one got this dewy-eyed look and made some lame pseudo-sentimental comment about "all the beautiful soldiers' children here in the front." Gag. Several were tooting their own horns about having gone to Kabul and/or Kandahar to visit the troops, saying they saw first-hand the hardships and tough terrain you were dealing with. First, please don't rub it in that you got to go to Afghanistan to see our soldiers! Many in the audience would have killed for the chance. Second, exactly how hard can life be for a soldier at one of those two places? Compared to what you're doing, it sounds like a day at the beach.

As you no doubt know by now, the next local OCS board is March 11. That means they might–if we are really, really lucky–send you home in time for it. This, of course, puts me in fits of excitement–what better birthday gift could I possibly ask for than you?! I'm trying not to get my hopes up, but, boy, wouldn't that be great?

I miss you like crazy.

Much, much love,
Alycia

KABOOM.

"Get off your asses! You heard the rockets, and you know the drill. Move it!"

"C'mon, Sarge. Nobody gives a fuck about rockets. Can't we just die in peace."

"No, Jimmy. I'm here to save you from yourself. Put your gear on you fat bastard and get the truck ready. QRF timeline says we gotta roll out in five."

Jimmy grinned, stretched, and put down his Gameboy. "Roger, Sergeant."

I knew Jimmy would be fine. He was a good dude. I walked outside and saw that Hogu was already lubing down the MK-19.

SHHEEEEEOOOOO

"All good, man?"

SHHEEEEEOOOOO

"You know, it, Pace. Another day at the fucking office."

SHHEEEEEOOOOO

"Yeah, I get it. Who gives a fuck, right?"

KABOOM.

"Pretty much. Stupid fucking rockets."

KABOOM.

"How far off do they sound to you?"

"I dunno, maybe five hundred meters? Probably in one of those fields down the road."

"Yeah. Probably so."

KABOOM.

"That last one seemed closer. Like, *in-the-base*, closer."

Hogu looked toward the gate where a plume of black smoke was rising.

"Yeah, maybe so. Whatever. Haji-ass motherfuckers can't hit shit."

I nodded. "Yeah. I don't know what they aim with, but it clearly sucks. Ah well, let's get this over with."

The rest of the section was out and getting the trucks started by now. I saw Sergeant Chadwick leaving the TOC and walked up to get the intel dump on where we were headed.

"We got a grid, Pace. Here it is. You got point today. Get us out there so we can verify there's nobody at the fucking POO site and get back in time for chow. Goddamn stupid rocket attack."

I scribbled down the digits. "Roger, Sergeant. Rolling out now."

I hopped in the truck and told Jimmy to head toward the gate. We only needed two trucks today for the counter-rocket patrol, so there wasn't much to coordinate.

I hollered up to Gardea as we held at the exit to the FOB. "We got the other truck?"

"Roger, Sergeant. We're good to go."

"Killer One, this is Killer One-Alpha. Rolling out. Time now, over."

"Roger, One-Alpha. I'll call it up to the TOC."

We rolled out of the gate, past the ever-present and always bored-looking Afghan guard, and turned right, looping around the base toward the point I'd marked on the map. The point wasn't far from the base–no more than 800 meters–which was weird. Also, the point wasn't in the mountains, it was out in the middle of a field. Who'd shoot rockets from the middle of a field?

My brain chewed on these details as we rolled toward the identified grid. As we neared it, we turned off of the main road and onto a farm track that led between two fields. Unlike in Sh'kin, the fields here didn't have rock-piles separating them; the roads seemed to fill that function.

"Alright, stop here."

Gardea responded from the turret. "You sure, Sergeant? This doesn't look like a POO to me."

"It's where they told us, man."

I climbed out of the truck and met Chadwick, who had just pulled up behind us.

"This is it, Sergeant."

"Nah, there's no way. Let me see your map."

We pulled it out and laid it out on the hood. Both of us re-plotted the grid.

SHHEEEEEOOOOO

"Ah, Jesus. More fucking rockets. Really?"

KABOOM.

The impact was less than a hundred meters from us. The rocket had landed square in the middle of one of the adjacent fields and set a pile of old brush on fire. My brain made a sudden connection.

"Hey, Sergeant. Any chance they gave us the wrong grid? This looks a lot more like a point of impact than a point of origin."

SHHEEEEEOOOOO

KABOOM.

Another impact. This one was close, too, but hit in the field on the other side of us. Chadwick looked at me for a second. The fire had started producing thick white smoke that was wafting toward us.

"Since when do rockets set stuff on fire? That ever happened before?"

"Not that I remember. No idea what's up."

Hogu called down to us from the other truck's turret. "Hey, Sergeants, come here for a second."

SHHEEEEEOOOOO

KABOOM.

We walked over to Chadwick's truck as another rocket impacted nearby.

"What's up, Hogu?"

"You guys smell that?"

I sniffed the air. Hogu was right. The smoke had a distinct smell. I couldn't immediately place it. Chadwick replied. "I smell something. No idea what it is, though. Something in those rockets smells funny."

"Yeah, I know what you're saying. The rocket smoke *does* smell weird. But I know what the rest of the smell is. Motherfucking Mary Jane!"

I sniffed again. It *was* dope. The rockets must have set somebody's marijuana crop on fire.

Hogu started to breathe exaggeratedly deeply. "This shit is like being back home, Sergeant. The Army can't even get me in trouble, 'cause this shit is in the line of duty."

Chadwick seemed perplexed for a minute.

SHHEEEEEOOOOO

KABOOM

But he recovered quickly. “Alright, you pothead fuck, that’s enough of that. Clearly we’re in the fucking impact area, so let’s get out of here. I don’t need any of you piece of shit druggies getting high.”

“Ah, c’mon, Sarge. Can’t we stay just a little longer? They aren’t going to hit us anyway."

“Fuck no, Hogu. Pace, load up and get us out of here.”

By the time I was back in my truck, the smoke from the rockets had gotten thick enough to make me cough. Jimmy looked over at me with a silly grin on his face. “Can’t we stick around for a few more minutes, Sergeant?”

“Nope, we sure can’t, Jimmy. Besides, it smells like there’s more than just pot in that smoke. Head back to the base.”

We turned the truck around and rolled back the way we’d come, through the thick smoke. All of us were coughing by the time we came out the other side, but the coughing stopped soon after the air cleared.

What a clusterfuck! How had we ended up in the impact area? Fucking intel. “Yeah,” my inner voice said, “but you could have checked that shit more thoroughly yourself, too.”

The drive back to base was short, and in a few minutes we were there. Romero met us at the tent.

“Chadwick, I followed along in the TOC. Describe that smoke to me in more detail.”

“It was white and thick as fuck, Sergeant. The rockets hit some brush pile that turned out to be dope and set it on fire, but that wasn’t the only source of the smoke. The other rockets were burning when they landed, too, and the smoke smelled bad.”

Romero’s brow knit as he nodded and listened. “Get your section to the aid station. That shit sounds like it might have been white phosphorus, and, if it was, you need to get your lungs checked out.”

“Roger, Sergeant.”

Romero left and Chadwick turned toward us. “Alright, all of you get your asses to the aid station so you can get your vaginas checked out since apparently we’re all broke-dicks now.”

“Can we grab chow first, Sergeant? Hot chow closes in twenty minutes.”

"Fuck no, you can't. You heard Romero. Your precious lungs are broken and you need to get checked out so you don't start bleeding out of your asses or something."

Grumbling, we all dropped our kit and headed to the aid station, which was a small mud-brick building across the base. None of us had ever been there before. Aside from the risk of missing hot chow, a visit to the aid station created the risk that you would be labeled a 'broke-dick.' And, in the infantry, there is nothing worse than being thought of as a broke-dick.[7]

As noted in the glossary, a broke-dick is someone who abuses the medical system to avoid work. Unfortunately, in the simple dog-eat-dog world of the infantry, there is often little distinction drawn between those who abuse the system and those who need to visit the medics for a legitimate reason, like having a broken leg. This means that anyone standing in the line for sick call when it's time for morning PT is likely to be tagged as a broke-dick by his squad mates. Consequently, even infantrymen with sprained ankles, stress fractures, or bleeding ulcers tend to avoid the aid station at all costs.

Not that we could avoid medical treatment today, of course. We'd been directly ordered to report to the aid station. Fortunately, we were all going to the aid station together, and it was in a remote enough location that none of the other platoons saw us enter it, so we had some chance of escaping without being tagged as broke-dicks.

When we arrived, we lined up and took turns knocking on the plywood door. The orderly called us one-by-one. When it was my turn, I walked through the door and was greeted by a bored-looking specialist wearing DCU bottoms and a t-shirt. My eyes narrowed as I surveyed him and his work space. He had a phone, a desk, and a pile of snacks. Motherfucking pogues. This is where all the good shit went! But my moment of infantry rage passed quickly. It wasn't his fault that he'd made better life decisions than I had, after all.

"Walk behind the curtain, take off your DCU top, and sit down on the bed, Sergeant."

"Roger."

I did as he asked, and in about a minute the doctor walked in. He was tall and older and wearing glasses.

"What's up, Sir? Am I going to die?"

"Eventually? 100%. Imminently? Probably not. We just need to check your lungs for

7. This is debateable, of course. Other rivals for the 'worst ever' title are 'chow thief,' 'surface shitter,' and, of course, 'buddy fucker.'

damage. White phosphorus inhalation is a big deal if you get too much of it. But since you were outside, you'll probably be fine."

The doc put a stethoscope to my chest and listened to me breathe a few times. Then he moved it to my back and did the same thing.

"You seem fine. I'll make a note in your records. You're good to go."

"Thanks, Doc. I appreciate it."

I left the aid station and turned to head back to the tent. As I was walking away, a Hilux pulled up, and a couple of Afghans got out and started unloading dead bodies off the back. Other than being dead, the bodies were dressed exactly like the two guys unloading them: man jammies, mushroom hats, and flip flops. The two guys handled their task in a leisurely way. They picked the bodies up one by one and laid them out in a neat row next to the aid station. I looked around. I'd been the last guy to get checked out, and I didn't see any other Americans in the area, so I was alone with the two Afghans. Was this normal? Did they always stack dead bodies here? Did I need to do something about this?

I watched for several minutes, fascinated by the weird scene. After they moved the last body, they stopped working and leaned against the truck, clearly waiting for something. I stared at them in confusion for longer than was socially appropriate. Eventually, they noticed me and stared back. We didn't speak. What was there to say? After a few more seconds of the weird stare-off, I realized hot chow was probably about to close and that I may have already missed it. This snapped me out of the moment and I turned to leave, giving the Afghans and their morbid project one last backward glance.

I walked back across the base to the Chow Hall and pushed inside just in time to see the last container of hot food vanishing out the other door. Hogu was still sitting at one of the tables eating.

"Ha! You just missed chow, Pace. Sucks to be you."

I considered my remaining options and settled on a handful of Nutrigrain bars and a couple of packets of peanut butter. As I sat down next to Hogu, I couldn't help but compare my dinner to the remnants of his, which had been surf n' turf. It's true that the steak was always overdone and the lobster tails were rubbery as hell, but that meal was still light years beyond what I was looking at.

"If you want, you can pick some of the meat out of this leftover tail."

"I appreciate that, dick. I'll keep your generosity in mind next time I'm putting together the guard roster."

Hogu laughed. "Fuck it, right? The deployment is almost over. It's definitely time to get the fuck out of this place."

"I completely agree. You heard when we might be flying yet?"

"Early March is the rumor, but who knows? I heard the replacement unit should be here in a few weeks."

I nodded, munching on a bar and washing it down with a mouthful of warm, shelf-stable milk.

"Man, they can't get here fast enough."

By late February, the rumors about us heading home had materialized into a plan about how redeployment would work. Of course, as lower enlisted folk we knew very little about the plan itself, but since it was an Army plan, the details were carved in butter anyway and subject to change with zero notice. This meant our lack of knowledge wasn't especially detrimental.

I discussed a few of the rumors I'd heard with the PL while we were working out at the gym. I was spotting him as he got ready for a set of bench press.

"Hey Sir, so what's the plan to get home?"

I lifted the bar off the rack for him, and he started pressing.

"Well, I think we're going to rotate out of here by sections. Once our replacements are on ground, First Section will head back to Bagram to wait for a flight home. Second Second will stick around here for one more week to hand over with the new guys, then they'll head up to Bagram, too."

He racked the bar and sat up. We switched places.

"That makes sense. Any idea on the timeline yet?"

"Nope. No idea. All depends on the weather, really. You know how delicate our rotary wing friends are."

"For sure. Wouldn't want to risk them getting stuck here for the night if the weather turned bad. They might have to sleep on a cot or miss movie night in Bagram or something."

"Yep. The cold reality is that one of those Chinooks is worth more money than our entire platoon with all its equipment. So, in the Army's mind, keeping the helicopter alive is definitely worth more than pretty much any amount of your time."

"Man, reality hurts, Sir. Oof, little help here."

I missed my last rep, and the PL helped me get the bar back on the rack. I sat up, and we switched places again.

"What I heard, Sir, is that we're actually going to do a ground convoy all the way back to Bagram so we can get the vehicles home."

"Where'd you hear that?"

"At the chow hall."

The PL thought for a moment as he worked through his set.

"I can't say for sure that's not true, but it's not what they're putting out in the TOC."

"Would you say, on average, that the TOC has put out more reliable information than chow hall rumor on this deployment?"

"Fifty-fifty, I think. Honestly, either option seems plausible enough."

He racked the weight again and stood up. Chadwick poked his head into the gym and interrupted our workout.

"Pace! Been looking for you. Get your team ready. There's some asshole on the ridge overlooking the FOB, and we gotta go deal with him. Romero's briefing us at the trucks in ten minutes."

"Roger, Sergeant." I turned to the PL. "Looks like you gotta finish without me, Sir."

"Go forth and do great things."

"You not coming, Sir?"

"Nah, I used to think I knew everything about combat and needed to plan it all myself, but ever since I started incorporating you guys into the planning process it's made our plans better. You guys got this."

I left the gym, grabbed my gear, and headed to the truck. The rest of the section was already assembled around the hood of Chadwick's truck with a bunch of guys from C Co. When I walked up, Romero started putting out guidance on the operation.

"Alright, homos, don't go smelling the barn on me just yet. There's still a war to fight. We've got a dickhead up on that ridge over there." He paused and pointed to the mountainside that overlooked the base. "One of the guard towers says he's up there with binos scouting the place out, and the headshed wants us to go pick him up for questioning. We can't get up there with trucks, so we'll take the road around to the back side of the ridge, then dismount and walk up to the top of the ridge to grab him. Any questions?"

Nobody had any. We loaded up the trucks and rolled out. The drive around to the back side of the ridge took around ten minutes. We parked the vehicles, then climbed out to get ready to move. Romero gave the team a quick update.

"Chadwick, you stay with the trucks and pull security. Everyone else, spread out and let's move up the hill. Pace, you stick with me."

"Roger, Sergeant."

We fanned out in a long line, with about ten meters between each individual. We had a squad from C Co, complete with a machine gun team, so there were thirteen of us altogether. I looked up to the top of the ridge. It didn't look that steep. I figured it would take about fifteen minutes to climb.

Fifteen minutes later, when we were about a quarter of the way up the ridge, I realized I was way wrong. I looked over at Romero, who was breathing as hard as I was and sweating heavily.

"Holy shit, Sergeant, this climb was harder than it looked."

"No kidding, Pace. That's why I brought you up here on this. Folks need to feel what climbing these mountains is like. You think you're sucking? Look at that guy." Romero pointed to the C Co machine gunner, who was thirty yards to the right of us. "You know how much that guy is carrying? The 240 is twenty-five pounds. He's got a starter belt of a hundred rounds, and another couple hundred rounds in his pack. That's another twenty pounds or so. That's on top of the armor, helmet, and whatever else he's got in his kit. Probably brings his total carrying weight up to eighty or ninety pounds. He's just like our trucks. Loaded down with so much shit that he can't function and will probably break. "

We stood there, catching our breaths, and watching the machine gunner and his assistant kneel for a second to recover. He looked spent as hell, and we had a lot of climb left.

"How maneuverable do you think that guy is in this terrain? Can he close with and destroy the enemy? Fuck no. He can barely make it up the damned mountain at a slow walk, even with nobody shooting at us." Romero looked up toward the top of the ridge. As he did so, a question crossed my mind.

"Hey, Sergeant. That guy's going to be long gone by the time we get up there, isn't he?"

"Of course he is. That dude was born and raised here, and he probably isn't wearing anything but some man jammies and a pair of flip flops. He was long gone before we even parked the trucks, and there's no way in hell we're catching him."

"Why are we doing this, then?"

Romero laughed. "To prove a point. Because we were told to. To see if that guy was dumb enough to leave something behind. All of that, I guess."

I thought about this the rest of the way up the climb. It was grueling and took us more than an hour. By the time we neared the top, I was drenched with sweat and exhausted, and the machine gunner looked like he was going to pass out.

We finally crested the ridge and were greeted with a view of the entire Orgun valley spread out in front of us. I could see every inch of the base, and if I were a bad guy watching a helicopter land right now, it'd be pretty easy to put a rocket into it. The machine gunner and his assistant immediately set up the gun and lay down to establish security. The rest of us scoured the area for any pocket litter the Afghan who'd been up here might have left behind. We found none.

Romero told the RTO to report our findings to the TOC and request permission to return to base. While we waited for the transmission and reply, we continued our conversation.

"So, if we aren't maneuverable enough to catch one dude who might or might not have been collecting information on the base to shoot rockets at it, what are the odds we've caught anyone else during our time here?"

"To be honest with you, I assess we've missed eighty percent of the bad guys we were going after. Think about it. If we head to a village to search it, how long does it take us to get there? Hours, right? Do you think any serious terrorist wouldn't have people looking out for us? I mean, he knows exactly where we're coming from. Why wouldn't he?"

He kicked a rock down the ridge line and continued. "And if he sees us coming, he's got *hours* to load up and get out of the area. Hell, he can have his wife go to the market, buy a goat, slaughter it, and make it into a picnic for him and the family, and he'll still be gone long before we show up. The only people we're catching are the dumbasses who aren't smart enough to leave town, or the jackasses idiotic enough to actually shoot at us. How long do you think it'll be before all of them are dead, though? What's left after that?"

"So, what do we do then?"

"To kill terrorists? We gotta get much faster. We need lift to actually surprise people, and we need to strip all this shit off." He flicked one of the straps on his armor. "So we can actually maneuver."

"Let's say we did. What then? Does killing these people even matter? How many of them are terrorists at all? Was this guy we were looking for today a terrorist, or was he just some dickhead up here with his goats?"

"Probably just some dickhead. Maybe he was up here making twenty bucks to collect information or maybe he was making five bucks just to walk around up here and see how we'd react. Or maybe neither one, and he was some random Haji on his morning constitutional. Who knows?"

"So what's the point, then? Why are we even over here?"

Romero stood looking out across the valley at Orgun.

"What would you do if you were them?"

"If I were an Afghan?"

"Yeah, if you were an Afghan living in Orgun. You grew up in the '80s when the Russians were here, and you remember them building a base outside of your village and dropping mines all around it. You remember them fucking with your family, killing your relatives, and generally causing problems. Then, a bunch of your cousins get together and started fucking with them back. Your cousins beat the Russians, who go home with their tails between their legs. And then, ten years later, another foreign army shows up and occupies the same exact base. Your cousins are pissed and come ask if you want to help get rid of these new foreign dickheads. What would you do?"

"I guess I'd probably help them out."

"Hell yeah, you would. And what if some priest at your church told you he had a friend who'd give you free weapons to do it. Would you take them?"

"I probably would."

Romero nodded. "Yeah. Who wouldn't?"

I thought for a minute, staring out into the distance and listening to the electronic beeps and squawks of the RTO's radio. "So, what's the way ahead then?"

He laughed. "Fuck, I don't know, Pace. Ask some officer that shit. That's their business. They tell me to kill people, so I kill people the best way I know how. Just because I see where the Afghans are coming from doesn't mean I don't like killing their dumb asses. I'd be alright with us staying around here killing people for a while, I guess. Beats peacetime bullshit. What I want to see change is how we're doing it. We have to get more maneuverable and be able to respond more quickly, but that means we have to accept more risk. Commanders are more concerned about losing a guy than winning, and we're never going to accomplish the mission with that attitude."

I thought about his response. I guess I understood it, sort of. Combat could be pretty exciting, for sure. Was that reason enough to kill a bunch of people, though? Was any of this worth losing anyone on our side?

I hadn't reached a conclusion by the time the RTO finished his call and reported to Romero that we were cleared to return to base. We picked ourselves back up, and after one last birdseye look at Orgun, we humped our way back down to the trucks and drove back to the base.

February 22, 2004
Fort Drum, NY

My love,

I'm gearing up psychologically for our reunion–I'm almost nervous about it! Irrationally, I'm worried I'll disappoint you somehow, although I can't picture how that could possibly be. I'm worried we'll fight over stupid stuff or that your return won't live up to our expectations somehow. I have a feeling that everything will be lovely, but these concerns do flit through my head from time to time.

But then, I probably have plenty of time to wait. It seems like you'd know by now if they were going to send you home in time for that OCS board on 3/11.

We had an FRG meeting this week–mostly we just got tax information. SFC Gilbert (who was there) has a really good attitude, and he gets around quite well. I saw him chasing several squealing children, all of whom were fascinated by his prosthetic leg. He seems happy just to be alive.

Melissa Butera had her baby shower today. I brought homemade fruit dip, vegetable dip, and large platters of cut fruits and veggies. It was nice to touch base with the other wives, and I've been enjoying the leftovers all evening. Win-win. It sounds like Brian will be sent home really soon, ideally in time for the baby's birth. I can't believe poor Melissa has been without him for her entire pregnancy. I really, really hope that never happens to us.

I said goodbye to Misty this week! She's gone from James Street for good now, and I keep wondering what's going to happen when Tyrone gets back to clear housing. It's going to be a total disaster for that poor guy. I hope he has a friend with a pickup, or he'll be hauling an awful lot of trash in his car.

Take care, and I hope to talk to you soon!!

Love,
Alycia

"Fuck," I thought, "Everyone on this plane is forty years old. They all must be generals or sergeants major or something." My spidey senses, finely honed from years of getting in trouble, were running wild at the possibility of getting the shit smoked out of me.

Butera and I were the only two guys from the 1-87 on this plane. The rest of the unit was still packing gear in Bagram. Butera was a friend from the mortar platoon who lived across the street from me on James Street. His wife, Melissa, was about to give birth. So, the First Sergeant had agreed to send him home ahead of the main body of redeploying soldiers. He'd cleared me to go home early, too, so that I'd be able to make the OCS board, although he'd only done so grudgingly and because Captain McCarron had insisted on it. The First Sergeant had been pissed at Romero for a week for convincing the PL to end-around him on that, though my impression was that Romero wasn't too worried about any consequences.

As with everything in the Army, our flights were slow and chaotic. Once we'd finally gotten out of Orgun on a Chinook, we'd been held up at Bagram for two days waiting on an airplane home. Bagram was amazing compared to our previous firebases. We were put up in a transient tent the battalion maintained for folks rotating in and out of Afghanistan, but since nobody else was in it, we had the entire thirty-man tent to ourselves. The thing was even climate controlled. Butera and I spent our first evening there marveling at the relative luxury of our surroundings.

"Man, Butera, this is a totally different way to deploy."

"Yeah, no shit. Air conditioned tent. Clean port-a-shitters. Chow hall with a salad bar. Movie night. Phones that work. Badass gym. Motherfuckers probably either get fat or jacked on their rotations."

"Right? Even if the food on Orgun hadn't sucked ass, I couldn't have put on any muscle anyway because I stayed so sick that I shat out every calorie I put in my mouth."

Butera laughed. "So, what do we do for the rest of our time here?"

"Eat until we're tired; sleep until we're hungry? Turn food into shit? Steal oxygen? What else? I think I saw a big stack of books in the MWR tent; maybe they have something worth reading."[8]

"I was thinking about heading out to the Haji mart and picking up some stuff for Melissa."

"That's a good idea, man. Let's scope it out after chow tomorrow."

I'd slept well that night in the clean, smoke-free, climate-controlled air of the tent. The next day, we walked out to the huge bazaar where Afghans sold everything from switchblades to Persian rugs. I bought a hat, some earrings for Alycia, and some Afghan currency, and we spent the rest of the day loafing around the base and trying our best to avoid the attention of anyone over the rank of sergeant.

We mostly succeeded in this until we boarded the plane to fly to Germany. Looking around, the first person I noticed was a wizened old dude with a gray Tom Selleck moustache. "Shit," I thought, "That dude's got to be at least a colonel."

He was already strapped into one of the plane's jump seats, so I shuffled past him and tried to find a place next to someone less august-looking. But the next open seat was next to another equally old guy. This one looked like George C. Scott. "What the fuck," I thought. "Is this some sort of command plane?"

The Air Force load master was giving me the 'hurry the fuck up' stink eye, so I had to sit somewhere. But I sure didn't want to spend eight hours holding in my farts so as not to piss off some four-star and his sergeant major. Hurriedly, I took one last look around, found another open spot, and flopped into it, hoping for the best. I stuffed my collection of bags and gear under my seat, looked up, and locked eyes with the oldest and most serious looking dude I'd seen so far. He stared at me hard, and I felt like he was pissed that I'd just taken up the space he was going to use for legroom.

"Fuck my life. He's probably like my division commander or something." I stuffed a set of ear plugs into my ears to drown out the drone of the plane's engines, pulled a book out of my bag, and pointedly started to read. I figured he'd eventually get bored with staring at me.

By the time we took off, my legs were already cramped from exaggeratedly minding

---

8. MWR, or Morale, Welfare and Recreation, tents were facilities set up to provide entertainment options for deployed soldiers. These tents varied in size and quality, but in 2004 the tent at Bagram had a giant pile of old books, a ping pong table, a foosball table, and a TV with a bunch of movies you could check out. It also had free phones.

the other guy's space. However, by unobtrusive side-eye recon, it seemed that he had fallen asleep, so I took the opportunity to take a quick peek at his collar to see what I was up against.

"A fucking specialist?" I was so surprised I said it out loud, and I had to look twice to make sure I wasn't seeing things. Was I missing something? Nope, there it was, plain as day. This old dude was a specialist. Emboldened by my discovery, I started looking around at the other passengers. Mr. August was a sergeant, and the George C. Scott lookalike was a friggin private first class. Was this a plane full of criminals? Had all these guys been busted down for DUIs or something? My confused brain scrambled for more data. My eyes landed on one of their unit patches. It was a bird head with ears.

What the hell unit is that? It wasn't the 101st, which was the only bird-headed unit patch I could think of. And what kind of bird has ears?[9] My mind churned through the possibilities. Old people. Too little rank for their age. Weird unit patch. Suddenly, the pieces fell into place and my brain leapt to the conclusion that these guys were National Guard.

I laughed out loud in relief. I wasn't on a command bird at all. I was on a plane full of the oldest squad of army privates on the planet. I immediately relaxed and stretched out my aching legs, bumping the sleeping guy across from me. Since we were equal competitors in the battle for legroom, I encroached on his space a bit to achieve parity.[10]

From that point on, the trip was no more uncomfortable than any other Air Force flight, and within a few hours I didn't feel the least bit bad when I had to jostle the shit out of everyone while stepping over them to take a leak in the C-130's weird urinal system.[11]

We arrived in Germany hours later, deplaned, and transferred through Customs to our next plane, which was the one that would take us back to Syracuse. It was only then, while the random Air Force guy was going through my paperwork, that the reality of

---

9. I later learned the image on the patch wasn't a bird head at all. The patch belonged to the Delaware National Guard, and the animal in question was a griffin.

10. He didn't wake up.

11. For those who've never used it, the restroom on a C-130 is nothing but a sort of funnel welded into the skin of the plane that you piss into. I suspect the system dumps the piss straight out of the side of the plane, but I have no data to back that up. I have no idea how one might use this system to take a crap. In twenty-three years of service, I never once saw anyone try.

what was about to happen hit me. I was going home. *The deployment was over.*

"Hey, Butera, I dunno why, but I just realized we're about to see our wives for the first time in nine months."

"Yeah, Pace. No shit."

"The gravity of that just sunk in. Dude, our combat trip is over."

Butera nodded, looking completely unimpressed. "Yeah, Pace. Again, no shit. Hadn't you already realized that? Where the hell did you think we were going when we left Bagram?"

"I guess I was so caught up in the Army jack-assery of the flight, I missed the significance of what we were actually doing."

Butera shook his head and laughed. "That college degree didn't make you any smarter, man. You've got your head right up your ass."

"Yeah. I guess I do."

I chewed on the reality of going home for the entire trans-Atlantic flight, and a knot started to form in my gut as I thought about it some more. I hadn't seen Alycia in nine months. We'd only been married for a little over two years. This deployment represented a huge chunk of that time. Plus, she'd been in grad school for half of it, having a great time and hanging out with dudes from all over the country. What if she'd found someone better than me? I didn't think she had, but what did I know?

I'd certainly seen plenty of guys get screwed and have nasty surprises waiting for them back home. One guy's wife had left him a month into the tour and he hadn't heard about it until after Christmas when the command had told him to get on a plane. It turned out that his wife had locked his dogs in the house with about two hundred pounds of dog food, divorced him, and left. The dogs were fine, but the house was covered in shit. It smelled so bad the neighbors called the cops, and the housing office had flipped out, so the guy had to fly home to deal with it. I was sure Alycia wasn't like that, but was I dumb to be as certain as I was? And even if she hadn't done anything crazy, we'd been apart for a long time; would our relationship be the same? I knew I was being an idiot, but I couldn't get rid of the fear that things weren't going to be the same as they were when I left.

Otherwise, the flight home was great. Back in 2004, when the Global War on Terror was still new and folks still remembered September 11th, a guy flying in uniform could expect an amazingly warm welcome pretty much everywhere he went. In Germany, we'd transitioned from a C-130 to a commercial jet, so instead of being strapped into canvas jump seats we enjoyed the relative comfort of a Delta airliner. I'm not sure the people on my immediate left and right were thrilled with the way I smelled, but the rest of the passengers and the crew ensured I had a great time.

The stewardesses were especially friendly. "Would you like a drink, Sir?"

"Absolutely, I would. Jack and Coke, please. What do I owe you?"

The stewardess smiled. "No charge. Thanks for your service."

The passengers were interested and complimentary. "Hey there, soldier. How was it over there?"

"Nothing we couldn't handle, Sir."

"Well, it's a hell of a thing y'all are doing over there. Thanks for what you do."

"No problem, Sir."

Even the airport staff were pleasant. "Attention in the terminal, we'd like to extend a special thank you to our soldiers in uniform who are flying home today."[12]

While it was a bit embarrassing, all of the praise and 'thanks for your services' went to my head a little and gave me a sense of bravado that was both out of character and a bit hazardous for a guy who hadn't had anything to drink for nine months. Technically, I shouldn't have been drinking in uniform at all, and I definitely shouldn't have been drinking liquor, but why not? I was coming home from combat, damnit.

I was buzzed and telling stories within a few hours of taking off from Germany, and the only thing that saved me from making a complete ass of myself was that the combined effect of the booze and transient lifestyle ensured that I fell asleep for most of the flight and didn't wake up until the plane landed.

By the time we touched down in Syracuse, it was evening, and the combination of bravado and jetlag turned into apprehension. I walked off the plane, turned a few corners, and suddenly saw Alycia standing there.

She was just as beautiful as I remembered, standing there in jeans and a purple sweater, with her long hair falling down around her shoulders. We both just stood there looking at each other stupidly for a second, and then we rushed into each other's arms.

"I'm so happy to see you!"

"I missed you so much!"

---

12. The sole exception to these positive interactions was my encounter with the German airport security personnel. They were in no way grateful for my service, and, instead, were quite unhappy that I had a knife in my pocket. After a very one-sided argument, they took it and did whatever Germans do with confiscated weaponry.

We kissed for longer than was socially appropriate, but nobody seemed to mind. Eventually we finished, and, flushed with happiness, we went to pick up my baggage and walk out to the car.

"Are you okay to drive?"

I thought for a minute. "Maybe you should, just for tonight. I haven't driven anything in months, and I definitely haven't followed any rules of the road or driven in the snow. I might be a little rusty, and I'm definitely jetlagged as hell."

She nodded, and we climbed into our 1997 Nissan Altima and drove away. We stopped at the Olive Garden in Syracuse for dinner, then drove up Interstate 81 toward Fort Drum and our house. Our conversation was easier than I'd worried it might be, and within a few minutes of being back together, Alycia and I were talking as freely and easily as we ever had. By the time we turned onto James Street and pulled into our driveway, it was late at night and I was freshly caught up on all the various scandals and drama of our little slice of the world. Alycia parked the car and I climbed out into the frigid air to look around the neighborhood.

The streetlights were shining brightly on the heaped piles of snow. Next door, Trey and Brandy's cars were still parked in their driveway. Tyrone's car was gone, but otherwise, the whole street appeared exactly like I remembered it. Everything was quiet and deceptively normal looking. The stolen car, the drug fire, the domestic violence and the many, many divorces all seemed to have been buried under the blanket of fresh white snow. There were no visible signs of all the craziness that had apparently taken place over the last nine months.

"Looking around this place right now, you wouldn't imagine how much drama is packed into this neighborhood. Where does all the crazy go at night, do you think? Or is it hibernating for the winter?"

Alycia shrugged. "Oh, it's definitely not hibernating for the winter. Tyrone was confined to the barracks, so that's probably why their house is quiet, but in general, I guess even crazy has to sleep sometime."

We stood looking around for a moment.

"It's good to be home, babe."

"Let's go inside. It's too cold out here to stand around bullshitting."

15

March, 2004: Fort Drum, New York

Back in the early days of the Soviet Union, the sequence of five year plans rolled out by Stalin exponentially increased the industrial capacity of that country by staggering amounts. Reality was somewhat different than reported, though. While the gains, expressed as percentages, appeared enormous on paper, critics often noted that the overall numbers weren't as large as they seemed–a three hundred percent increase in machine shops is substantial, but if you only had one shop to begin with, and your rival has four thousand, it's not all that great. Still, that sort of growth promoted a hubris in the Soviet people, and many of them felt great pride in their achievements.

That's exactly what I felt like walking around the PX with a Combat Infantryman's Badge on my chest, nine months of combat pay in my pocket, and a fresh raise from E-4 to E-5.[1]

In 2004, a Combat Infantryman's Badge, or CIB, was a rare and wonderful thing in the Army, particularly at Fort Drum, where the 1-87 was the first unit to earn the badge since the 10th Mountain had deployed to Somalia in the '90s. The CIB marked its wearer as a combat veteran. Moreover, it marked him as a combat veteran who'd been in direct fire contact with the enemy, and in the mind of the average infantryman, that

1. In actual terms, upon getting promoted to sergeant, my gross pay had increased from $1,726 a month to $1,901.10, which, viewed through 2025 eyes, appears a little ridiculous.

also signified that he was better than pretty much anyone who didn't have one.[2]

This was my attitude as I carried my dry cleaning through the sliding glass doors of the Post Exchange. I was as rich as Croesus and mighty as Leonidus, and everyone I saw I judged accordingly.

Sergeant first class with a Ranger tab but no CIB? Pussy.

Lieutenant colonel with airborne wings and an EIB? Pussy.[3]

Private first class with a combat patch but no CIB? Gray area, but probably still a pussy.

The judgment continued as I turned right and headed for the cleaners. It might not seem like it, but dry cleaning was an important aspect of my new mindset. Before the deployment, when I'd been a broke specialist, dry cleaning had been an unaffordable luxury. But I still needed a freshly pressed uniform every Monday. So Alycia had dutifully ironed my uniform every weekend and often cut my hair, too. While my wife is many things, a barber she is not, and starching my uniform was one of her least favorite chores. Consequently, now that I was a rich sergeant, we allocated fifteen bucks a week to pay someone else to take care of my hair.[4]

Our new affluence allowed for a few other lifestyle improvements, too. Alycia finally got an engagement ring, and I'd been able to splurge on a long-handled ice scraper with a brush that made de-icing the car substantially easier. We were also able to drink more often, with enough cash in our pockets to afford beer all the way up to the night before payday. Videos from Blockbuster weren't as tightly rationed as before, and occasionally we could even see a movie in the theater.

The other guys in the unit splurged similarly. Barret worked a deal with some guy for a motorcycle frame. He paid with a money order, but he was scammed, so he ended up with neither the frame nor the cash. Undeterred, he went out and picked up a crotch rocket instead. The Hawk invested heavily in a hamster breeding scheme he ran out of

2. Plenty of CIBs were earned in questionable circumstances, including indirect fire attacks on firebases where the fire was in no way well aimed or dangerous. Deep in every infantryman's soul, he knows whether or not he actually earned his CIB.

3. Albeit a grudgingly saluted pussy. My hubris hadn't made me suicidal.

4. I still polished my own boots, though. I was never that bougie.

his garage in on-post housing.[5] Other guys, like Ross, Williams, Raimond and Gardea, splurged in a more traditional fashion and went on a tattoo rampage. Some of these looked good. Others, like the full back infantry logos, or full-sleeve, sixty-four point 'freedom isn't free' tatts were likely to age less well.

We even had a sentimental set of spenders, whose cash went to enrich or surgically enhance the various working girls at the strip clubs of Watertown, Kingston, and Syracuse. Over the weeks that followed our return, a few of these beneficiaries even ended up joining the Army community as wives.

While our tastes varied, in the wake of the austerity and discomfort we soldiers of the 1-87 experienced over the last nine months, we all indulged in some kind of extravagance. It isn't fair to say we were *nouveau riche*, because we weren't *riche* at all. But for a bunch of previously broke-ass twenty-five year olds, we were definitely living scratch-off-lottery-ticket-winner large.

Unfortunately, as Robert Frost once wrote, nothing gold can stay, and neither did our boomtown wealth last. The bust started slowly and subtly, with barracks rats who'd been enthusiastic partiers one day earlier quietly refusing to join their comrades at the strip club or on out-of-town, weekend benders. Over the next months, the signs of trouble grew and became more noticeable. The battalion parking lot, which had been replete with shiny new cars, trucks, and bikes, began to thin out as one new vehicle after another, purchased with a ridiculously high-interest loan, was hauled back to the dealer to be sold to another sucker. On James Street, a similar story played out with rented furniture, with the added wrinkle that Rent-A-Center had to actually catch the person at home to repossess its stuff. Our neighbor, Brandy, played this cat and mouse game with the company for months, with Alycia serving as the unenthusiastic informant/mediator for both parties.

"Morning, ma'am. She home today, or is she hiding out in there and avoiding us?"

"I don't believe so. She's been gone for a while."

Once the repo truck was safely out of sight, Brandy popped her head out of the door.

"Hey, Alycia. Did the Rent-A-Center guys ask if I was home?"

"Yeah. I told them you were out."

---

5. His intent had been to raise and sell the rodents to scientific research labs. The effort had gone well until he found out the hamsters had to be certified to be used in research, so his were unfortunately useless for that purpose. In frustration, he'd given up on the project and released the hamsters into his yard, much to the frustration of his neighbors and to the delight of the neighborhood cats.

"Awesome, thanks! They gotta catch me if they want to take this motherfucking big screen back!"

Both sides appeased, Alycia went back to digging into the flowerbed.

The only people who really didn't suffer from the collapse in spending were the body modification junkies. For better and for worse, ink lasts forever. So, while a tattoo might not have been the best long-term investment, it at least wasn't repossessable. Sure, the dude didn't make anything off the investment, but at least he couldn't really lose his principal. The boob jobs guys bought for their significant others were often an even better deal. In fact, they might have been the most financially responsible purchases made during our post-deployment boom. Not only were fake tits permanent–so much so that they often outlasted the relationship that spawned them–but since a good number of the recipients were strippers, the augmentation often increased their earning potential.[6] And, of course, as Brandy was fond of pointing out while dramatically grabbing her boobs, "They can't repo these babies!"

The spending spree and its aftermath got so bad that Romero had me teach a class on personal financial management to the platoon.[7] The privates of AT platoon dutifully sat through my presentation, and they even (mostly) didn't fall asleep, although that might have been because I made the first guy I caught dozing koalify for ten minutes.[8] At the end of the class, E-brake asked to keep the butcher-block I'd made so he could review it further, which I took to be a good sign, and most of the rest of the platoon continued to spend themselves into oblivion, which I accepted as less good.

In the post-deployment period, people's personal finances weren't the only thing in turmoil. Relationship chaos prevailed as well. Guys finalized divorces that had occurred (with or without their knowledge) while they were deployed, cleaned up the messes left in the houses when their ex-wives abandoned them, and settled disputes with the dudes

6. They're even tax deductible!

7. Having a college degree made me the default choice for any task that involved typing, reading, or anything even vaguely intellectual, like writing all the platoon's award citations.

8. Koalifying is when a person is made to wrap his arms and legs around a tree or telephone pole and suspend himself that way for a period of time. Koalifying can be done either head up or head down.

who'd screwed their significant others while they were at war.[9] Some of the disputes were settled face-to-face in a drunken fist-fight. Others were settled through the Army chain of command by turning in the revenge photos their mocking wives had mailed them in Afghanistan. In the days before deep fakes, it was hard to deny photographic evidence, and several guys ended up facing UCMJ for adultery after a cuckolded husband sent a photo of the culprit in the act to his commander.

Some of the relationship craziness even hit James Street, and it wasn't long before I found out how the longtime drama between Misty and Tyrone was finally resolved. After yet another big fight with Misty, Tyrone had been confined to the barracks for months. From Alycia's letters, I knew that Misty had stayed in the house all winter, doing her best to trash it, and that her crowning achievement had been to avoid taking out the trash entirely, instead leaving it to freeze in a giant heap in the garage. The rest of the story remained a mystery, though, until I walked into the house one day and found out what happened.

Alycia ambushed me before I even got my boots off.

"Oh my gosh, you won't believe what happened! Tyrone finally came home and opened the garage!"

"You're kidding, really? What happened? Was it amazing?"

"Oh, was it ever! And I was accidentally in the perfect position to see the whole thing."

Alycia continued excitedly. "So, I was cleaning the bedroom this morning, and I happened to look out of the window just as Tyrone pulled up and parked in his driveway. I was so excited, I stopped what I was doing and sat down on the bed to see what would happen."

"Didn't you feel embarrassed? That's so voyeuristic."

"I know, I know. It was a guilty pleasure, for sure. I couldn't look away, though, you know?"

"Yeah, I get it. Go on."

"So, he gets out of the car, and everything seems fine. He's wearing his BDUs."

I interrupted. "Yeah, he's probably still not authorized to wear civilian clothes.

9. One of the most colorful divorces was Sergeant Skerret's. His philosophy on marriage, expressed in his thickly accented English one night over too many beers, had always seemed questionable. "The first time, I marry a good looking girl, and she cheat on me. This time, I marry an ugly girl, so she no cheat on me." After the deployment, his wife, who had never appreciated the philosophy much, cheated on him, got into an enthusiastic fistfight with him in the parking lot of a bar, and eventually divorced him anyway.

Normal part of the punishment process."

She continued. "He walks up to the garage and gets ready to open it. You've got to understand that I've been watching this drama play out for months, and I know exactly what's about to happen. I've pictured it in my head a hundred different ways and wondered how it was going to actually happen, so I can't describe how anxious I was to see it."

"I get it, for sure. Stop beating around the bush!"

"Okay, okay. So Tyrone walks up to the garage and bends down to grab the handle. Then, as he throws the door open, a *mountain* of garbage falls out. It's like a tidal wave of half-frozen garbage bags. There are ripped bags, full bags, small bags, big bags. This garbage goes *everywhere."*

"Oh my God, what did he do?"

"He just started swearing. Wild, colorful swearing. Swearing that was so loud I could hear it distinctly across the street. He called Misty every name in the book, and some that I don't think are. Then he started hopping up and down in pure rage. He was literally *hopping mad.* I've never seen that!"

"I can see why!"

"Oh, for sure. Me too. Anyway, so he just stood there, jumping and swearing for several long seconds. It could have been a whole minute."

"And you're just watching all of this?"

"Shamelessly. Staring at it like an old lady watching a soap opera. I couldn't look away."

"Well, what happened next?"

"Eventually, he seemed to calm down a bit and stopped shouting. He kicked the trash a little with his foot, shaking his head, and then seemed to sort of check out of the situation and walked over to the front door. He opened it, walked inside, and then I lost sight of him for a bit."

"So, that was the end of it?"

"Definitely not. He wasn't inside for five minutes before I heard more shouting. You remember all the shit I told you about what Misty did to the inside of the house, right?"

"How could I forget? Rocky the dog eating the trim off the doors. Misty drilling holes through the walls so she could run cable to the bedrooms. And who knows what else?"

"Exactly. He must have seen all that, because he was walking around shouting for a good while before he calmed down and walked outside."

"Poor guy."

"Right? The funny thing is, I don't think Misty even thought she was doing anything serious. I think she thought of it all as a sort of giant revenge joke."

"It's not going to be funny when housing docks Tyrone's pay for the next six months."

"No. No, it really isn't. He already lost a bunch of rank."

"I guess that's what happens when you marry a crazy bitch."

"Hey! She's my friend. She wasn't any crazier than he was, and they helped me out a lot while you were gone."

"Alright, that's fair. Still."

"Yeah, I get it. I don't think you can live on the emotional edge like that for very long without something or someone exploding."

After Alycia finished her story, we continued our now-very-normal seeming evening of a simple home cooked dinner and a high-dollar new release from Blockbuster.[10]

As we sat on the couch watching *Secondhand Lions,* Alycia looked down at her new engagement ring.

"You know, I really didn't need to get this. It seems extravagant."

"No, you're wrong about that. You did need it. It was all well and good to get married the way we did, but it's important to me to make sure you know I'm invested in this for real. We can't brush everything aside just because it's expensive. Sometimes you have to spend cash on the things that really matter. It's only money, after all."

She squeezed me in an affectionate hug. "I love you."

"I love you, too, babe!"

As we fell asleep that night, my brain started churning on that idea–that you should spend money on the things that really matter. Surely, all the guys at work thought that what they'd spent money on mattered, too: What was the difference?

The best I could come up with before I fell asleep was that the most important differentiator was whether you could actually pay for the thing without going into debt. Living for the now was fine, but when you had to mortgage the future to do it, it probably wasn't. Especially when that mortgage rate was at eleven percent.

---

10. New releases were $3.25, whereas old movies were usually a dollar.

I had been uncomfortably sitting in the grey sterile room for about an hour. I'd already filled a blue book with a bunch of hand written answers to questions about how I'd handle this ethical dilemma or that leadership challenge, and now I was waiting for the board to call me in for the oral question and answer session.

I thought back to the last time I'd been in this situation–during the sergeant's board. There were similarities, like the line of people and our collective anxiousness, but it was the differences that really attracted my attention. The other people, for example, clearly weren't infantrymen. Two were older looking NCOs. One was a staff sergeant, and the other was a sergeant first class. Neither had a combat patch or CIB, but other than their lack of infantry insignia, they looked about like the NCOs I expected the Army to produce–hard-bitten and cynical professionals who appeared to be carved out of boot leather. The third person had none of these qualities. She was a pretty young girl wearing sergeant's rank. Her hair was impeccably done, her nails were long and well-manicured, and she had a slick looking leather planner. She looked smart and pulled together. "Hell," I thought, "She even *smells* pulled together."

Because we were both sergeants, I assumed we'd joined around the same time. I reflected on the last four years. I thought about the cold nights, the bleeding feet, and the oh-so-much shouting and anger. I thought about the raging shits and the vomit and the laughing and the terror and the death. That was the Army I knew. That was everything I thought the Army was.

Had this girl been in that Army? I looked her over again, taking in the clean uniform and the spit-shined jump boots. I couldn't imagine she had. And if she hadn't, which Army *had* she been in, and what had her experience there been? Was *that* the Army I was walking into now–an Army with polished, professional people who had the time and desire to ensure they smelled nice before they came to work? I looked down at my own uniform. It was freshly pressed, but it had seen three years of hard use, and it looked like it. The colors were faded, and there were stitched up spots where I'd had to repair gashes in the fabric. My boots were in similar condition. They were nicely polished (I thought), but they'd been resoled once, and the leather had scratches that no amount of shoe polish could cover up.

Next to Ms. Perfect, I looked pretty ratty, and as I sat and waited to see whether or not I'd be selected to become an officer, it was hard to imagine an Army in which we both had a place. Which one of us was the sort of person the Army was looking for?

A uniformed staff sergeant stepped through the door and into the board room, interrupting my stream of thought. "Sergeant Pace, please report to the president of the board."

"Roger, Sergeant."

I stood up and walked across the room, past the two crusty sergeants and Ms. Perfect. I briefly stood outside the door to the board room and brushed a speck of something off my jacket. Despite being absolutely certain that this experience would be less unpleasant than 99% of what I'd experienced in the Army thus far, I was still anxious. What would happen if I failed? What if they weren't looking for dirty grunts at all? What then? What the hell would my life look like if I walked back to the unit in disgrace as a non-select?

"Ah, fuck it," I thought. "What's the worst they can do to me? Make me burn shit for a living?"

I checked my uniform one last time and walked through the door.

The room was as gray and plain as the one I'd just left, with standard government fluorescent lighting overhead and thin commercial-grade carpet underfoot. There was a long metal-and-wood-veneer table in front of me, and behind that table sat four uniformed officers. Unlike me, they were dressed in their Class A's.

I walked into the center of the room, stopped a few feet in front of the table, and saluted.

"Sergeant Pace reporting to the president of the board."

While I held my salute and waited to be acknowledged, I took in the members of the board. I was shocked to see that the president, the guy I was standing directly in front of, was a fat old man. What remained of his hair was gray, and his face was creased. He had to be at least forty, which made him a Methuselah in Army terms. He had a thick stack of ribbons, but neither a CIB nor a combat patch. To an infantry sergeant, fresh out of the 'Stan, he looked about as combat-ready as my grandmother.

The guys to his left and right were younger, but still old. I suspected none of them were under thirty-five. They also each had a stack of ribbons, but, again, none of them had a CIB or a combat patch.

"At ease, Sergeant Pace."

I transitioned from the position of attention to parade rest.

"Sergeant Pace, the purpose of this board is to determine whether or not to recommend you for selection as an officer candidate. Do you understand that?"

"Yes, Sir, I do."

"Good. We'll be asking you a series of questions to assess your fitness to serve as an officer in the United States Army. Based on those answers and the overall strength of your packet, we will then give our recommendation to the Army. Is that clear?"

"Yes, Sir."

"We'll begin. Captain Johnson has the first question."

One of the officers to the president's right spoke up: "Sergeant Pace, what would you say the most important quality in an Army leader is?"

I thought for a few seconds before responding. A million data points from the various team leaders, squad leaders, and platoon leaders I'd served under raced through my mind. What had the good ones done right? What had the bad ones done wrong? "Sir, I believe the most important quality an Army leader can possess is the confidence to make a decision, but that confidence has to be balanced by the humility to recognize when he's wrong."

The members of the board scratched on the paperwork in front of them with their pens. Was my answer right? Could it have been better?

Other questions followed from each of the other members of the board. They asked questions related to military ethics, they presented hypothetical situations and asked what I'd do in them, and they asked how I'd handle various leadership challenges. Some of the questions were easy, and I answered them without even really thinking about what I said. Others were stickier, and by the end of some of my answers I felt like I'd tied myself in logical knots. After ten minutes or so of this, the president of the board asked the final question.

"Sergeant Pace, why do you want to be an officer?"

Of all the questions the board asked, this was one of the simplest, but it stumped me, and it was several long seconds before I replied. Why *did* I want to be an officer? Hell, why did I want to stay in the Army at all? Why was I going through all of this trouble just to stick around an organization that seemed hell-bent and determined to make my life miserable? Images flashed through my head.

Brock and I were sitting in our truck in Uzbekistan, watching the LT impale bees and eating MREs for the sixtieth meal in a row as we stared out across the airfield. Neither of us had any more idea about what we were doing or why we were there than a trained orangutan would have.

The platoon was sitting in the arms room scrubbing weapons at 1630 on a Friday afternoon. Nobody in charge was anywhere in sight. None of us knew what was going on. All of us were irritated at what we perceived to be a complete waste of time.

Sobon and I, staring at that car full of approaching Afghans and wondering if we should kill them all or not.

An answer suddenly congealed in my brain and started dribbling out of my mouth:

"Sir, I want to be an officer because during my time as a soldier in the Army, I have felt many times that I didn't know why we were doing whatever it was we were doing. That lack of understanding, or lack of purpose, meant that even when I was doing something that ended up being important, I never did it as well or with as much motivation as I would have if only someone had explained why it mattered."

The dribble increased into a stream, as I warmed to the subject.

"During the initial invasion, our platoon sat in Uzbekistan for weeks with no idea whatsoever about what we were doing. That, combined with poor living conditions, led to guys becoming demotivated and caused disciplinary problems that wouldn't have happened if we'd understood the importance of what we were doing in K-2. Similarly, during my next combat deployment, there were dozens of situations where we made life or death decisions with no real understanding of the potential impact of what we were doing. We were told what to do, at least sometimes, but rarely why we were doing it, which meant our decisions often seem short-sighted when I look back on them.

"As an officer, I want to change that. I want to ensure that I always understand the 'why' behind what I'm doing, and I want to impart that sense of purpose to my guys. I want them to always understand the context in which they're operating and what their actions are ultimately trying to result in, because I think that's the key to unlocking their potential as soldiers."[11]

My brain should have stopped there. A couple of members of the board were nodding along and writing things down, and nobody seemed to think my answer had been dumb. But then my mouth leaked out the last part.

"Also, I'd really like to make more money."[12]

At this comment, one of the board members coughed, and the president looked up from his writing and cracked the faintest hint of a grin.

---

11. This version of the answer benefited from all the polish editing a written product can provide. The original was a bit more of a jumblefuck.

12. I've often felt like I was an outside observer listening to the things that came out of my mouth, rather than the person responsible for producing the comments. This was one of those times.

"Thank you, Sergeant Pace. That's all the questions we have for you today. You may return to the waiting room, and you'll receive guidance from there."

"Roger, Sir."

I saluted, waited for the president to return the salute, then turned on my heel and left the room. As I walked out, the two other NCOs and Ms. Perfect looked up inquiringly, probably trying to see if they could glean any information from my demeanor about what to expect. I pointedly ignored them and turned toward the staff sergeant who'd called me into the room a few minutes earlier.

"What's next, Sergeant?"

"Your unit will be notified of the results of the board in a few weeks. Until then, no action is required. Best of luck to you."

"Thanks, Sergeant. Is that all, then? Am I free to go?"

"Yep. Don't talk to any of the other applicants about the board process, though."

"Don't worry about that. I wouldn't want to deny them the fun of experiencing that themselves."

The staff sergeant nodded. I gathered my things, vaguely gestured 'good luck' to the other applicants, and left.

As I drove back to the company building, I realized I was sweaty. I must have been more nervous than I thought during the interview. The board had taken longer than I thought, too. When I'd walked out, I figured I'd been answering questions for no more than a few minutes, but the clock suggested it had been closer to fifteen. I thought about my answers. Had they made sense? Were they coherent? Were they what the board was hoping to hear? I didn't know, and I supposed I wouldn't find out until my packet came back. Until then, there was plenty of work to do back at AT platoon, and Romero was going to jump my shit if I didn't get back to doing it.

Grochowski, a clean-cut new guy with a college degree, was repositioning the sand bags around the Ma Deuce's tripod so the feet wouldn't shift when he started firing. We were on a machine gun range in the Fort Drum training area, and the platoon was once again running through our training task list to get the new soldiers up to speed on their infantry skills before the unit headed back to combat. As I watched him smash the bags into place, I was irritated.

Why the fuck were we training with sand bags? Why was the gun mounted in a fixed position? We'd never once used either in combat. For that matter, why were we practicing stationary gunnery at all? Every time I'd fired the Ma Deuce in combat, I'd been on a truck, and most of that time that truck had been moving. However, this was the way we'd always run ranges, and it was the way the training manual I'd found in the platoon office had said ranges should be run. Still, it irritated me. We were training guys wrong.

"Stop what you're doing, Private."

Grochowski looked up at me. "Roger, Sergeant. What's up?"

"We're going to change the table. Ramsey, go get one of the trucks and bring it over here."

"Roger, Sergeant."

I looked up at the range control tower where our new platoon sergeant, Sergeant First Class Jones, was monitoring the training. Or at least I thought he was. I hadn't seen him all morning, and the range control tower windows were mirrored, so I wasn't really sure what he might be doing–or if he was there at all. Would he mind? Did I care? Romero had vanished into a special operations pipeline a few weeks prior, and Jones had appeared in the platoon command post a few days later. He was quiet and hadn't said much of anything so far.

When the truck arrived, I had the team mount the gun on the turret and told Grochowski to get ready to fire. I then turned to Raimond and Ramsey. "Alright, you two, get in the driver's seat and the back seat, and lock and load your rifles."

"Roger, Sergeant."

"Now, listen up. Here's the drill. You can all see the targetry. There's the vehicle," I pointed in one direction at a large, wooden truck target, "and there's the two groups of dismounted infantry." I pointed at two groups of four man-sized silhouettes. "When I say 'fire'," you will all engage your targets until they're destroyed."

Ramsey had a question. "You want us all to shoot at the same time, Sergeant?"

"Yep. Everyone at once. And Ramsey, since you're the driver, once you hit your first target, I want you to get the truck ready to move and bound up to that point," I pointed at another firing position on the range, "while the other guys keep shooting."

"Roger, Sergeant."

There was a pause while the team processed the guidance. Raimond raised his hand.

"Hey Sergeant Pace, is that alright? Aren't we breaking the range rules or something? We've never done it that way before."

He was probably right. I'd never been on a range where we did anything like this. Still, I was a sergeant and a combat veteran, right? Who knew better than me how to train guys for war?

"That's my problem, man. If the PSG doesn't like it, he can come straighten me out. We gotta get ready for war, though, right?"

The team nodded. "Roger, Sergeant."

"Then let's get to it."

The team locked magazines into their rifles, climbed into the truck and shut the doors. I stepped behind one of the rear doors of the truck and shouted, "*Fire!*"

BOOM BOOM BOOM

The familiar blast of the Ma Deuce rocked the truck. The two front doors popped open, and Raimond and Ramsey leapt out. Ramsey leaned against the door to use it for cover, and Raimond ran up to brace himself on the hood of the truck so he could fire more accurately. The sounds of their rifle fire soon joined the .50 cal's blasts.

BOOM pop pop pop BOOM BOOM pop pop pop

I looked downrange. The targets were taking plenty of fire. I looked at the team. They were shooting and shouting at one another. It looked about right, but something was still a little off. What was it?

Ramsey hit his target and shouted for Raimond to get in the truck so they could move. While Grochowski kept shooting, Raimond climbed back in, and Ramsey maneuvered the truck twenty yards up to the next firing position. When the team got there, they climbed back out and started firing again. Grochowski was doing well. His fire was accurate and he was controlling the gun. It looked easier than I remembered. Suddenly, a switch flipped in my brain. *That* was what was missing!

I scooped a handful of .50 cal brass and links off the ground and jogged over to where the team was still firing. I took careful aim, and slung one of the links at Grochowski's head. It hit him in the neck, causing him to jerk the gun and fire wide. He looked around in confusion and finally saw me.

"*What the fuck are you looking at?* Keep firing! You think nothing's going to hit you in combat?!"

"Roger, Sergeant!"

I repeated the process several more times, hurling pieces of brass and machine gun links at the team as they tried to both engage accurately and look out for flying debris. After thirty seconds or so, all the targets were down, and I hollered, "*Cease fire, cease fire, cease fire!*"

The team stopped firing, Ramsey started to clear his weapon.

"Nope, don't clear that shit. Keep it live. You'll carry it that way in combat, right?"

"Roger, Sergeant."

"Alright, the targets are down. How do y'all think you did?"

Grochowski spoke up first. "I think the engagement went well, Sergeant."

"Did it? I thought you were slow getting the gun on target, and when you got hit with that first piece of brass you lost focus, and it took you a while to start hitting anything again."

Grochowski nodded, "Roger, Sergeant."

"Shit's going to get wild when you're in a firefight, man. It's not going to be like this. It's not going to start or stop in a defined way, and the targets aren't going to sit there and wait for you to shoot them. Sometimes, you won't be able to see them at all. And you're going to get hit by stuff. Debris. Shrapnel from indirect. Brass from your buddy's gun. The ground when your truck rolls over. All of that's going to happen, and you'll need to be able to focus even when it does and stay focused on what you're doing."

The team nodded again.

"Now, help me get this reset for the other guys. We're going to do it again."

"Roger, Sergeant."

We reset the drill and ran it again for the other team. Then we ran it a half dozen more times. Each time, the teams were forced to get their rifles and gear out of the truck, get into a good firing position, and dodge or deal with the debris I threw at them. By lunch time, they were getting better at putting it all together, even when I started adding in extra challenges like making them pass ammo up to the gunner, or adding a fourth person to the team. By the middle of the afternoon, I thought we were actually doing pretty well.

We'd just started an iteration that had the dismounted guys run forward to take cover while the .50 fired over them, when a Humvee I didn't recognize pulled up behind us. I ignored it while we finished the current run.

I flung a piece of brass at one of the dismounts and shouted at him. "You! You're dead." Over the booms of the Ma Deuce, I shouted at his buddy, "You! Get him back to the truck. He's hit!"

"Roger, Sergeant!"

Ramsey grabbed his compadre and started dragging him by his gear toward the vehicle, while Grochowski kept firing at the targets down range. Ramsey had just managed to wedge Raimond into the back seat when I heard, "*CEASE FIRE, CEASE FIRE, CEASE FIRE!*" from behind me. The firing stopped, and I turned around, annoyed, to see who was fucking up my training. Our new first sergeant was standing in front of me, and he looked furious.

"Hey, First Sergeant, what's up?"

The first sergeant's face was red, and his jaw was clenched. "Sergeant Pace, come with me. The rest of you, stand at ease."

"Roger, First Sergeant."

He led me back toward his truck in silence. What was up, I wondered? Why was he mad? Why was he even here? Had somebody in the company lost a weapon or something? The first sergeant hadn't been with the unit for very long. He was an older guy who'd just arrived from the Old Guard in D.C, and our limited interactions had been neutral so far. I looked up at the tower. I didn't see any sign of the platoon sergeant.

As soon as we got to the trucks, the first sergeant whirled on me. I reflexively snapped to parade rest.

"Sergeant Pace! What in God's name kind of training are you running here?"

I was confused. "Machine gun training, First Sergeant."

"Why are the other soldiers firing rifles, then. And why is the gun mounted on the truck? And why are you maneuvering and firing? Have you cleared any of this through range control?"

"No, First Sergeant. We're just training the way I thought made sense."

My face must have relayed some of the confusion I felt, because his tone somewhat softened.

"The way you thought made sense? Do you see those two checkerboard signs?"

I looked at the two edges where he was pointing. The large checkerboards represented the left-and-right most limits of the range. I nodded. "Roger, First Sergeant, I do."

"And do you realize that your team was firing outside those limits?"

"What do you mean, First Sergeant? They were only firing at the targets."

"Yes, but the targets are designed to be engaged deliberately, from the established positions. If you change your position, the line of fire changes, too, and the bullets can hit outside the confines of the range."

I looked at what he was showing me. He was right. The angle changed when we moved the truck, so some of the bullets were probably hitting outside the range limits.

"Roger, First Sergeant. That makes sense. Is that a big deal?"

His eyes went wide. "Of course, it's a big deal! What if there was another unit training over there?"

"Downrange? Is that possible?"

"Well, I mean, they wouldn't be downrange, but if there was a ricochet, it's possible someone could be hit."

I thought about it for a minute. I remembered Sergeant Major Watson once lecturing me on the 'little bullet, big sky' theory, which explained why it was statistically very unlikely for someone to be hit in combat. I thought about all the bullets that had been fired at me.

"But, won't there be shots fired at us in combat First Sergeant? Don't the guys need to be ready?"

The first sergeant stared at me. His face wasn't angry anymore. It even looked a bit bewildered. He took a solid several seconds before he spoke again.

"Sergeant Pace, have you been to the range safety class yet? Are you range control qualified?"

"No, First Sergeant."

He closed his eyes and nodded, thinking.

"Where is your platoon sergeant?"

"He's in the tower, First Sergeant."

He nodded again. "Tell the guys you're done training for now. Break for chow. I'll have new guidance for you before the nightfire this evening."

"Roger, First Sergeant."

"Carry on, Sergeant Pace."

The first sergeant turned to walk toward the tower.

"Hey First Sergeant, did I do something wrong? Isn't this what training is supposed to look like?"

He paused briefly and said over his shoulder, "I'll get you updated guidance before the nightfire this evening, Sergeant Pace."

I nodded and walked over to where the platoon had been pretending to clean weapons as they struggled to hear what sort of ass chewing I was receiving. As I walked up, Raimond broke the ice. "What's First Sergeant pissed about, Sergeant Pace?"

"I dunno, man. Maybe we filled out some paperwork wrong or something."

We cleaned up the brass and wiped down the vehicles, then broke out a box of MREs for chow. A half hour passed before the first sergeant and the platoon sergeant walked out of the tower and down the stairs. The first sergeant climbed into his truck and drove away, and the platoon sergeant walked over to talk to us.

"Alright, Sergeant Pace, here's how tonight is going to run. There will be no maneuver, and only one soldier will fire at a time. You will engage only the targets in your lane, and you will engage only when instructed. There will be no evacuation of casualties, and you will not throw anything at one another. Am I clear?"

"Roger, Sergeant. WILCO."

The platoon sergeant nodded, grabbed an MRE, and walked back to the tower. I watched him climb the stairs and thought about what he'd said. I'd always heard the cliche that you should train as you fight. Well, I'd been to war, and this was how we'd fought. So, what was the problem?

Across the Army in 2004-2005, this same source of friction–old Army processes grinding up against fresh ideas from combat–surfaced a hundred times in a hundred different units, and the results of these tensions had profound impacts on the organization as a whole. When a unit rotated back from a combat deployment, the young soldiers who still had time left on their contracts got promoted to sergeant and took over junior leadership positions. Simultaneously, the senior leaders they'd served under during the deployment rotated out for new assignments and were generally replaced by guys who'd been serving in training or headquarters billets and hadn't seen combat yet. These senior leaders were experienced, but their experience was from the peace-time Army of the 90s, and they generally wanted training to run the way they'd always run it. After all, generations of combat experience had produced the current training guidelines, so why change them? Newly promoted leaders like me, on the other hand, had rarely practiced things 'the old way' enough times to understand why they were the way they were. My generation of combat vets only saw a lot of risk-averse Army nonsense that seemed to focus more on looking polished and keeping soldiers busy than on actually preparing guys for combat.

In the structured world of the Army, generational tension normally sorts itself out very quickly and one-sidedly. What made this situation different, however, was that many senior leaders were insecure in the strength of their positions. Their subordinates, after all, *had been to war*, had seen the truth of combat and had first-hand information on what worked and didn't work. When it came to developing realistic training, which was worth more: a career spent pretending to fight or a year spent actually fighting?

The truth, as it so often is, was somewhere in the middle. Senior leaders were often risk averse and hesitant to adapt training standards to match the demands made by the new war. Conversely, young leaders like me were often too quick to throw the baby out with the bathwater when reworking training scenarios. We hadn't experienced the casualties that could occur when training is poorly planned or recklessly executed, so we didn't value the wisdom that was baked into many of the seemingly stale and outdated Army processes.

In the best cases, units were able to identify a golden mean and refine established Army training methodologies by incorporating fresh observations from the war. Units that did this successfully often returned to combat much better prepared than they had been on their first rotation. Unfortunately, units that didn't adapt in those ways often suffered from poor morale and low retention as irritated seniors sought to hammer their rebellious subordinates into compliance, and the subordinates themselves subsequently voted with their feet and left service. During future rotations, I worked alongside units like this and saw them struggle with disciplinary issues and suffer from a lack of junior leader initiative.

As for the rest of our training that night, it was by-the-book and dull. I had the guys zero their lasers, then we engaged all the targets with the Ma Deuce from a sand-bag-reinforced tripod on the ground, just as I remembered doing when I'd first joined the platoon and totally unlike I'd ever done in combat. The targets even had chemlights on them so that they'd glow in the dark–just like Afghan terrorists didn't–to prevent anyone from accidentally firing at the wrong thing and causing a safety violation.

As I stretched out my sleeping bag that night, I was mildly irritated by the whole scenario. What a bunch of bullshit. Why was everyone so averse to realistic training? The only thing all this chemlights-on-the-target crap would do was get guys used to something that would never happen in real life. We weren't honing our skills; we were building training scars.

My irritation flared up but then faded as I crawled into my fart sack and got ready for a few hours of sleep before I had to get up and check on the guards. Since I'd joined the Army, there'd always been something to be outraged about, and now that I was staying in for at least another couple of years, this would always be the case. But once the day was over and it was time to settle down for some well-earned rest, none of that mattered all that much. The Army was what it had probably always been, and, for the most part, that was a pretty good thing. There was plenty I didn't like about it, but that was life, wasn't it?

"After all," I thought, "it'll buff out."

Epilogue: May, 2005: Fort Drum, New York

Every new beginning comes from some other beginning's end.

Semisonic, "Closing Time"

"What the hell are you shitbags doing over there?! Get off your asses and get to work!"

My asshole puckered a little as I heard the NCO yelling at someone. I looked around reflexively to see if he was berating me before my brain snapped into gear, and I realized he wasn't. I was a lieutenant now, and there's no way in hell that ass chewing had been directed my way. But old habits die hard.

My eyes finally landed on the source of the sergeant's ire, a group of five privates who were sitting outside the company area with their backs against the wall, smoking and joking. I couldn't help but laugh a little as I watched the sergeant stomp across the grass shouting at them. The privates leapt to their feet, put out their cigarettes, and snapped to parade rest.

"Pick up those brooms and unfuck yourselves, yardbirds, or I'm going to stick my foot so far up your ass you'll taste fucking boot polish."

I stood fifteen feet away on the sidewalk, a fascinated observer watching the familiar scene play out. I'd been the recipient of similar ass chewings a thousand times, and I'd even doled out a few myself. This time, though, I wasn't a gorilla anymore–now I was

Dian Fossey.[1] I was still kind of a member of their world, but now I was in a different caste. My existence touched theirs sometimes, but otherwise operated in a parallel, and slightly more pleasant, plane.

I stood watching until the sergeant finished squaring away his guys and walked off, glowering. His path took him past me, and his expression changed entirely as he approached, switching from anger to professional neutrality in a flash as he raised his hand in a crisp salute.

"To the top, Sir."

I was momentarily caught off guard. *Receiving* salutes was still an uncomfortable and surprising thing. I shuffled, came to attention, and returned the salute.

"All the way, Sergeant."

My reply had been a little too late and a bit too disjointed to be quite right, and the sergeant had doubtless recognized that, but he wasn't going to say anything about it. I was only a second lieutenant, and thus not much of an officer. Still, I was an officer.

The sergeant walked past me and turned a corner, and I watched as the privates, who'd been working more or less diligently while he was in their line of sight, started slacking almost immediately. I laughed to myself and walked into the company headquarters building to find the commander. I needed to get him to sign my paperwork so that I could leave. I was only back on Fort Drum for a few days to clear the post, and I needed to get everything done quickly so I could drive down to Fort Knox, Kentucky, where Alycia was getting our apartment settled.

OCS, or Officer Candidate School, had been a little painful. The PL's report about the course had been correct, and I'd spent twelve weeks at Fort Benning marching around in an ascot, organizing the pencils in my desk drawer, and trying not to fall asleep in sweaty classrooms. Every night that we weren't in the field we had to line up in the hallway, chug a canteen of water, and sing the OCS song at the tops of our lungs to be granted permission to go to bed.

As a former infantryman, I found none of the course material challenging. But since most of the other candidates were folks who'd joined straight out of college, the instructors had to spend most of their time teaching them the basics of Army life. For those of us who'd been in for a few years–particularly those of us who'd served in

---

1. The author of *Gorillas in the Mist*. She'd written the book after spending years living in and around gorilla society in Africa.

the combat arms–OCS was just a silly block that needed to be checked in order to get commissioned. Adapting to the schoolhouse and its seemingly infinite number of nitpicky rules had been difficult, though. Consequently, I'd spent many an evening marching around the OCS perimeter on a punishment tour for running my mouth or blowing off some requirement I'd deemed a waste of time.

Eventually the course had ended. I'd graduated and was assigned to be an armor officer in Fort Riley, Kansas. Before moving to Kansas, though, I had to complete the Armor Officer Basic Course at Fort Knox. And before doing *that*, I had to clear Fort Drum. Which meant I had to go back to my old unit as a lieutenant.

I walked into the platoon headquarters where Sergeant Chadwick was irritatedly typing something into our platoon's single computer.

"Man, fuck this piece of shit computer. It's all been fucking downhill since we stopped writing shit down on a board in the hallway and started this digital, New Age hippy bullshit."

He stopped pounding on the keys and turned to look at me.

"Hey Sir, what can I do for...Pace? Is that you?"

"What's up, Sergeant? I'm back!"

"Holy shit! Lieutenant Dan! Sir! Man, the Army must be hard up to make your ass an officer."

"I guess so. I tried to talk them out of it, but nobody would listen. I'm just passing through to get my clearing papers signed by the commander, then I gotta get on the road."

"That's incredible, man. *Sir*, I mean."

I laughed. "C'mon. I'm the same turd I always was."

He shook his head in disbelief. It was weird. Overnight, our relationship had changed entirely. He'd been my boss when I left, and, now, in the general military sense, I was his.

"Where is everyone? Any of the guys around?"

"Nah, not really. There's an FTX coming up, and everyone else is down in the motor pool getting all our shit ready. Even the new PL is down there. They only left my ass up here to update our PERSTAT in this stupid fucking computer."

"Clearly they put our most tech-savvy on the job."

"Right? Your college-degree-having ass should be doing this, not me. I had Grochowski doing it for a while, but he needs to learn other shit, too. So I sent him down with the rest of the guys."

"It looks pretty fun. I'm jealous."

"You won't have to be for long, motherfucker. Guess what lieutenants do? This paperwork shit right here."

"That's a good point."

I looked down the hallway toward the commander's office. McCarron had left, but it looked like the new guy was there, and I needed to catch him before he left for lunch or something.

I turned back to Chadwick. "Well, I gotta run, man. Best of luck to you."

He stood up and embraced me. "It's been a pleasure serving with you, *Sir.* I'll see you around."

"Yeah, you too. Later, Sergeant."

I walked down the hall to the company headquarters. Nobody was around, but there were clearly a couple of people talking in the commander's office, so I leaned on the desk to wait until they finished.

I'd only been down to this end of the hall a handful of times. This was the first sergeant's and commander's territory, and no sane private wanted to hang out here, ever. To do so risked uncomfortable questions or worse from folks who outranked your boss's boss. Though that wasn't my problem anymore. As a lieutenant, and a lieutenant who was leaving the unit at that, I had zero concerns.

The conversation in the office concluded, and I heard the shuffle of chairs that indicated people inside were about to exit. I stood up to meet the new company commander, and was instead surprised to see Lieutenant Colonel Howard, the battalion commander, step out of the office, followed by a captain I didn't know. I immediately straightened to the position of attention.

"Sir."

LTC Howard looked at me for a moment, brows knitted, then said, "Who are you, Lieutenant? I don't know you. Did you just get to the battalion?"

"No, Sir. I just graduated OCS, and I'm getting ready to leave Drum for the basic course."

His face opened up a bit. "So, you were one of my guys overseas then?"

"Yes, Sir. AT Platoon. We were in Sh'kin, then Orgun."

"That's wonderful. Your platoon did a really good job over there, and I appreciate it."

"Thanks, Sir."

The commander shook my hand. "Well, thanks again for what you did, and congratulations on earning your commission. Good luck to you, son."

"Thanks, Sir."

The battalion commander left, leaving me standing with the captain, whom I didn't know.

"I overheard your conversation, LT. It sounds like you just need a signature to leave?"

"Yes, Sir. That's it."

The captain signed my paperwork and handed it back to me.

"Good luck to you."

"Thanks, Sir."

I turned and walked past the platoon headquarters, where Chadwick was still typing, and left the building. When I walked outside, the privates had dropped their brooms and were smoking again. As I walked past them, one of the group fumbled with his cigarette and roughly said, "Group, attention!" He then saluted. "To the top, Sir."

"All the way, gentlemen. Carry on!"[2]

I continued down the sidewalk, laughing quietly to myself, got to the parking lot, climbed in my car, and drove off post for the last time. On a whim, I decided to swing past our house on James Street before I left to say goodbye to the place where Alycia and I had made so many memories. When I pulled up in front of our house, I got out and stood in the yard to look around. The house didn't have new occupants yet, and with our Altima parked in the driveway, it almost looked like we still lived there.

I glanced across the street, where a car I didn't recognize sat in Tyrone and Misty's driveway. A new family must have moved in. I was sure housing had plastered over all visible signs of what had occurred inside, but what about the invisible remnants? Had Misty and Tyrone left their mark on the house's spirit if there was such a thing?

And what about Brandy and Trey's place? They'd moved out just before we left, and the last thing we heard about them was when a strange woman knocked on our door a few months back and asked if they still lived there.

"No," Alycia had told her. "They moved out last week."

"WHAT?! Those motherfuckers owe me ten grand! I loaned it to them to open a

2. An important part of OCS was learning to stop talking like a protagonist from a Quentin Tarantino movie. It's been a tough habit to break, and I still swear reflexively when telling Army stories even today.

bar."

I wondered again if the particular brand of crazy that was such an integral part of Army family life stained the *Qi* of the houses where those families lived. If so, what sort of chaotic energy accumulated over the years as family after family rotated through the buildings? And what about the positive aspects of Army family life? Did all of that warmth, openness and energy linger around the neighborhood, too? What weird mixture did it yield, and what effect did it have on the future occupants? Did it doom them to reenact the same soap opera-ish stuff as the previous occupants, or did every family get a clean slate? As I left the neighborhood and drove toward the interstate, I pondered all of this and more.

How much had my time at Fort Drum captured the Army experience as a whole, and what was in store for me at my next assignment? Would I be able to enact the changes I thought were necessary to fix whatever I found there, or was everything too stale to evolve? Would I go back to war again? If so, what would it be like to do so as an officer? The future seemed wide open and exciting as I turned onto Interstate 81 and headed south. Whatever was coming, I was looking forward to figuring it out.

Afterward

But all the miles that separate,
disappear now when I'm dreaming of your face.
Three Doors Down, "Here Without You"

There was something that reprised the wonder of childhood about being a private in the Army. The experience was so unlike anything I'd encountered previously, that it really did feel as though I'd fallen through the looking glass: everything I encountered had such a bizarre quality to it that it frequently left me standing in stunned silence. Or stunned silence until whatever the thing I'd encountered required me to do something awful, uncomfortable, or ridiculous.

Despite my years at Fort Drum being the most physically uncomfortable period of my life, though, I remember that time fondly, and none of the eighteen years of Army life that followed have matched that time for sheer richness of experience. Moreover, I've never been closer to anyone since as I was to the friends I made during those four years. Maybe that was just due to the sheer volume of time we spent together, but I think there was more to it. I think there was actually something unique about the whole experience of being wartime privates, and because we–the Brocks, the Sobons, the Hogus, and The Hawks–experienced it together, we will be forever bonded in a way that can't be reproduced.

Because you can never go back. And you can never do it again. There are some experiences, like war, that can't be unseen, and as is said about things that are seen that can't be unseen, the first time is always transformative. Like Neo's first trip into the real world in *The Matrix*, once your perspective is changed, it can never return to the way it was, and that creates a separation between you and everyone else who hasn't had a similar experience.

I remember how the hippies talked about Woodstock when I was a kid. I never really understood the feeling they were trying to describe–I couldn't understand because I hadn't been there. It irritated me at the time, but I think I get it now. Afghanistan was my Woodstock. *And* it was my red pill. Now I can't help but see the world in a new light.

Speaking of seeing things in a new light, when I wrote this, I drew heavily on the letters Alycia and I wrote throughout the period in which this book is set.[1] Aside from being incredibly cringeworthy, one of the things that shocked me most rereading them is how little I understood about what was going on when I wrote them. It's as though Specialist Pace saw everything through a soda straw.

Rereading my hand-scribbled words, one would think no combat mission we conducted and no training we endured had any purpose at all other than to inconvenience us and make us miserable. That's certainly how I felt at the time; that was my lived experience. With twenty-plus additional years of service (and a good deal more time at war) under my belt, I now have the perspective to put much of what we did in context, but providing that context wouldn't serve the purpose of telling it like it was, when purposeless confusion defined what it was like to be a private at war.

Having said that, when I zoom out a bit, it's also funny to think about the number of things I still *can't* put in context. Like what we were doing there at all. While toppling the Taliban in the wake of September 11th still seems (to me) as though it was a worthwhile endeavor, I don't think having the 10th Mountain Division drive around Paktika province in 2003 and 2004 was doing much of anything but pissing off the people who lived there. Years later, as a retired Colonel, Captain Lopez captured this sentiment well when he said, "We did a lot, but I'm not sure we got a lot done." He said this when we were musing about how the war ultimately ended, though.

To *Specialist* Pace and the men of AT Platoon, the greater context of what we were doing was mostly unimportant. While we did occasionally hunger for the 'why' behind what we were doing, particularly toward the end of the 03-04 rotation, that was usually further up Maslov's hierarchy than we managed to get. Instead, chow, Imodium, our buddies, whether our sergeants were pissed off, when the mail would come, and when

1. If you've never re-read old letters you wrote when you were young and dumb, I highly recommend it.

the next guard shift was scheduled, were the things our worlds revolved around. I now know from years of reading that these simple things, or their historical equivalents, have always been the primary concerns for infantrymen at war.

For me, this simplistic outlook on the world is what makes the infantryman, who is so often surrounded by and immersed in horror, such a likable figure. His (generally) guileless pursuit of the basics we all look for in our daily life–food, sleep, and a touch of comfort–make him relatable in a way that so much of war really isn't. Literary critics have said that Tolkien used the hobbits in his books to help the reader relate to the world-shaping events and larger-than-life characters in *The Lord of the Rings.* I believe the recorded accounts of infantrymen, particularly when those accounts include their thoughts about what they've seen and dialogue with their peers, serve a similar function in the real world. These stories provide readers with a guided tour of a world they wouldn't otherwise (and probably wouldn't want to) experience themselves. Aside from being entertaining, this service is inherently valuable. By allowing people to better understand war and the impact it has on its frontline participants, these stories help ensure society neither takes war too lightly, nor shies away from it when it's necessary. At a time when a shrinking percentage of the population has front-line combat experience, I believe the risk that either decision will be made in ignorance is increasingly likely. I hope that the story of my friends and me in Afghanistan can narrow this perspective gap, and by doing so, ensure that while the junior enlisted members of the 1-87 may not have always understood why we were doing whatever it was we were tasked to do, what we did should still have value to the country we all fought so hard to protect.

Acknowledgements

This book wouldn't have been possible without the help of a great number of people. Many of them were present during the described events. Their assistance in triangulating the mish-mash of fuzzy memories, letters, and soda-straw perspectives we've all been left with after more than twenty years into an accurate and compelling story has been invaluable. Others are friends and associates who provided other sorts of help, like editing, reviewing, and telling me when things weren't quite coming together very well. One is married to my dumb ass, and has been since this silliness began all those years ago.

Their names are here, listed in no particular order: Dave Martineau, Brett Hamilton, Dan Romero, Jack Horn, Casey Sobon, Rob Ritchie, Adam Everley, Randy and Laura (the Story Ninjas), Scott Chadwick, Buddy Epting, Rhys Williams, The Hawk, Joey Hogu, Chad Woody (the cover artist), Eric Lopez, the amazing Anna Simons, and, of course, the love of my life and fellow James Street veteran, my wife Alycia.

About the Authors

Daniel and Alycia Pace grew up in Texas, where they met in 1990. They were married in 2001 after Daniel enlisted in the Army, and they moved 12 times over the next twenty-two years to places as varied as Fort Riley, Kansas; Miami, Florida; and Stuttgart, Germany. Daniel retired from the Army as a Special Forces Lieutenant Colonel in 2023, and he and Alycia now reside in Northwest Florida with their five children.

It's hard for a book to get noticed these days. Please consider leaving a review on Amazon to help me continue writing.

https://www.danielvpace.com/

www.ingramcontent.com/pod-product-compliance
Ingram Content Group UK Ltd.
Pitfield, Milton Keynes, MK11 3LW, UK
UKHW022027041026
15591UKWH00012B/534

9 798990 128484